S0-BWX-175

Colloid and Interface Science

VOL. III

Adsorption, Catalysis, Solid Surfaces, Wetting, Surface Tension, and Water

EDITED BY

MILTON KERKER

Clarkson College of Technology
Potsdam, New York

Academic Press Inc.

New York San Francisco London 1976

A Subsidiary of Harcourt Brace Jovanovich, Publishers

ACADEMIC PRESS, INC.
111 Fifth Avenue, New York, New York 10003

United Kingdom Edition published by
ACADEMIC PRESS, INC. (LONDON) LTD.
24/28 Oval Road, London NW1

Library of Congress Cataloging in Publication Data

International Conference on Colloids and Surfaces,
50th, San Juan, P.R., 1976.
Colloid and interface science.

CONTENTS: v. 2. Aerosols, emulsions, and surfactants.–v. 3. Adsorption, catalysis, solid surfaces, wetting, surface tension, and water.–v. 4. Hydrosols and rheology. [etc.]
1. Colloids–Congresses. 2. Surface chemistry –Congresses. I. Kerker, Milton, II. Title.
QD549.I6 1976 541'.345 76-47668
ISBN 0–12–404503–0 (v. 3)

PRINTED IN THE UNITED STATES OF AMERICA

Contents

CONTENTS

CONTENTS

List of Contributors

G. Patricia Angel, Department of Chemistry, Emory University, Atlanta, Georgia 30322

M.P. Aronson, Lever Brothers Company, Research Division, 45 River Road, Edgewater, New Jersey 07020

R.F. Baddour, Department of Chemical Engineering, Massachusetts Institute of Technology, Cambridge, Massachusetts 02139

Fredrick S. Baker, Amherst College, Amherst, Massachusetts 01002

Stuart S. Barton, Department of Chemistry and Chemical Engineering, Royal Military College of Canada, Kingston, Ontario K7L 2W3, Canada

Ralph A. Beebe, Amherst College, Amherst, Massachusetts 01002

H. Bilinski, Institute "Ruder Boskovic," Zagreb, Yugoslavia

Raghupathy Bollini, University of Nebraska, Lincoln, Nebraska

M. Branica, Institute "Ruder Boskovic," Zagreb, Yugoslavia

Chester V. Braun, Jr., Laboratory for Water Research, Department of Chemistry, University of Miami, P.O. Box 249148, Coral Gables, Florida 33124

Z. Calahorra, Department of Chemistry, Technion, Israel Institute of Technology, Haifa, Israel

J.V. Calara, Department of Mining, Metallurgical and Fuels Engineering, 412 Mineral Science Building, The University of Utah, Salt Lake City, Utah 84112

P. Canesson, Groupe de Physico-Chimie Minerale et de Catalyse, Universite Catholique de Louvain, Place Croix du Sud 1, B-1348 Louvain-la-Neuve, Belgium

T-H. Chiu, Medical Group, Avco Everett Research Laboratory, Inc., Everett, Massachusetts 02149

David M. Clementz, Department of Agronomy, Life Science Building, Purdue University, West Lafayette, Indiana 47907

H. Lawrence Clever, Department of Chemistry, Emory University, Atlanta, Georgia 30322

H.D. Cochran, Department of Chemical Engineering, Massachusetts Institute of Technology, Cambridge, Massachusetts 02139

George L. Cohen, Bristol-Myers Products, Department of Physical Chemistry, 1350 Liberty Avenue, Hillside, New Jersey 07207

Mary Ellen Counts, Department of Chemistry, Virginia Polytechnic Institute and State University, Blacksburg, Virginia 24061

D.G. Crosby, Department of Environmental Toxicology, University of California, Davis, California 95416

Alvin W. Czanderna, Physics Department, Clarkson College of Technology, Potsdam, New York 13676

J.P. Damon, Laboratoire de Chimie Generale, Université Scientifique et Medicale de Grenoble, BP 53 38041 Grenoble Cedex, France

Burtron H. Davis, Associate Professor, Chemistry, Potomac State College, West Virginia University, Keyser, West Virginia 26726

F.D. Declerck, Groupe de Physico-Chimie Minerale et de Catalyse, Université Catholique de Louvain, Place Croix du Sud 1, B-1348 Louvain-la-Neuve, Belgium

C. Defossé, Groupe de Physico-Chimie Minerale et de Catalyse, Université Catholique de Louvain, Place Croix du Sud 1, B-1348 Louvain-la-Neuve, Belgium

B. Delmon, Groupe de Physico-Chimie Minerale et de Catalyse, Université Catholique de Louvain, Place Croix du Sud 1, B-1348 Louvain-la-Neuve, Belgium

T.J. Dickinson, Department of Physics, Washington State University, Pullman, Washington 99163

R.G. Donnelly, Department of Chemical Engineering, Massachusetts Institute of Technology, Cambridge, Massachusetts 02139

Clifton W. Draper, Department of Chemistry, The Pennsylvania State University, University Park, Pennsylvania 16802

W. Drost-Hansen, Department of Chemistry, Laboratory for Water Research, University of Miami, P.O. Box 249148, Coral Gables, Florida 33124

David W. Dwight, Department of Chemistry, Virginia Polytechnic Institute and State University, Blacksburg, Virginia 24061

W.P. Ellis, University of California, Los Alamos Scientific Laboratory, Los Alamos, New Mexico 87545

Frank M. Etzler, Department of Chemistry, Laboratory for Water Research, University of Miami, P.O. Box 249148, Coral Gables, Florida 33124

John L. Falconer, Department of Chemical Engineering, University of Colorado, Boulder, Colorado 80302

R.S. Farinato, Department of Chemistry, University of Massachusetts, Amherst, Massachusetts 01002

R.M. Fitch, Department of Chemistry, Institute of Materials Science, University of Connecticut, Storrs, Connecticut 06268

M. Fletcher, Marine Science Laboratory, Menai Bridge, Gwynedd, Great Britain LL595EH

M. Folman, Department of Chemistry, Technion, Israel Institute of Technology, Haifa, Israel

Tomlinson Fort, Jr., Department of Chemical Engineering, Carnegie-Mellon University, Schenley Park, Pittsburgh, Pennsylvania 15213

Helga Füredi-Milhofer, Institute "Ruder Boskovic," Zagreb, Yugoslavia

E.L. Fuller, Jr., Chemistry Division, Hollifield National Laboratory, Oak Ridge, Tennessee 37830

C. Gajria, Department of Chemistry, Institute of Materials Science, University of Connecticut, Storrs, Connecticut 06268

R.B. Gammage, Health Physics Division, Oak Ridge National Laboratory, P.O. Box X, Oak Ridge, Tennessee 37830

Robert J. Good, Department of Chemical Engineering, State University of New York at Buffalo, Buffalo, New York 14214

Flemming Yssing Hansen, Fysisk-Kemisk Institut, Technical University of Denmark, DK 2800 Lyngby, Denmark

A.H. Hardin, Energy Research Laboratories, CANMET, Ottawa, Canada

Y. Harnoy, Department of Mechanical Engineering, University of Toronto, Toronto, Canada M5S 1A4

Brian H. Harrison, Department of Chemistry and Chemical Engineering, Royal Military College of Canada, Kingston, Ontario K7L 2W3, Canada

Douglas Henderson, IBM Research Laboratory, Monterey and Cottle Roads, San Jose, California 95193

Marcel Hennenberg, Fysisk-Kemisk Institut, Technical University of Denmark, DK 2800 Lyngby, Denmark

Vladimir Hlady, Institute "Ruder Boskovic," Zagreb, Yugoslavia

Frank J. Holly, Eye Research Institute of Retina Foundation, 20 Staniford Street, Boston, Massachusetts 02114

H.F. Holmes, Health Physics Division, Oak Ridge National Laboratory, P.O. Box X, Oak Ridge, Tennessee 37830

L.E. Iton, Argonne National Laboratory, 9700 South Cass Avenue, Argonne, Illinois 60439

Karl F. Keirstead, Reed Paper Ltd., Pulp & Paper Group, Chemical Division, Lignin Products, P.O. Box 2025, Quebec, Quebec, Canada, G1K 7N1

Arthur S. Kesten, Kinetics and Environmental Sciences, United Technologies Research Center, East Hartford, Connecticut 06108

A.D. King, Jr., Department of Chemistry, University of Georgia, Athens, Georgia 30602

Riki Kobayashi, Department of Chemical Engineering, Rice University, George R. Brown School of Engineering, Houston, Texas 77001

S. Kozar, Institute "Ruder Boskovic," Zagreb, Yugoslavia

Joel L. Lebowitz, Yeshiva University, New York, New York 10033

D.M. Lederman, Medical Group, Avco Everett Research Laboratory, Inc., 2385 Revere Beach Parkway, Everett, Massachusetts 02149

Richard S. Lemons, Department of Chemistry, The Pennsylvania State University, University Park, Pennsylvania 16802

A.J. Léonard, Groupe de Physico-Chimie Minerale et de Catalyse, Université Catholique de Louvain, Place Croix du Sud 1, B-1348 Louvain-la-Neuve, Belgium

S. Levine, Department of Mathematics, University of Manchester, Manchester, M13 9 PL, England

Alvin C. Levy, School of Chemistry, Georgia Institute of Technology, Atlanta, Georgia 30332

N.-J. Lin, Department of Chemical Engineering, State University of New York at Buffalo, Buffalo, New York 14214

G.I. Loeb, Ocean Sciences Division, Naval Research Laboratory (Branch 8354), Washington, D.C. 20375

D.G. Loffler, Department of Chemical Engineering and Materials Science, University of Minnesota, Minneapolis, Minnesota 55455

Philip F. Low, Department of Agronomy, Life Science Building, Purdue University, West Lafayette, Indiana 47907

A. Lubezky, Department of Chemistry, Technion, Israel Institute of Technology, Haifa, Israel

Thomas R. McGregor, Dow Badische Company, Anderson, South Carolina 29621

R.J. Madix, Department of Chemical Engineering, Stanford University, Stanford, California 94305

J. Adin Mann, Jr., Chemical Engineering Department, Case Western Reserve University, Cleveland, Ohio 44106

Pierre J. Marteney, Kinetics and Environmental Sciences, United Technologies Research Center, East Hartford, Connecticut 06108

R. Massoudi, Department of Chemistry, University of Georgia, Athens, Georgia 30602

T. Matsushima, Department of Chemistry, The University of Texas at Austin, Austin, Texas 78712

F.J. Micale, Center for Surface & Coatings Research, Lehigh University, Bethlehem, Pennsylvania 18015

J.D. Miller, Department of Mining, Metallurgical and Fuels Engineering, 412 Mineral Science Building, The University of Utah, Salt Lake City, Utah 84112

M. Modell, Department of Chemical Engineering, Massachusetts Institute of Technology, Cambridge, Massachusetts 02139

B.A. Morrow, Chemistry Department, University of Ottawa, Ottawa, Canada

Richard S. Myers, Delta State College, P.O. Box 3255, Cleveland, Mississippi 38732

G. Neale, Department of Chemical Engineering, University of Alberta, Alberta, Canada

A.W. Neumann, Department of Mechanical Engineering, University of Toronto, Toronto, Canada M5S 1A4

E. Nyilas, Medical Group, Avco Everett Research Laboratory, Inc., 2385 Revere Beach Parkway, Everett, Massachusetts 02149

Robert A. Pierotti, School of Chemistry, Georgia Institute of Technology, Atlanta, Georgia 30332

F. Philips Pike, College of Engineering, University of South Carolina, Columbia, South Carolina 29208

H.M. Princen, Lever Brothers Company, Research Division, 45 River Road, Edgewater, New Jersey 07020

Frederick A. Putnam, Department of Chemical Engineering, Carnegie-Mellon University, Schenley Park, Pittsburgh, Pennsylvania 15213

A.V. Rapacchietta, Department of Mechanical Engineering, University of Toronto, Toronto, Canada M5S 1A4

P. Reed, Department of Mathematics, The University of Manchester, Manchester M13 9PL, England

Miguel F. Refojo, Eye Research Institute of Retina Foundation, 20 Staniford Street, Boston, Massachusetts 02114

L. Rodrique, Groupe de Physico-Chimie Minerale et de Catalyse, Université Catholique de Louvain, Place Croix du Sud 1, B-1348 Louvain-la-Neuve, Belgium

G.J. Roebersen, Rijksuniversiteit Utrecht, Van't Hoff Laboratorium voor Fysische-en Colloidchemie, Padualaan, Utrecht 2506, The Netherlands

Paul Rolniak, Department of Chemical Engineering, Rice University, George R. Brown School of Engineering, Houston, Texas 77001

Allan H. Rosenberg, Bristol-Myers Products, Department of Physical Chemistry, 1350 Liberty Avenue, Hillside, New Jersey 07207

Gerd M. Rosenblatt, Department of Chemistry, The Pennsylvania State University, University Park, Pennsylvania 16802

P.G. Rouxhet, Groupe de Physico-Chimie Minerale et de Catalyse, Université Catholique de Louvain, Place Croix du Sud 1, B-1348 Louvain-la-Neuve, Belgium

R.L. Rowell, Department of Chemistry, University of Massachusetts, Amherst, Massachusetts 01002

Hugo A. Ruiz, Department of Agronomy, Life Science Building, Purdue University, West Lafayette, Indiana 47907

T. Salman, Department of Mining and Metallurgical Engineering, McGill

University, P.O. Box 6070, Station ‘A’, Montreal, Quebec, Canada H3C 3GI

H. Saltsburg, Department of Chemical Engineering, University of Rochester, Rochester, New York 14627

Steven B. Sample, Southern Illinois University at Edwardsville, Edwardsville, Illinois

A. Sanfeld, Pool de Physique et Chimie Physique II, Batiment de Physique, Campus Plaine, Niveau 5, Université Libre de Bruxelles, Bd. du Triomphe, Bruxelles 1050, Belgium

Joseph J. Sangiovanni, Kinetics and Environmental Sciences, United Technologies Research Center, East Hartford, Connecticut 06108

L.D. Schmidt, Department of Chemical Engineering and Materials Science, University of Minnesota, Minneapolis, Minnesota 55455

Malcolm E. Schrader, Code 287, David Taylor Naval Ship Research and Development Center, Annapolis, Maryland 21402

P.O. Scokart, Groupe de Physico-Chimie Minerale et de Catalyse, Université Catholique de Louvain, Place Croix du Sud 1, B-1348 Louvain-la-Neuve, Belgium

Robert D. Shoup, Corning Glass Works, Corning, New York 14830

James A. Singmaster III, Agricultural Experiment Station, University of Puerto Rico, Rio Piedras, Puerto Rico 00928

Torben Smith Sørensen, Fysisk-Kemisk Institut, Technical University of Denmark, DK 2800 Lyngby, Denmark

D. Stanga, Department of Mechanical Engineering, University of Toronto, Toronto, Canada M5S 1A4

A. Steinchen, Pool de Physique et Chimie Physique II, Batiment de Physique, Campus Plaine, Niveau 5, Université Libre de Bruxelles, Bd. du Triomphe, Bruxelles 1050, Belgium

T. Takaishi, Institute for Atomic Energy, Rikkyo University, Yokosuka, Japan 240-01

John A. Tallmadge, Department of Chemical Engineering, Drexel University, Philadelphia, Pennsylvania 19104

T.N. Taylor, University of California, Los Alamos Scientific Laboratory, Los Alamos, New Mexico 87545

Chandrakant R. Thakkar, College of Engineering, University of South Carolina, Columbia, South Carolina 29208

C. Vayenas, Department of Chemical Engineering, University of Rochester, Rochester, New York 14627

A. Vrij, Rijksuniversiteit Utrecht, Van’t Hoff Laboratorium voor Fysische-en Colloidchemie, Padualaan, Utrecht 2506, The Netherlands

P.G. Wahlbeck, Department of Chemistry, Wichita State University, Wichita, Kansas 67208

Eduard O. Waisman, INTI, Buenos Aires, Argentina

E.J. Watson, Department of Mathematics, The University of Manchester, Manchester M13 9PL, England

Charles B. Weinberger, Department of Chemical Engineering, Drexel University, Philadelphia, Pennsylvania 19104

J.M. White, Department of Chemistry, The University of Texas at Austin, Austin, Texas 78712

James P. Wightman, Department of Chemistry, Virginia Polytechnic Institute and State University, Blacksburg, Virginia 24061

D. Ying, Department of Chemical Engineering, Stanford University, Stanford, California 94305

R.H. Yoon, Department of Mining and Metallurgical Engineering, McGill University, P.O. Box 6070, Station 'A', Montreal, Quebec, Canada H3C 3GI

Preface

This is the third volume of papers presented at the International Conference on Colloids and Surfaces, which was held in San Juan, Puerto Rico, June 21–25, 1976.

The morning sessions consisted of ten plenary lectures and thirty-four invited lectures on the following topics: rheology of disperse systems, surface thermodynamics, catalysis, aerosols, water at interfaces, stability and instability, solid surfaces, membranes, liquid crystals, and forces at interfaces. These papers appear in the first volume of the proceedings along with a general overview by A. M. Schwartz.

The afternoon sessions were devoted to 221 contributed papers. This volume includes contributed papers on the following subjects: adsorption, catalysis, solid surfaces, wetting, surface tension, and water. Three additional volumes include contributed papers on aerosols, emulsions, surfactants, hydrosols, rheology, biocolloids, polymers, monolayers, membranes, and general subjects.

The Conference was sponsored jointly by the Division of Colloid and Surface Chemistry of the American Chemical Society and the International Union of Pure and Applied Chemistry in celebration of the 50th Anniversary of the Division and the 50th Colloid and Surface Science Symposium.

The National Colloid Symposium originated at the University of Wisconsin in 1923 on the occasion of the presence there of The Svedberg as a Visiting Professor (see the interesting remarks of J. H. Mathews at the opening of the 40th National Colloid Symposium and also those of Lloyd H. Ryerson in the Journal of Colloid and Interface Science 22, 409 412 (1966)). It was during his stay at Wisconsin that Svedberg developed the ultracentrifuge, and he also made progress on moving boundary electrophoresis, which his student Tiselius brought to fruition.

The National Colloid Symposium is the oldest such divisional symposium within the American Chemical Society. There were no meetings in 1933 and during the war years 1943–1945, and this lapse accounts for the 50th National Colloid Symposium occurring on the 53rd anniversary.

However, these circumstances brought the numerical rank of the Symposium into phase with the age of the Division of Colloid and Surface Chemistry. The Division was established in 1926, partly as an outcome of the Symposium. Professor Mathews gives an amusing account of this in the article cited above.

The 50th anniversary meeting is also the first one bearing the new name

Colloid and Surface Science Symposium to reflect the breadth of interest and participation.

There were 476 participants including many from abroad.

This program could not have been organized without the assistance of a large number of persons and I do hope that they will not be offended if all of their names are not acknowledged. Still, the Organizing Committee should be mentioned: Milton Kerker, Chairman, Paul Becher, Tomlinson Fort, Jr., Howard Klevens, Henry Leidheiser, Jr., Egon Matijevic, Robert A. Pierotti, Robert L. Rowell, Anthony M. Schwartz, Gabor A. Somorjai, William A. Steele, Hendrick Van Olphen, and Albert C. Zettlemoyer.

Special appreciation is due to Robert L. Rowell and Albert C. Zettlemoyer. They served with me as an executive committee which made many of the difficult decisions. In addition Dr. Rowell handled publicity and announcements while Dr. Zettlemoyer worked zealously to raise funds among corporate donors to provide travel grants for some of the participants.

Teresa Adelmann worked hard and most effectively both prior to the meeting and at the meeting as secretary, executive directress, editress, and general overseer. She made the meeting and these Proceedings possible. We are indebted to her.

Milton Kerker

EFFECTS OF SURFACE WATER ON THE ADSORPTION OF INERT GASES*

R. B. Gammage and H. F. Holmes
Oak Ridge National Laboratory

ABSTRACT

Molecular water on mildly outgassed open oxide surfaces can, under special circumstances, significantly reduce the monolayer capacity for inert gas molecules. Non-porous thoria and cubic europia are two examples. The special proviso is that the surface be sufficiently uniform to permit the formation of a well structured, ice-like surface with non-polar character. This type of behavior reaches an extreme for a non-porous ground calcite; surface water converts a Type II isotherm for krypton into a Type III. These structured layers of adsorbed water adsorb nitrogen in a manner very similar to that found for various types of ice. For open oxide surface in a more thoroughly outgassed condition, the BET specific surface area does not generally vary with outgassing temperature at and above 150°C. The area occupied by an inert gas molecule does not change, therefore, with varying hydroxyl ion content of the surface.

I. INTRODUCTION

Even the serious reader of the literature on inert gas adsorptions by oxides and other polar solids is likely to be baffled by claims made for the effects of residual water. Karasz et al. (1) were among the first to demonstrate that molecular water on standard anatase surface significantly lowers the uptake of nitrogen and argon; adsorption of argon onto a sample of silver iodide, however, was not affected by the presence or absence of molecular water.

There is also confusion as to whether or not the concentration of surface hydroxyl groups can influence the apparent specific surface area. The few available examples given in Table 1 suggest that sometimes it does, sometimes it doesn't; in each instance the author claims that the deviations in specific surface, where they occur, arise from a changing occupancy area of the adsorbed nitrogen molecule.

---

*Research sponsored by the U. S. Energy Research and Development Administration under contract with Union Carbide Corporation.

TABLE 1
Effect of Outgassing Temperature on the Specific Surface Area by Nitrogen or Argon Adsorption for Several Different Oxides

	Specific Surface (M^2/g)									
Outgas. Temp. (°C)	Silica Gel KSK-2 (Ref.2)		Silica SB (Ref.3)	Silica SL (Ref.3)	Rutile (Ref.5)		Lunar Fines Apollo 16 (Ref.4)	ThO_2 Sample I (Refs.6,7)		Cubic Europium Oxide
	N_2	Ar	N_2	N_2	N_2	Ar	N_2	N_2	Ar	N_2
25					32	32	0.42	2.39	2.38	5.36
50					32	32	0.43			
100			334	593	32	32		2.74	2.52	
150	320	316			35	34				
200			341	611	36	36	(0.39)	2.70	-	
250					37	36				
300			337	637			0.50	2.70	-	5.72
350					37	37				
400	320	316	339	646			0.56			
450					37	36				
500							0.50	2.70	2.69	
600	320	-			17	17	0.50			
700					10	10	0.45			
800	314	319								

In studying polar solids [thoria (7), europia (8), lunar fines (4) and calcium carbonate (9)] we have been faced with the problem of deciding whether or not, and by how much, water or the elements of water are affecting the adsorption of nitrogen, argon and krypton. The purpose of this paper is to shed more light on these problems and pinpoint the circumstances under which surface water alters the adsorption characteristics. Several new, and several previously published, results are given in order to achieve this purpose.

II. EXPERIMENTAL

The oxides and calcium carbonate examined in this study were all low, or relatively low-area (< 10 m^2g^{-1}), non-porous solids. The criteria for non-porosity were a reversible inert gas adsorption isotherm with rapid kinetics of adsorption and t-plots which extrapolated to the origin. In the case of the thoria sample I, heat of immersion data in water confirmed the non-porous nature (10).

The thoria, I-ThO_2 (11), originated from thorium oxalate and was calcined at 1200°C. The cubic europia was obtained from the Molybdenum Company of America in high purity form. The lunar fines 63321 were returned from the Apollo 16 mission. They were rich in feldspar (∿ 30% Al_2O_3) and had "matured" through the combined effects of solar wind radiation damage and meteorite bombardment (12). The calcium carbonate sample was an Iceland spar ball-milled for 1000 hours, whose properties have been extensively studied (13)(14).

Measurements of the adsorption of nitrogen and argon were made on microbalance systems suited to the handling of low area solids (15), while the sorption onto calcium carbonate was measured volumetrically using ^{85}Kr - labeled krypton (16).

III. RESULTS AND DISCUSSION

A. Molecular Adsorbed Water

The adsorption of nitrogen and argon onto standard anatase is increasingly reduced by the first, second and third layers of preadsorbed water (1). This section of the present paper is addressed to ionic adsorbents outgassed at, or slightly above, room temperature. Can molecular surface water not removed in the mild outgassing significantly reduce the adsorption of an inert gas compared to adsorption on the well outgassed surface? It will be shown that the answer is yes, in special circumstances, but that the magnitude of the effect depends on the nature of the adsorbent and the structural perfection of the adsorbed water.

We have experienced two unambiguous examples of drastic effects of surface molecular water on the magnitude and energetics of inert gases adsorbed on two different kinds of non-porous, ionic materials.

The first example is a ground calcite which had been structurally degraded in plastic flow processes (13). Although extremely reactive, because of a high density of imperfections including 2% by weight of buried water taken up during the milling for 1000 hours, the flowed grains had been rendered non-porous to inert gas molecules (14). The hitherto unpublished isotherms of krypton shown in Fig. 1 were measured on the same specimen of this ground calcite. One isotherm was measured immediately after, the other 48 hours after cessation of prolonged outgassing in vacuo at 150°C. The effect of the time delay of 48 hours on the subsequent adsorption of krypton was depression of the BET c constant from a value of 800 to 2 (Type II → Type III isotherm). A similar, but less

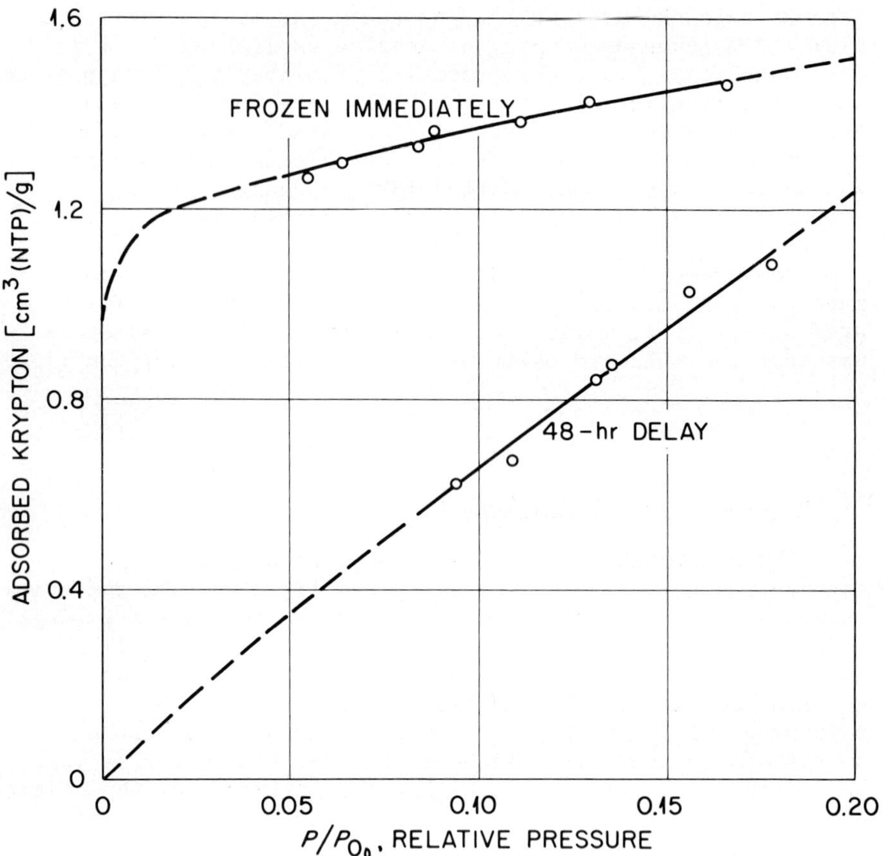

Fig. 1. Krypton adsorbed at 77°K on a single sample of calcite ground for 1000 hours; the solid was frozen either immediately after, or after 48 hours at room temperature, following outgassing at 150°C.

pronounced, change in the adsorption, Fig. 2, was observed with an identically outgassed sample of precipitated calcite with 3% by weight of occluded water. The c constant this time was reduced to 21. In both instances the most probable cause of the changes in adsorption behavior is water diffusing from the bulk to the surface after cessation of the outgassing at 150°C with structuring to present an annealed, or partially annealed, ice-like exterior to the adsorbing krypton.

In general, for various specimens of calcium carbonate containing water, unless the well outgassed solid was frozen immediately to prevent diffusion of water, anomalous adsorption behavior and erroneous and varying specific surface

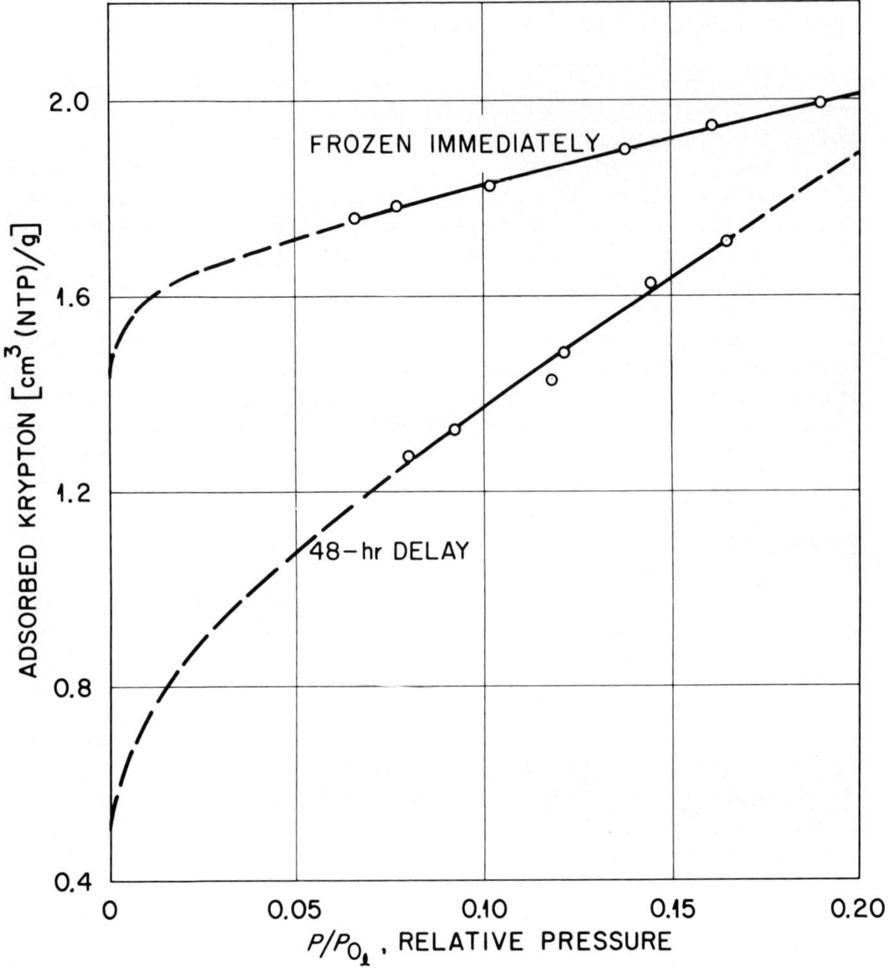

Fig. 2. Krypton adsorbed at 77°K on a single sample of precipitated calcite; the solid was frozen either immediately after, or after 48 hours at room temperature, following outgassing at 150°C.

measurement would ensue. However, adsorption of krypton on ten aliquots of the ground calcite (1000 hours) in the well outgassed condition yielded a specific surface of 6.92 m^2/g, constant within a standard deviation of 3%.

Adsorptions of butane at 0°C have already been reported (13) for the ground calcite (1000 hours). In very much the same way as for krypton adsorption, there was a conversion from a Type II to a Type III isotherm. At that time the resemblance to the isotherm for alkanes adsorbed on the surface of liquid water (17) was noted.

The second example concerns high-fired (1200°C), non-porous, low specific surface thorium oxide (sample I of reference 7) possessing a considerable degree of surface homogeneity. In addition to reversible, stepped isotherms of nitrogen and argon (7), net differential heats of adsorption, derived from heat of immersion measurements, indicate two distinct layers of molecular water bound to the fully hydroxylated surface with net enthalpy values of -8.6 and -1.8 kcals/mole (10). The formation of these discrete layers of water occurs very rapidly in aqueous media (10). Reaching the equilibrium state through adsorption of water vapor is, in contrast, a very laborious process requiring many sorption cycles to relative pressures approaching saturation (18). The apparent specific surface, determined by nitrogen adsorption, is 2.39 m^2/g after carefully equilibrating with water vapor and outgassing at 25°C (7) compared to 2.70 m^2/g for the surface outgassed at 500°C, a 13% increase. Slightly lesser amounts of irreversibly bound and less well-structured molecular water give intermediate values for the specific surface; 2.51 m^2/g was an earlier published result (19) where less effort was made to ensure the equilibrium condition.

The n value plots for nitrogen and argon adsorbed on the most carefully equilibrated water-thoria system are shown in Fig. 3. The isotherms are striking for their overlap; the slight divergence below a P/P_0 of 0.4 is associated with the lower BET c constant of 22 for argon compared to 33 for nitrogen. That one is, in effect, measuring inert gas adsorption on ice is confirmed by inclusion in Fig. 3 of the nitrogen isotherm measured by Adamson *et al.* on a non-annealed ice, designated 2A (20), for which the BET c constant was a comparable value of 39. The coincidence is nearly exact. It was, in fact, Adamson and Dormant (21) who first recognized the similarity between isotherms of nitrogen for ice and water covered anatase (1) mentioned previously.

Fully annealed ice samples studied by the Adamson group (20)produced Type III isotherms of the sort obtained for krypton adsorbed on water-covered, ground calcite (Fig. 1). The degree of ordering of the molecular water on the polar surface (oxide or carbonate) seems to control the extent to which the inert gas adsorption isotherm is changed. To bring the specific surface of the water covered thoria (2.39 m^2/g) into line with the "real" specific surface of 2.70 m^2/g, obtained after high temperature outgassing, requires a value

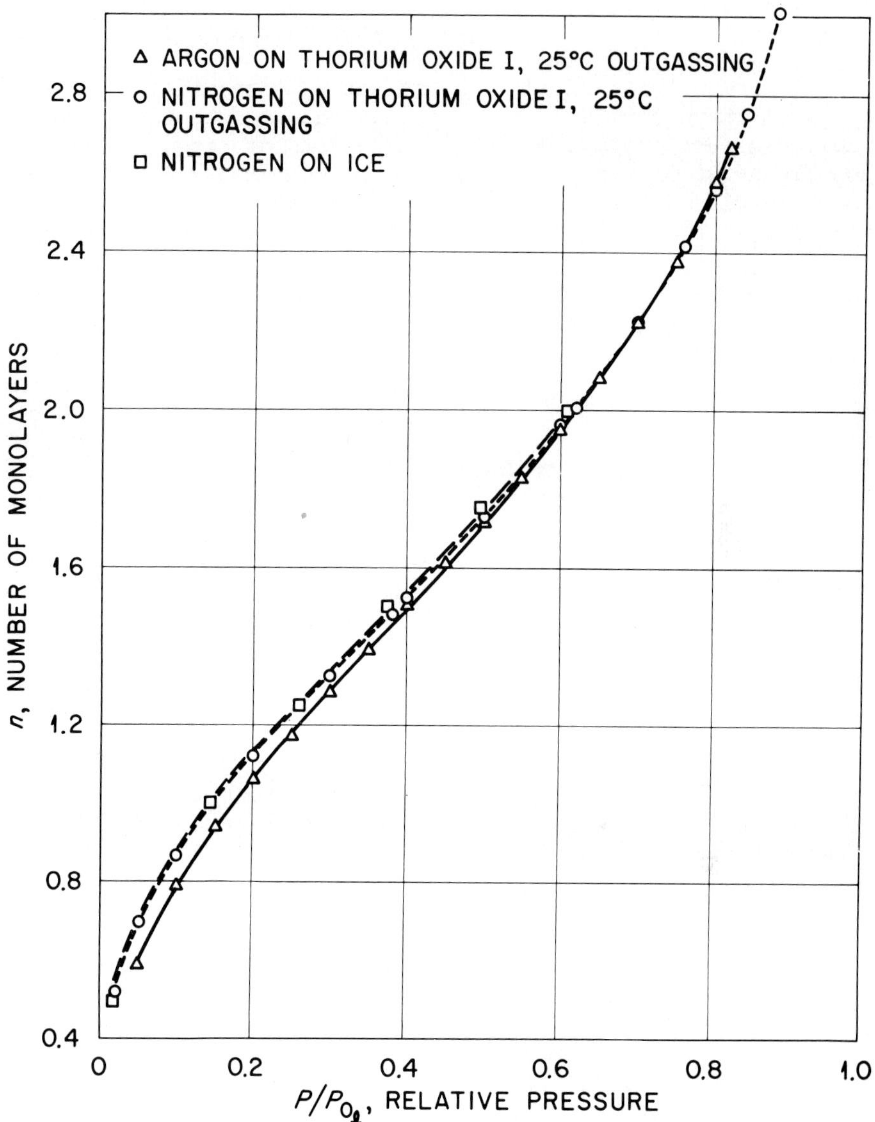

Fig. 3. n-plots of nitrogen and argon adsorbed at 77°K on uniform, non-porous thoria (Sample I) with an equilibrium amount of surface water for the 25°C outgassed condition, together with an n-plot for nitrogen on an ice which exhibits a like BET c constant (ice 2A of ref. 20).

of 18.3 Å^2 per nitrogen molecule. Another example of larger than the normal cross-sectional area of 16.2 Å^2 for nitrogen is the well known 20 Å^2 for graphitized carbon (22).

To demonstrate vividly the effect on adsorption of changing from the non-polar, ice-like substrate to the highly polar thoria surface outgassed at 500°C, n-plots of argon and nitrogen are shown in Fig. 4, and should be compared against those shown in Fig. 3. The slight step structure in the nitrogen isotherm we attribute to surface homogeneity. Reasons for

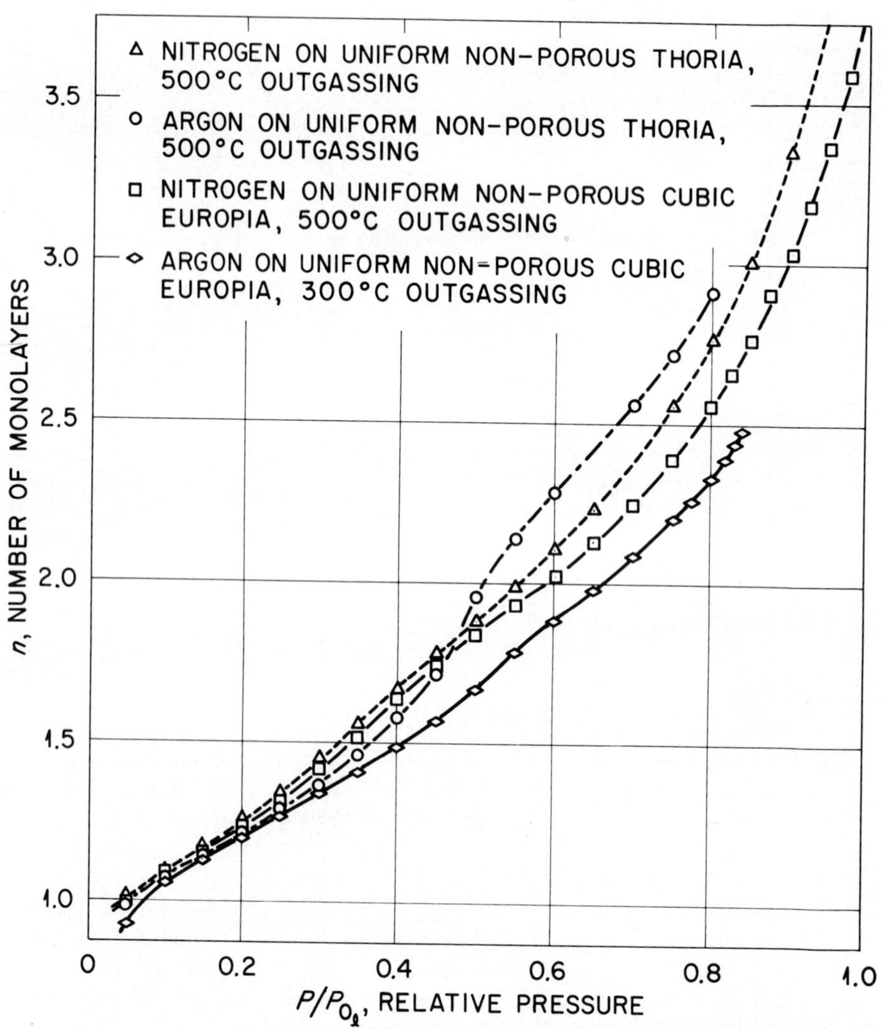

Fig. 4. n-plots of nitrogen and argon adsorbed at 77°K on uniform, non-porous thoria and cubic europia outgassed at 500°C.

the stepped structure of the argon isotherm lie beyond the scope of this paper.

Another example of inert gas adsorption being influenced by adsorbed molecular water is the absorption on a non-porous, cubic europium oxide, again with some homogeneous surface character. The effects, however, were less pronounced than those with the thorium oxide sample I. Increase in outgassing temperature from 25° to 300°C produced an apparent increase in specific surface of 7% (from 5.36 to 5.72 m^2/g) compared to 13% for the thorium oxide example which has been discussed. One surmises that the degree of oxide surface homogeneity and degree of structural perfection in the adsorbed water layer is less for the europia than the thoria.

Confirming the non-porous character of the europia, the isotherms of the nitrogen and argon were reversible over the entire range of relative pressure. The n-plots of nitrogen and argon are reproduced in Fig. 4. The validity of the steps and the reproducibility were checked in repeated isotherm measurement. The europia had been subjected in a prior experiment to adsorption of water vapor at 25°C. This was followed by "clean-up" in oxygen at 300° or 500°C. The similarities to the n-plots obtained for the thorium oxide are quite striking. In spite of these characteristics and the apparently firm indication that surface molecular water is reducing the monolayer capacity for nitrogen, we choose not to confer the same high degree of "unambiguousness" that we attach to the thoria and ground calcite examples. This is because a quite massive interaction with adsorbed water vapor had taken place with irreversible binding of water to the extent of 70 OH/100 $Å^2$ on (100) faces of the cubic europium oxide (23). The issue is clouded because of possible real area changes occurring in the extensively hydrated oxide substrate upon increase of the outgassing temperature.

The lunar fines also fall into the category of being less than totally satisfactory in exemplifying definite effects on apparent specific surface caused by molecular water. There are three reasons. The specific surface of the fines 63321 is about 0.5 m^2/g and the anticipated maximum variation with outgassing temperature of about 10% is only 0.5 m^2/g. Since the experimental error is ± 0.02 m^2/g the changes we seek are barely discernible from experimental inaccuracy. The BET c constant is about 150 for surface outgassed at 25°C, which hardly approaches the value of 2 for the ground calcite with its annealed, ice-like layer of surface water. The layer of adsorbed water on lunar fines is, in all probability, insufficiently ordered to produce a more open than normal monolayer of adsorbed nitrogen molecules. Finally, lunar fines are rich in buried solar wind, principally hydrogen in the form of small gas bubbles (12). This gas diffuses out of the grains at > 150°C and poses the possibility of real surface area changes taking place due to structural alteration.

The data in Table 1 for rutile (5) are also regarded as unsatisfactory for explanation in terms of well-structured water layers. The acknowledged porosity leaves a possibility that micropores may be present which can remain blocked by water in an outgassing at 100°C. The ∿ 15% increase in apparent specific surface after outgassing at 200°C would then come about from an opening up of these micropores.

It is of value to discuss these various findings with respect to the now commonly used standard or universal isotherm of nitrogen on non-porous oxides. In the general case of a high specific surface, oxide having a heterogeneous but still non-porous surface, it would be unlikely for firmly bound molecular water to become sufficiently ordered and non-polar to affect significantly the standard nitrogen isotherm. This, in our judgment, is one of the reasons for the successful operation of the α_s (24) and t-plot methods of analysis (25). However, as a general principle, it is advisable to use well-outgassed surfaces for obtaining reference isotherms. There is then no possibility of well-structured water layers being present to give unusual behavior in the adsorption of nitrogen.

B. Surface Hydroxyl Groups

Molecular water strongly adsorbed in the micropores of a finely pored oxide blocks the entry of adsorbed inert gas molecules and reduces the apparent specific surface area. To drive off this molecular water completely requires outgassing temperatures ranging from about 200°C for microporous samples of silica (26), calcium suphate (27) and alumino-silicates of lunar origin (28), to about 400°C for a microporous thorium oxide (29). Where any increase in specific surface of an oxide is noted one cannot attribute the increase to changes in the specific interactions between hydroxyl groups and nitrogen molecules unless one is certain there are no micropores to complicate the issue. We now consider the data of Table 1 with this proviso in mind, especially the data relating to outgassing temperatures of 150°C or more. At 150°C all, or a major fraction, of strongly bound surface molecular water will be removed from the open surfaces of these oxides. The surfaces will still have essentially their full complement of surface hydroxyl groups.

HIGH PRESSURE ADSORPTION OF METHANE ON 5A MOLECULAR SIEVE ZEOLITE BY ELUTION CHROMATOGRAPHY

Paul Rolniak and Riki Kobayashi
William Marsh Rice University

I. ABSTRACT

The adsorption of methane on Linde 5A molecular sieve zeolite pellets has been studied by gas chromatography from 344 to 8963 kPa at 288, 298, and 308 K. The gas chromatographic technique yields both the Gibbs, or differential, adsorption and the absolute adsorption. The intercrystalline gas volume was determined by helium perturbation of a carbon dioxide carrier gas. The data is correlated by the Dubinin-Astakhov equation using a saturation fugacity corresponding to a gas phase density equal to the limiting adsorption density. Using this correlation, the heats of adsorption and the adsorbed phase entropies were calculated as functions of the amount of adsorption. In the range of pressures studied, the adsorbed phase behaves like a highly compressed gas.

II. INTRODUCTION

Molecular sieve zeolites are used as adsorbents in the natural gas processing industries. Low pressure, pure component isotherms have been extensively studied (1,2). However, processing pressures are generally far above the low pressure conditions where most data are taken. As part of a continuing effort at this lab in high pressure adsorption, the adsorption behavior of methane on 5A molecular sieve pellets has been studied.

At high pressures, differences in the definitions of adsorption can be seen. The Gibbs definition of adsorption, a differential adsorption, is the excess of the adsorbed species over that which would be in the adsorption volume if the adsorbent had no intermolecular force. This definition of adsorption is a natural result of a thermodynamic analysis of the system. The absolute adsorption is the total amount of the adsorbed species in the adsorption volume. This definition of adsorption is generally the type of adsorption described by adsorption correlations.

Elution chromatography is perfectly suited to high pressure adsorption studies because it yields both types of adsorption. In this technique, an adsorbable carrier gas, e.g. methane, is flowed through a column packed with the zeolite, brought to equilibrium at constant temperature and pressure, and then pulsed with an injection of radioactive methane. From the retention time of the radioactive peak, the adsorption properties can be calculated.

The basic chromatographic equation used in this work can be found by solving the appropriate partial differential equation describing the process for the first chromatographic moment (3,4). The result is

$$A_{abs} = \left[\frac{t_r \dot{V} T_c Z_c}{T_p Z_p} - V_g\right] \frac{P}{Z_c R T_c m} \tag{1}$$

$$A_{Gibbs} = \left[\frac{t_r \dot{V} T_c Z_c}{T_p Z_p} - V_G\right] \frac{P}{Z_c R T_c m} \tag{2}$$

III. APPARATUS

A simplified schematic of the apparatus is shown in Fig. 1. Central to the design of the apparatus is the use of a tandem proportioning Ruska pump of 1000 cc total volume. This pump was used to generate the very stable, low flow rates required in high pressure work. The flow rate capability of the pump varies from 2.5 to 560 cc/hr. Two chromatographic valves from Valco Instruments were mounted inside the constant temperature bath. One, a ten port valve, was used as a double injection valve. One side of this valve was used for helium injections; the other, for tritiated methane injections. The other chromatographic valve was a six port valve used as a switching valve between the packed column and the by-pass column. The empty by-pass column is used as a reference to subtract out system dead volume and detector response time effects.

A Gow Mac 470 thermal conductivity cell was used to detect concentration perturbations of helium and neon. Signals from the tritiated methane injections were detected with a 2.34 ml ionization chamber (5) using a Cary 401 electrometer. Typically, the magnitude of the radioactive signal was in the 10^{-14} to 10^{-15} ampere range.

The carrier gases employed were Linde UHP methane, Linde Coleman grade carbon dioxide, and Bureau of Mines high purity

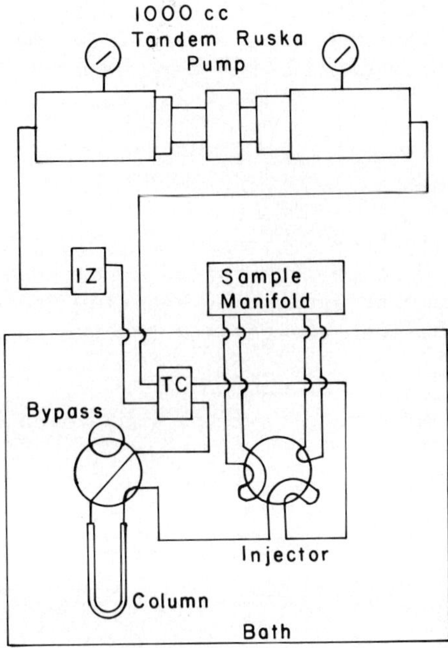

Fig. 1. Simplified schematic of chromatographic apparatus. TC - thermal conductivity cell. IZ - ionization chamber.

helium. These gases were passed over a bed of activated 13X pellets to further dry them. Matheson research grade neon was used as a perturbation gas. The adsorbent was Linde 5A molecular sieve zeolite pelletized with about 20 percent clay binder. Physical properties for the methane were calculated from the BWRS equation of state (6).

IV. PROCEDURE

The molecular sieve pellets were crushed and sieved to 40-60 mesh and packed into a 0.47 cm ID by 30.48 cm long stainless steel column. After determining the activated weight of the adsorbent, 3.100g, the column was installed on the apparatus and reactivated at 400°C. and less than 10^{-4} torr for more than eight hours. The system was then filled with helium carrier gas and the total free gas volume in the column was determined by perturbations of non-adsorbed neon. Next, carbon dioxide was used as the carrier gas with helium perturbations to determine the extra crystalline volume (carbon dioxide occupies all the internal crystalline volume).

Finally, after reactivation of the zeolite to remove the carbon dioxide, retention times of tritiated methane perturbations of a methane carrier gas were used to calculate the methane adsorption isotherms at 15, 25, and 35°C.

V. THERMODYNAMICS

The following thermodynamic analysis follows that of Hill (7). Assuming that the adsorption system can be treated as a pseudo one component system, by equating the change in chemical potential between the gas and the adsorbed phases, then

$$d\mu_1 = -S_s\, dT + v_s\, dP + \frac{N_a}{N_1}\, d\phi = d\mu_g = -S_g\, dT + v_g\, dp \qquad (3)$$

$$(S_g - S_s) = (v_g - v_s)\left(\frac{\partial p}{\partial T}\right)_\phi \qquad (P = p) \qquad (4)$$

$$q_\phi = (S_g - S_s)T = v_g\, T\left(1 - \frac{v_s}{v_g}\right)\left(\frac{\partial p}{\partial T}\right)_\phi \qquad (5)$$

$$q_\phi = \left(1 - \frac{v_s}{v_g}\right) v_g T \left(\frac{\partial p}{\partial f}\right)_{T,\phi} \left[\left(\frac{\partial f}{\partial T}\right)_\phi + \left(\frac{\partial f}{\partial p}\right)_{T,\phi} \left(\frac{\partial p}{\partial T}\right)_{f,\phi}\right] \qquad (6)$$

$$q_\phi = \left(1 - \frac{v_s}{v_g}\right)\left[-R\left(\frac{\partial\, \ln f}{\partial\, 1/T}\right)_\phi + H_g - H_g^*\right] \qquad (7)$$

At constant temperature, ϕ can be calculated,

$$d\phi = \left(1 - \frac{v_s}{v_g}\right) v_g \frac{N_1}{N_a}\, dp \qquad (8)$$

$$\phi = \int_0^f (N_1 - V_s\, C_g) \frac{RT}{N_a}\, d\, \ln f \qquad (9)$$

The quantity, $\frac{1}{N_a}(N_1 - V_s\, C_g)$, is the Gibbs adsorption isotherm, and so by calculating ϕ, then q_ϕ and S_s can be found. It should be noted that q_ϕ goes to zero as the gas phase molar volume approaches that of the adsorbed phase.

VI. RESULTS

The total gas volume of the column, V_G, was found to be 4.14 cc. The extra-crystalline volume, V_g, was found to be 3.53 cc. The resulting intracrystalline volume, 0.196 cc/g, agrees well with the data of other investigators (8) for pure crystals of 5A zeolite when the correction for the 20 percent binder is taken into account.

The absolute adsorption isotherms of methane on 5A molecular sieve pellets is shown in Fig. 2. The corresponding

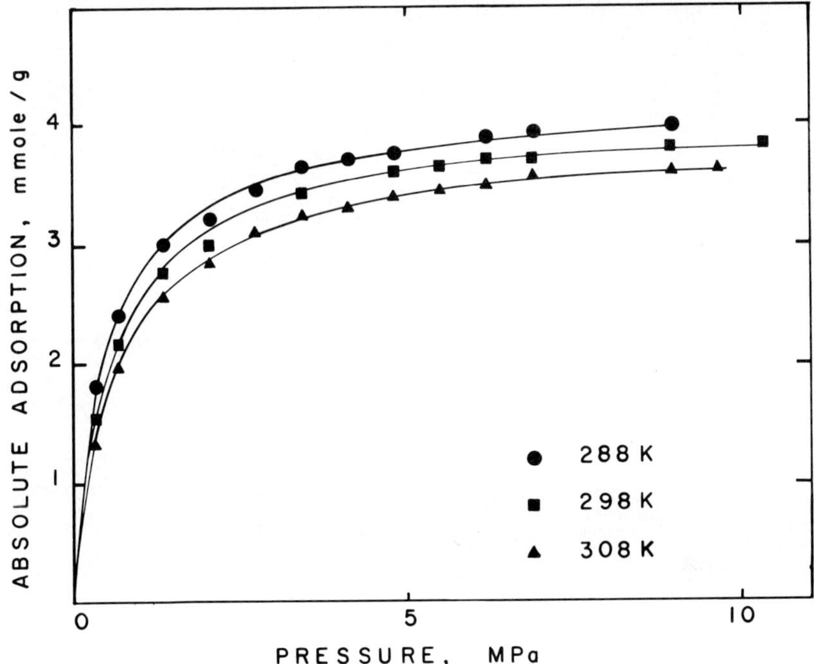

Fig. 2. Absolute adsorption isotherms for methane-5A molecular sieve system. Solid curves drawn according to Equation 10.

Gibbs adsorption isotherms are shown in Fig. 3. A comparison of the absolute and Gibbs adsorption is shown in Fig. 4 along with data taken with a microbalance (9). It is important to note that at high pressures the two types of adsorption diverge.

A useful correlation often used is the Dubinin-Astakhov equation (10).

$$A_{abs} = A_o \exp -\left(\frac{RT}{E} \ln f_s/f\right)^{\eta} \tag{10}$$

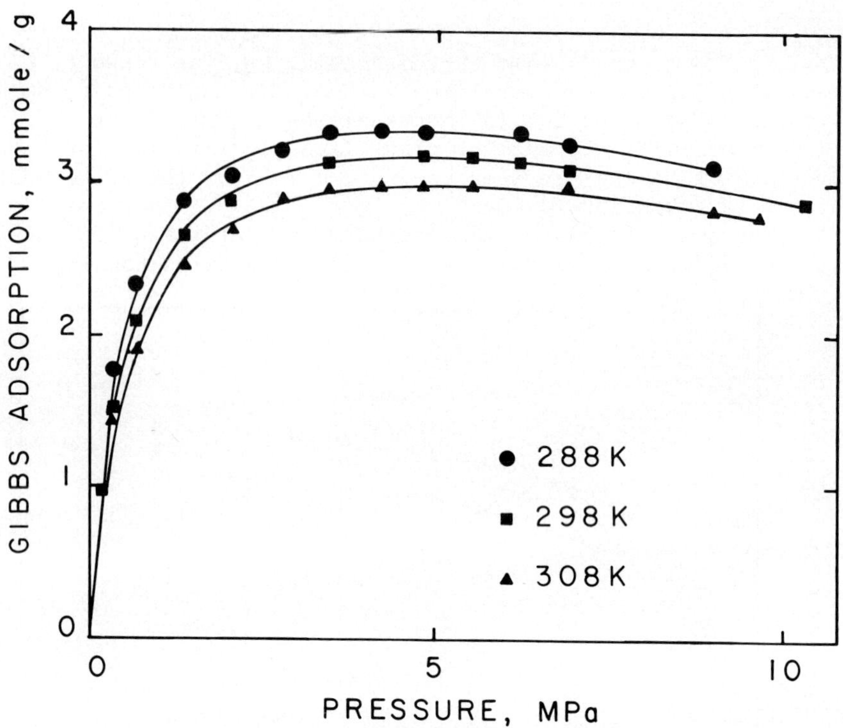

Fig. 3. Gibbs adsorption isotherms for methane-5A molecular sieve system.

In applying equation (10) to adsorption of gases far above their critical, the definition of f_s becomes troublesome. Previously, f_s was found by extrapolating the vapor pressure curve into the supercritical region (10). At that point, however, the gas phase properties bear no relation to the adsorbed phase properties. This situation is conveniently taken care of by defining f_s as occurring at the pressure where the gas phase density equals the adsorbed phase density at that particular temperature. Fig. 5 is a plot of the characteristic curve for methane adsorption on 5A molecular sieve. The parameters for the correlation are given in Table 1. Also, the solid curves in Fig. 2 and 3 are drawn according to this correlation.

The integration of equation (9) was carried out numerically using the correlation. The corresponding heats of adsorption, q_ϕ, and entropies, S_s, were then calculated and are plotted in Fig. 6 and 7 respectively.

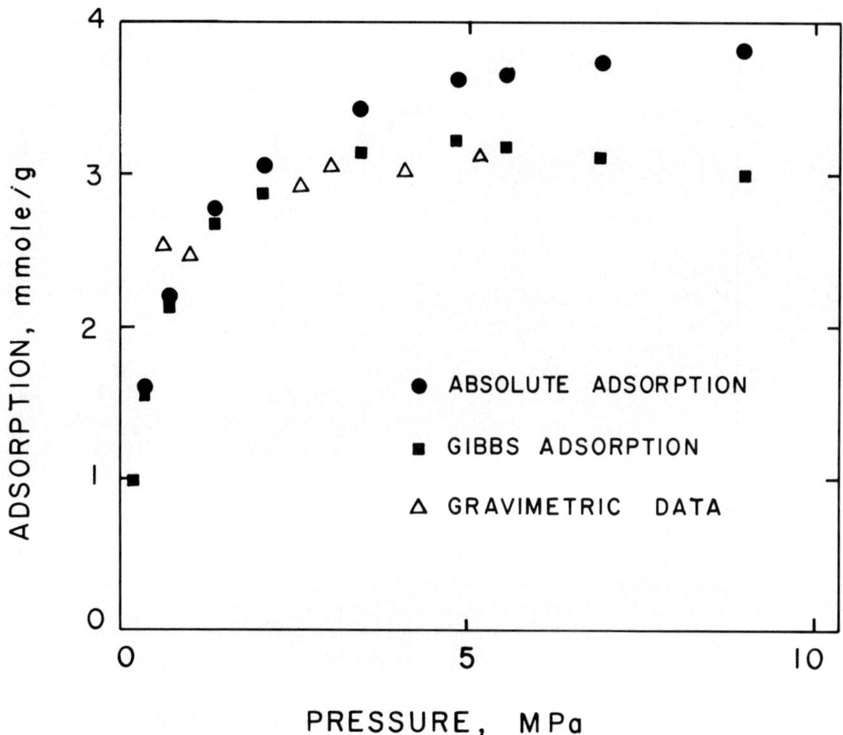

Fig. 4. Comparison of absolute and Gibbs adsorption at 298K for methane-5A molecular sieve system.

TABLE 1

Parameters of Dubinin-Astakhov equation to describe methane adsorption on 5A molecular sieve pellets

T, K	A_o, *mmole/g*	f_s, *psia*
288	4.206	11230
298	4.009	10560
308	3.800	10100

E = 3240 cal/gmole, η = 3.33

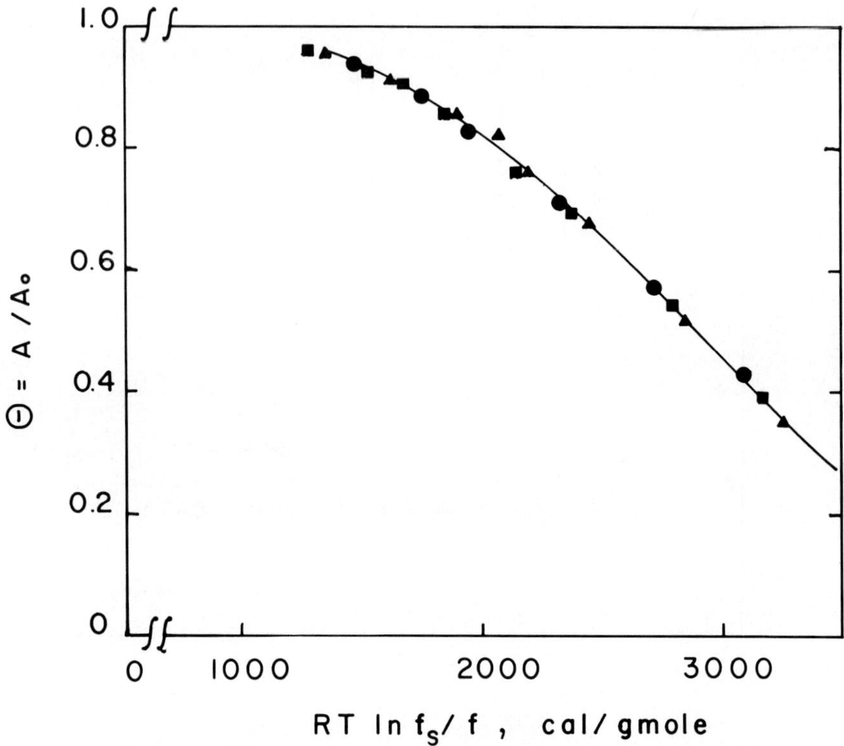

Fig. 5. Dubinin plot of methane-5A absolute adsorption data. Solid curve drawn according to Equation 10.

VII. DISCUSSION

The recent statistical thermodynamic approach of Stroud et al. (2) to methane adsorption on 5A synthetic zeolite met with only limited success. The statistical thermodynamic model of Ruthven and Loughlin (11) performs well up to about half saturation. This leads to interesting speculation of the character of the adsorbed phase above the critical temperature. From Table 1, the limiting adsorption values, A_o, certainly represent liquid-like densities, 19.38 to 21.45 mole/l, and are temperature dependent. The adsorbed phase entropies represent the entropies of the adsorbate molecules in the intense potential fields inside the 5A cavities. From Fig. 7, these entropies follow the gas phase behavior fairly closely -- only at a lower value. At this temperature, these values of density and entropy correspond fairly closely to highly compressed methane gas in the 2,000 to 10,000 psia region, which is also represented in Fig. 7. Uncertainties in the

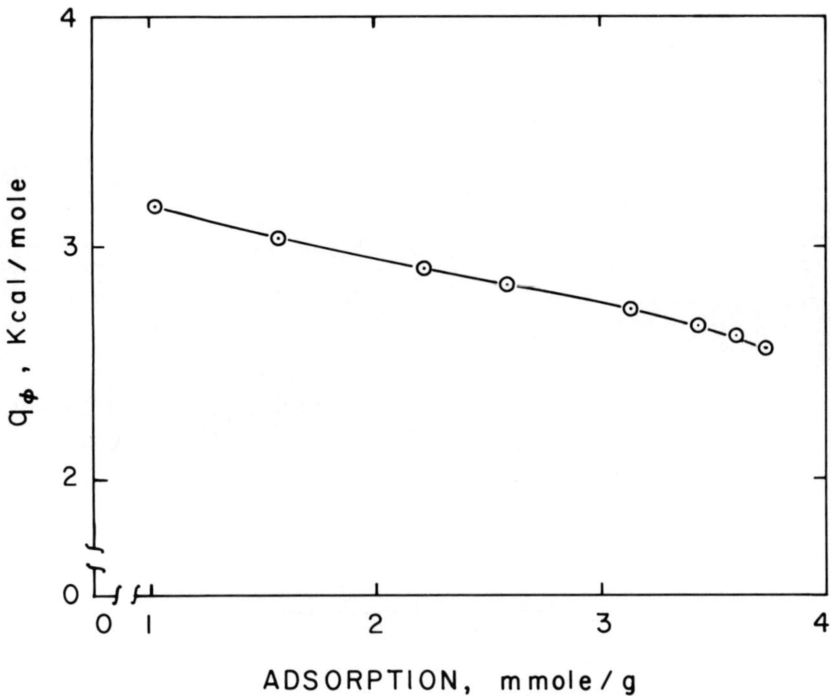

Fig. 6. Heat of adsorption of methane on 5A molecular sieve.

exact adsorbed phase density and also the effect of the adsorbent potential field limit the accuracy of the comparison.

VIII. CONCLUSIONS

High pressure elution chromatography with radioactive tracers yields useful data in systems where normal gravimetric and volumetric methods cannot be conveniently applied. With a slight modification, the Dubinin-Astakhov equation has been found to be quite useful in correlating absolute adsorption data for methane adsorption on 5A molecular sieve. The adsorbed phase at the conditions studied shows properties similar to a highly compressed gas.

IX. LIST OF SYMBOLS

A = adsorption amount
C = concentration

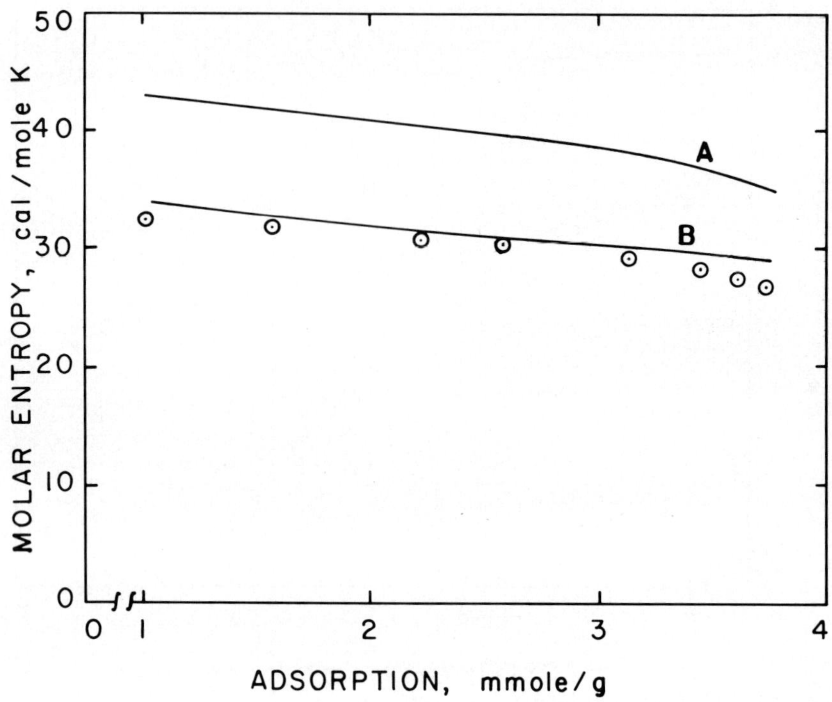

Fig. 7. Comparison of molar entropy of adsorbed methane phase (⊙) with equilibrium gas phase (curve A) and with highly compressed gas at the same temperature and density as the adsorbed phase (curve B).

E = characteristic energy
f = fugacity
H = molar enthalpy
m = mass of adsorbent
N = number of moles
p = gas phase equilibrium pressure
P = adsorbed phase equilibrium pressure
q_ϕ = heat of adsorption at constant ϕ
R = gas constant
S = molar entropy
t_r = retention time of tritiated methane peak
T = absolute temperature
v = molar volume
V = volume
Z = compressibility factor for gas
ϕ = change in chemical potential of adsorbent
μ = chemical potential

η = exponent of equation (10)

Subscripts

a = adsorbent
abs = absolute
c = column
g = gas phase
G = total gas phase
$Gibbs$ = Gibbs or differential
P = pump
o = saturation state
1 = adsorbate

Superscripts

$*$ = low pressure ideal gas reference state

X. REFERENCES

1. Breck, D. W., "Zeolite Molecular Sieves," John Wiley, New York, 1974.
2. Stroud, H. J., Richards, E., Limcharoen, P., and Parsonage, N. G., J. Chem. Soc., Fara. Trans. I, **72** (4), 942 (1976).
3. Horn, F. J. M., AIChE J. **17** (3), 613 (1971).
4. Hattaway, D. R., Ph.D. thesis, Rice University, October 1973.
5. Chu, T. C., Ph.D. thesis, Rice University, August 1972.
6. Starling, K. E., "Fluid Thermodynamic Properties for Light Petroleum Systems," Gulf Publishing Company, Houston, Tx., 1973.
7. Hill, T. L., J. Chem. Phys. **18** (3), 246 (1950).
8. Breck, D. W., and Grose, K. W., Advan. Chem. Ser. **121**, 319 (1973).
9. Lederman, P. B., Ph.D. thesis, University of Michigan, March 1971.
10. Bering, B. P., Dubinin, M. M., and Serpinsky, V. V., J. Colloid Interface Sci. **38** (1) 185 (1972).
11. Ruthven, D. M., Loughlin, K. F., and Holborrow, K. A., Chem. Eng. Sci. **28**, 701 (1973).

The National Science Foundation provided financial support for this research.

MECHANISM FOR THE DESORPTION PROCESS FOR CESIUM AND CESIUM HALIDES FROM NICKEL SURFACES

by P. G. Wahlbeck
Wichita State University

Desorption kinetics for Cs, CsCl, and CsI adsorbed on Ni single crystal surfaces have been determined by using a pulsed molecular beam technique. The temperatures of the Ni surfaces were varied to study the temperature dependence of the mean adsorption lifetimes (reciprocals of first order rate constants). A mechanism for the desorption process is proposed which utilizes motion of adsorbed species through the sequence of sites: kink site, ledge site, terrace site, desorption site, and the gas phase. As the experiments were performed, the species initially were placed principally on terrace sites. The rate determining step is the motion from the terrace site to the desorption site involving surface migration. The desorption site is a site at which the binding energy of the adsorbed species is small (perhaps an impurity site or dislocation). The proposed mechanism is in accord with existing data.

I. INTRODUCTION

The process of desorption of adsorbed species is an important process in the area of gas-surface interactions and is particularly significant in catalysis. If species are truly adsorbed on the surface, the expected thermal accommodation coefficients would be unity, and conservation of momenta from the original gaseous state would not be expected to be important (directed scattering would not be expected). The desorption process of adsorbed species from a crystal surface is related to the evaporation process of atoms from a bulk crystal surface; the differences are associated with the identity and binding of the species to the solid surface.

The desorption process for small fractions of surface coverage would be expected to follow first order kinetics. The reciprocal of the first order rate constant is identified as the mean adsorption lifetime, τ. The temperature dependence of τ is given by

$$\tau = \tau_0 \exp (\Delta H^*/RT) \qquad [1]$$

where τ_0 is a pre-exponential factor, ΔH^* is the activation enthalpy, R is the gas constant, and T is absolute temperature of surface. By statistical mechanical arguments, τ_0 is identified with the reciprocal of the vibrational frequency of the adsorbed species perpendicular to the surface. Thus, τ_0 would be expected to be *ca.* 10^{-13} s.

Chemical kinetics studies have been performed using pulsed molecular beam techniques. The experimental technique was initiated by Knauer(1) and Starodubtsev(2) and was developed by Scheer and Fine(3). A review of literature was given by Huang and Wahlbeck(4). Extensive studies have been performed on the desorption kinetics of ions from metal surfaces by Scheer and Fine and others; these studies have yielded large values of activation enthalpies of 2 - 3 eV and small τ_0 values of *ca.* 10^{-13} s. Mean adsorption lifetimes of neutral metal atoms on W and oxygenated W have been studied by Shelton and Cho(5) and by Hudson and Sandejas(6). These studies have yielded ΔH^* values of 1 - 4 eV and τ_0 values of 10^{-10} - 10^{-17} s. The adsorption and desorption of CO from Ni surfaces has been studied by Degras(7), by Madden *et al.*(8), and by Tracy (9).

Experimental research on desorption kinetics performed in this laboratory has been reported by Huang and Wahlbeck(4) and by Liu and Wahlbeck(10). A summary of their results is given in Table 1.

TABLE 1
Summary of Experimental Data

Adsorbent	*Surface*	*ΔH^*/kcal*	*τ_0/s*
Cs[a]	Ni(111)	3.3±0.4	(1.7±1.6)10^{-3}
CsCl[b]	Ni(111)	5.04±0.41	(1.22±0.31)10^{-4}
	Ni(100)	7.99±0.77	(2.38±1.18)10^{-5}
CsI[a]	Ni(111)	13.1±0.9	(3.20±1.84)10^{-6}
	Ni(100)	14.2±1.2	(4.17±2.81)10^{-6}

a. For details, see M. B. Liu and P. G. Wahlbeck(10).
b. For details, see L. W. Huang and P. G. Wahlbeck(4).

From the data in Table 1, one concluded that (1) ΔH^* values are all less than 0.6 eV, (2) ΔH^* values increase with the adsorbed species Cs, CsCl, and CsI, (3) ΔH^* values do show a surface structure dependence, particularly for CsCl, and (4) τ_0 values are much larger than the expected 10^{-13} s value. Lin and Wahlbeck(11) have performed related experiments using a different, but similar, apparatus in which the angle of incidence and the angle of detection of desorbing species may be changed. Lin and Wahlbeck(11) have found that in cases of desorption of CsCl from Ni single crystal surfaces and polycrystalline surfaces that the cosine law of restitution is valid. They also found ΔH^* values in excellent agreement with those observed by Huang and Wahlbeck(4) and Liu and Wahlbeck(10) as reported in Table 1.

Desorption is related to evaporation. Knacke and Stranski(12) proposed a mechanism for the evaporation process in which the following steps were proposed:

$$\begin{array}{ccccc} \text{X(kink site)} & \underset{1'}{\overset{1}{\rightleftharpoons}} & \text{X(ledge site)} & \underset{2'}{\overset{2}{\rightleftharpoons}} & \text{X(terrace site)} \\ & & & & 3 \downarrow\uparrow 3' \\ & & & & \text{X(g)} \end{array} \quad . \qquad [2]$$

The types of surface sites indicated in equation [2] may be briefly described. Terrace sites are sites occuring on the flat planes of a crystal. A ledge occurs when a new atomic layer is added to a terrace to give upward contour to the crystal; a ledge site is an atomic position along a ledge. A kink position occurs when a new atomic row is added to a ledge giving an in-out contour; a kink site is an atomic position at a kink. The atoms progress in steps following a decreasing order of binding energies. The rate determining step was proposed to be Step 1; see Hultman and Rosenblatt(13). The relationship between the evaporation mechanism given by equation [2] and the desorption process will be presented.

II. PROPOSED MECHANISM

A satisfactory mechanism must be able to explain the experimental kinetic data for the process. In the present case of the desorption process, the proposed mechanism would have to have a rate determining step with a ΔH^* and τ_0 in accord with kinetic data reported in the literature and in Table 1. Data which are reported in Table 1 all have large values of τ_0 compared to the expected value of 10^{-13} s.

The mechanism which is proposed for the desorption process is a modification of the mechanism originally proposed by Knacke and Stranski (12) for the vaporization of solid crystals, see equation [2]. The proposed mechanism is

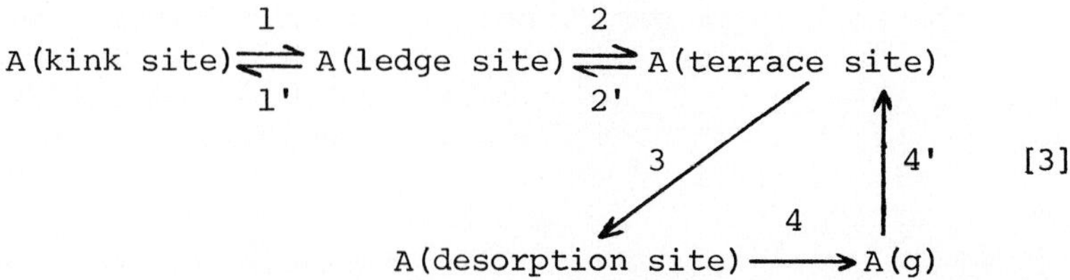

The symbol A is used to represent an adsorbed species on the crystal surface in contrast to the Stranski mechanism for bulk vaporization where the species X was from the solid crystal. The step involving surface migration to a desorption site, Step 3, was not in the Stranski mechanism.

Each of the steps in the proposed mechanism involves a change of position on the crystal surface except for Steps 4 and 4'. Each of these steps will have a time constant associated with it given by a variation on equation [1]; *i.e.*,

$$\tau = \tau_0^0 \; N \exp\;(\Delta H^*/RT) \qquad [4]$$

where τ_0^0 will be assumed to be 10^{-13} s which is the approximate time for one vibrational motion and N is the number of events in the step. It is assumed that steps which require multiple, N, repeated events will have a multiplier, N, to determine the time constant. The experimentally observed τ_0 value should correspond to $N\tau_0^0$ of the rate determining step, and similarly the experimentally observed ΔH^* should be the ΔH of the rate determining step. In order to estimate τ values for the different steps in the process, one must estimate ΔH and N values for each step.

A reasonable choice for the rate determining step to fit data given in Table 1 is Step 3. This is the only step which would have a large enough number of surface migration steps so as to give a value of τ_0 which would be appropriate for the experimental data. Thus, in the estimation of ΔH and N for Step 3, the experimental data have been used.

Values for ΔH for the steps of the mechanism have been estimated on the basis of the number of nearest neighbors (coordination number) following an approach similar to that of Hirth and Pound (14). Values of binding energies for the various types of sites are indicated in Table 2. These binding

TABLE 2
Binding Energies of Adsorbed Species

Type of Site	*Binding Energy*[a]*/kcal*
Kink	55
Ledge	50
Ledge-bridged	38
Terrace	40
Bridged	35
Gas	0

a. Reference state is the gas state. Data are designed for the case of CsCl adsorbed on Ni(111).

energies are graphically represented in Figure 1. The binding

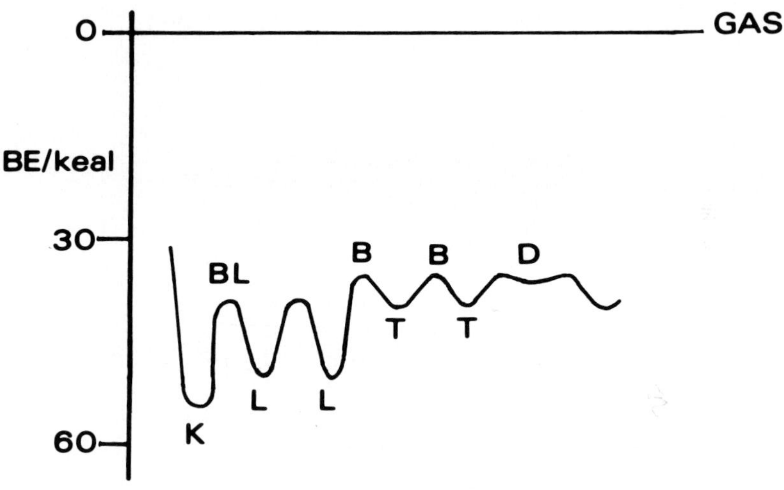

Fig. 1. Plot of binding energy (BE) for different types of positions of CsCl adsorbed on a Ni(111) surface. Meaning of symbols: K, kink site; L, ledge site; T, terrace site; LB, ledge bridged site; B, bridged site; D, desorption site. Energy of the gas phase is also indicated. See Table 2 for values of the binding energies.

energy of the kink site is assumed to be 55 kcal which is 10 kcal larger than ΔH°_{subl} for CsCl(s)(15). Movement of an absorbed species from a kink position to a ledge position involves the breaking of one bond. Movement from the ledge position to a terrace position involves the breaking of two bonds. The movement from one terrace site to a bridged position involves breaking one bond; as previously indicated this step is believed to be the rate determining step and ΔH is the experimental ΔH^*, 5.04 ±0.41 kcal. For adsorbed CsCl on Ni(111), the energy of one bond is assumed to be 5 kcal. The binding energy of the ledge-bridged position was assumed to be slightly greater than the bridged position. For the steps of the process in the mechanism, values of ΔH are obtained by using the binding energies of the initial site and the intermediate site; see Table 3.

When two types of events may occur, it is assumed that the relative probabilities of the events will be given by

$$P_1/P_2 = \exp(-\Delta H_1/RT)/\exp(-\Delta H_2/RT) \quad . \qquad [5]$$

For example, a molecule in the ledge position may move to another ledge position requiring a one-dimensional random walk or may move to a terrace position directly. Thus,

$$P_{Ledge \to Ledge}/P_{Ledge \to Terrace} = \frac{\exp(-12 \text{ kcal}/R\ 1000\ K)}{\exp(-15 \text{ kcal}/R\ 1000\ K)}$$

$$= 4.5 \quad , \qquad [6]$$

and

$$P_{Terrace \to Terrace}/P_{Terrace \to Gas} = \frac{\exp(-5 \text{ kcal}/R\ 1000\ K)}{\exp(-40 \text{ kcal}/R\ 1000\ K)}$$

$$= 4.5 \times 10^7 \quad . \qquad [7]$$

The N values of Table 3 have been estimated using calculations indicated by equation [5]. The N values which are set at unity are for steps in which a change in state occurs by a single motion. For example, a single motion is required for an adsorbed species to move from a kink position to a ledge position.

For Step 3, surface migration has been assumed where an average surface migration distance of r would be given by

$$r = \sqrt{N\ d^2/3} \quad , \qquad [8]$$

TABLE 3

Data for Steps of the Mechanism

Step #	*Process A(initial)* $\rightarrow A^{\dagger} \rightarrow$ *A(final)*	*N*	ΔH/*kcal*	τ/s (*)
1	A(kink) → A(ledge-bridged) → A(ledge)	1	17	5.2×10^{-10}
2	A(ledge) → A(ledge-bridged) → A(ledge)	10	12	4.2×10^{-10}
	A(ledge) → A(bridged) → A(terrace)	1	15	1.9×10^{-10}
3	A(terrace) → A(bridged) → A(terrace)	10^8	5	1.2×10^{-4}
	A(terrace) → A(bridged) → A(desorption)	1	5	1.2×10^{-12}
4	A(desorption) → A(g)	1	35	4.5×10^{-6}
1'	A(ledge) → A(ledge-bridged) → A(ledge)	10	12	4.2×10^{-10}
	A(ledge) → A(ledge-bridged) → A(kink)	1	12	4.2×10^{-11}
2'	A(terrace) → A(bridged) → A(terrace)	4000	5	4.9×10^{-9}
	A(terrace) → A(bridged) → A(ledge)	1	5	1.2×10^{-12}
4'	A(g) → A(terrace)	$\dot{Z} = \alpha_{trap} P/\sqrt{2\pi m k T}$		

(*)τ values are calculated using equation [4] with T = 1000 K.

where N is the number of jumps and d is the jump distance. For an N value of 10^8, the average surface migration distance would be

$$r = \sqrt{10^8 \cdot 1.43\ \text{Å}^2/3} = 8.2 \times 10^3\ \text{Å} \quad , \qquad [9]$$

where 1.43 Å results from an atomic radius of Ni atoms of 1.24 Å. Assuming circular symmetry, the number of desorption sites would be $5 \times 10^7\ cm^{-2}$.

The number of surface migration steps involved in Step 2' was estimated by assuming that the ledge to ledge distance was *ca.* 100 Å. For an average travel distance of 50 Å, the number of required steps would be 3600 according to equation [8].

In Table 3 are calculated values of τ for each step in the proposed mechanism. For the case of CsCl adsorbed on Ni(111) with estimated ΔH and N values, the rate determining step in the mechanism is observed to be Step 3 with a time of 1.2×10^{-4} s. For weakly adsorbed materials, the rate determining step is the step in which surface migration occurs with the adsorbed species moving on the surface until it reaches a site where the binding energy is less.

From the above and from entries in Table 3, one would conclude that the adsorbed species have multiple encounters with the ledge positions. Since the time constants are short for events associated with Steps 1 and 2, ledge and kink encounters can be viewed as "mirror" encounters in which the adsorbed species returns to the terrace to continue the random walk-surface migration.

If one uses a single bond energy of *ca.* 15 kcal or larger for setting the binding energies, the rate determining step would be expected to change to Step 1 for the desorption mechanism.

III. DISCUSSION

The proposed mechanism has as one of its important steps the surface migration of adsorbed species on the terraces until desorption occurs from a desorption site where the binding energy would be less. There are several pieces of information supportive of the proposed mechanism. These facts will be discussed in the following paragraphs.

Surface migration is a known means by which atoms can be transported across a surface; observations of surface migration have been made in studies with a field ion microscope by Ehrlich and coworkers; see Graham and Ehrlich(16).

Contaminants which give lower binding energies can serve as desorption sites. When contaminants have been accumulated on the surface by aging the surface for long times under ultrahigh vacuum conditions, the τ values and τ_0 values have decreased(11). These effects are to be expected if the contaminants serve as additional desorption sites.

Lin and Wahlbeck(11) observed that when the target surface temperature was made sufficiently small so that condensation of CsCl occurred, τ_0, became very small, *ca.* 5×10^{-11} s, and ΔH^* became *ca.* 25 kcal. When CsCl deposits on the surface, it is reasonable that the CsCl accumulates initially at kink sites on the Ni surface. The CsCl deposit will develop its own surface structure having kink, ledge, and terrace sites as described previously. With the increase in binding energy expected in this case, the rate determining step in the desorption process appears to become Step 1 where CsCl leaves a CsCl kink site.

The desorption mechanism, equation [3], has the feature that the rate determining step can shift from Step 3, surface migration, to Step 1, motion from a kink site to a ledge site as the binding energies increase. Thus, the mechanism can explain the data observed earlier for ion desorption as well as the data of Huang and Wahlbeck(4) and Liu and Wahlbeck(10).

In the Introduction, previous data were presented, much of this data was for ion desorption from metal surfaces. The data of Scheer and Fine(3) are typical giving a very small τ_0 and ΔH^* value of *ca.* 45 kcal. When contaminants are added to the surface, ΔH^* values decreased and τ_0 increased. In this case, it is reasonable that the mechanism is one in which the rate determining step is probably motion from a kink site to a ledge site. The binding of ions to the metal surface can be viewed as involving coulombic forces; contaminants serve as a dielectric material weakening the attractive forces.

The role of desorption sites in equation [2], the mechanism for evaporation of solid crystals, may have been overlooked. In these cases, the steps following Step 1 are all rapid which may have masked the presence of effects from desorption sites.

ACKNOWLEDGMENTS

The US National Science Foundation is thanked for financial support of the research which led to this paper. The support of Wichita State University is gratefully acknowledged.

REFERENCES

1. F. Knauer, Z. Phys. 125, 278 (1948).
2. S. V. Starodubtsev, Zh. Eksp. Teor. Fiz. 19, 215 (1949).
3. M. D. Sheer and J. Fine, J. Chem. Phys. 37, 107 (1962); 38, 307 (1963).
4. L. W. Huang and P. G. Wahlbeck, J. Phys. Chem. 80, (1976).
5. H. Shelton and A. Y. Cho, J. Appl. Phys. 37, 3544 (1966).
6. J. B. Hudson and J. S. Sandejas, J. Vac. Sci. Technol. 4, 230 (1967).
7. D. A. Degras, Adv. Vac. Sci. Technol., Proc. Int. Congr., 3rd, 1965, 2, 673 (1965); Vacuum 18, 122 (1968).
8. H. H. Madden, J. Küppers, and G. Ertl, J. Chem. Phys. 58, 3401 (1973).
9. J. C. Tracy, J. Chem. Phys. 56, 2736 (1972).
10. M. B. Liu and P. G. Wahlbeck, J. Phys. Chem. 80, (1976).
11. F. N. Lin and P. G. Wahlbeck, Unpublished data.
12. O. Knacke and I. N. Stranski, Prog. Metal Phys. 6, 181 (1956).
13. C. A. Hultman and G. M. Rosenblatt, Science 188, 145 (1975).
14. J. P. Hirth and G. M. Pound, J. Chem. Phys. 26, 1216 (1957).
15. W. D. Treadwell and W. Werner, Helv. Chim. Acta 36, 1436 (1953).
16. W. R. Graham and G. Ehrlich, Thin Solid Films 25, 85 (1975).

ADSORPTION AND DENSITY PROFILE OF A GAS NEAR A UNIFORM SURFACE

Douglas Henderson
IBM Research Laboratory

Eduardo Waisman* and Joel L. Lebowitz*
Yeshiva University

Densities profiles and adsorption isotherms are calculated, by means of the mean spherical approximation, for hard spheres near a uniform hard surface when there is an exponentially decaying attractive interaction between the wall and hard spheres.

I. INTRODUCTION

Recently, Henderson, Abraham, and Barker (1) have suggested that the problem of a fluid in contact with a wall can be regarded as the limiting case of a mixture in which one of the components becomes infinitely dilute and then infinitely large. In particular, they obtained the appropriate form of the Ornstein-Zernike equation,

$$h_{12} = c_{12} + \rho\, h_{21} * c_{11} , \qquad [1]$$

where the asterisk denotes a convolution, $\rho = N/V$ is the bulk density of the fluid in contact with the wall, $h_{ij}(r) + 1 = h_{ji}(r) + 1$ $g_{ij}(r)$ are the *radial distribution functions* (RDFs), i.e., $\rho_i g_{ij}(r)$ is the density of molecules of species i a distance r from the center of a molecule of species j and $c_{ij}(r) = c_{ji}(r)$ are the *direct correlation functions*. In equation [1], the particle of species 2 is the particle which is assumed to be infinite in size.

The Ornstein-Zernike equations are just the definition of the $c_{ij}(r)$ and become useful when supplemented by some approximate equations for the $c_{ij}(r)$. One of the most useful approximations is the *mean spherical approximation* (MSA) of Lebowitz and Percus (2). For systems for which the intermolecular potential is

*Supported by a grant from the Petroleum Research Foundation and by AFOSR Grant 73-2430 B

$$u_{ij}(r) = \infty \quad , r < \sigma_{ij} ,$$
$$= w_{ij}(r) \quad , r > \sigma_{ij} , \qquad [2]$$

where $\sigma_{ij} = (\sigma_{ii} + \sigma_{jj})/2$ and σ_{ii} is the molecular diameter, the MSA is

$$c_{ij} = - \beta w_{ij}(r) \ , \ r > \sigma_{ij} \qquad [3]$$

which together with the *exact* condition, $h_{ij}(r) = -1$, $r < \sigma_{ij}$ permits a solution of [1]. When $\beta = 0$ (i.e., infinite temperatures), equation [3] becomes the Percus-Yevick (3) approximation for hard spheres.

Henderson, Abraham, and Barker (HAB) have obtained the Percus-Yevick (PY) result for the density profile $\rho(x) = \rho g_{12}(\sigma_{12}+x)$ from [1] and the analytic expressions of Lebowitz (4) for $c_{12}(r)$. Similar results have been reported by Percus (5).

Subsequently, we (6) have obtained the solution of the MSA when $w_{11}(r) = 0$ and $w_{12}(r)$ is a Yukawa function, i.e.,

$$w_{12}(r) = - \varepsilon^{*} e^{-zr}/r \ , \quad r > \sigma_{12} \qquad [4]$$

In the limit $\sigma_{12} \to \infty$, equation [4] becomes

$$w_{12}(\sigma_{12}+x) = - \varepsilon \ e^{-zx} \ , \ x > 0 \ , \qquad [5]$$

where $x = r - \sigma_{12}$ and $\varepsilon = \varepsilon^{*} \ e^{-z\sigma_{12}}/\sigma_{12}$. Thus, the exponential function is the limit of the Yukawa potential which is appropriate to the wall-molecule interaction. Interestingly if the wall is assumed to consist of individual molecules which interact with the fluid by Yukawa intermolecular potentials and if the interaction of a fluid molecule is computed by integrating the interactions with all the molecules in the wall, the wall-molecule interaction is an exponential function with the same decay constant z as that of the individual wall-molecule fluid-molecule interaction. Thus, in both senses the exponential function is the natural extension of the Yukawa potential to wall-molecule interactions.

We have used (6) this solution to obtain the density profile of hard spheres near a hard wall by regarding $\beta\varepsilon$ not as a thermodynamic variable, but as an adjustable parameter which can be chosen by thermodynamic considerations. In this work we apply our results to the case where $w_{12}(r)$ is the attractive interaction between the molecules and the wall.

II. DENSITY PROFILE

As we have mentioned, the solution of [1] gives $h_{12}(r)$. The density profile of the molecules near the wall is just $\rho(x) = \rho[h(\sigma_{12}+x) + 1]$. We have plotted $\rho(x)$ in Fig. 1 for three thermodynamic states.

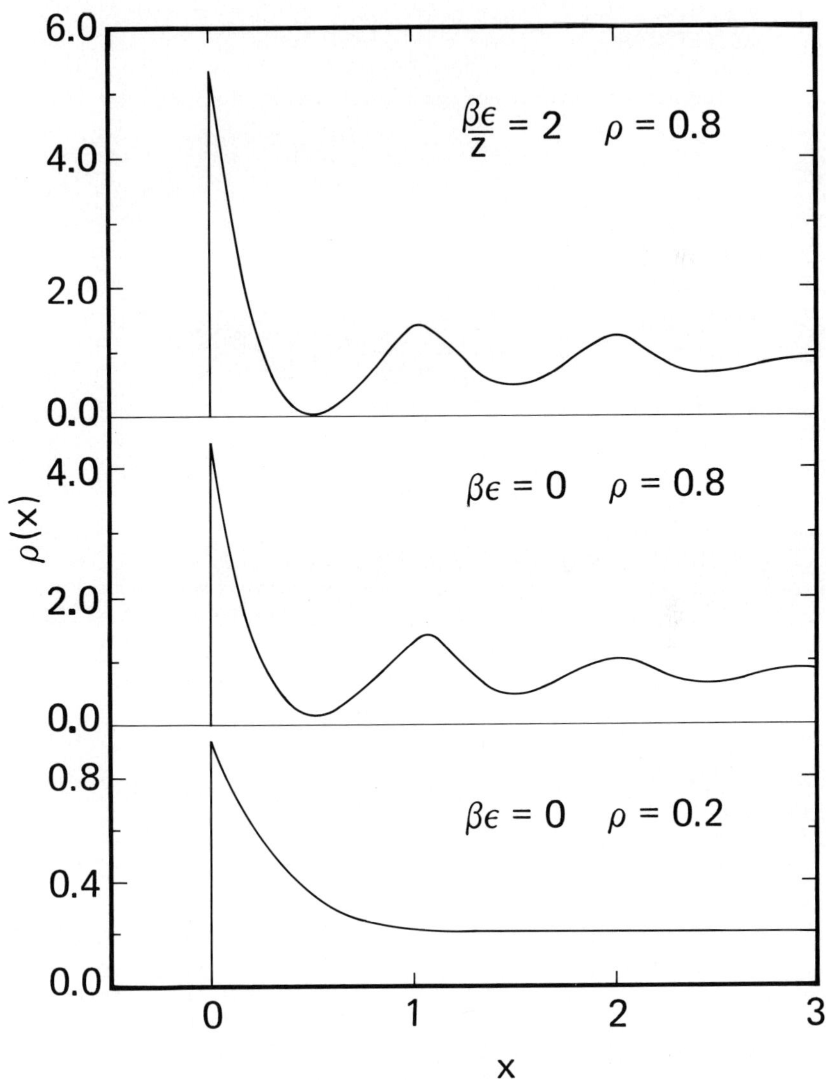

Fig. 1. Density profiles of gas near a wall

At low densities, the density profile is strongly affected by the attractive potential $w_{12}(r)$. However, at high densities, the density profile at $\beta\varepsilon/z = 2$ is nearly equal to that at $\beta\varepsilon = 0$. This means that structure of the fluid near the wall is dominated by the structure of hard spheres near a hard wall. This should mean that a perturbation treatment of fluids near a wall will be as successful as for bulk fluids.

III. ADSORPTION

The number of molecules per unit area absorbed by the wall is

$$N_a = \int_0^\infty [\rho(x) - \rho]dx \ . \qquad [6]$$

For the case when $w_{12}(x)$ is given by equation [5], we have shown (6) that this integral may be obtained analytically. Our result is

$$N_a = \rho \left[\frac{3\eta}{2(1+2\eta)} + \frac{\beta\varepsilon}{z} C \frac{(1-\eta)^4}{1+2\eta} \right] , \qquad [7]$$

where $\eta = \pi\rho\sigma^3_{11}/6$ and

$$C = \frac{z^3 e^z}{L(z) + S(z)e^z} , \qquad [8]$$

$$L(z) = 12\eta \left[(1+\eta/2)z + 1 + 2\eta \right] , \qquad [9]$$

$$S(z) = (1-\eta)^2 z^3 + 6\eta(1-\eta)z^2 + 18\eta^2 z - 12\eta(1+2\eta) \ . \qquad [10]$$

If $z \to \infty$, equation [7] becomes

$$N_a = \rho \left[\frac{3\eta}{2(1+2\eta)} + \frac{\beta\varepsilon}{z} \frac{(1-\eta)^2}{1+2\eta} \right] . \qquad [11]$$

The van der Waals limit, $z \to 0$, is of more interest. For this case

$$N_a = \rho \left[\frac{3\eta}{2(1+2\eta)} + \frac{\beta\varepsilon}{z} \frac{(1-\eta)^4}{(1+2\eta)^2} \right] . \qquad [12]$$

The second term in equation [12] gives the contribution of the attractive forces. It is of interest to note that it is damped at high densities because of the term $(1-\eta)^4/(1+2\eta)^2$ which is just the PY or MSA approximation for $kT\ \partial\rho/\partial p$ for bulk hard spheres. Thus as the bulk hard sphere fluid becomes incompressible, the contribution of the attractive forces becomes small. This is similar to the perturbation theory concepts of Barker and Henderson (8) for bulk fluids.

The adsorption, N_a, is plotted in Fig. 2

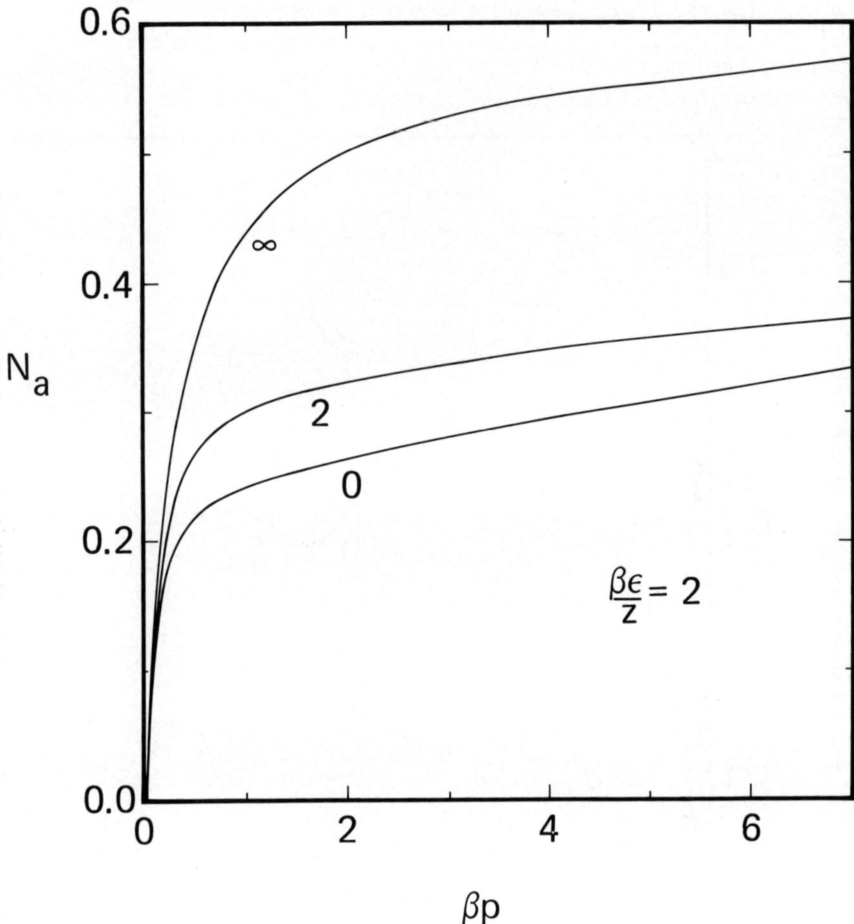

Fig. 2 Adsorption of gas for various values of z at $\beta\varepsilon/z = 2$.

for z = 0 and ∞ and for z = 2 which is a value appropriate for argon. The pressure of the bulk hard-sphere fluid is, in the PY (or MSA) approximation ,

$$\beta p/\rho = \frac{1+\eta+\eta^2}{(1-\eta)^3} \qquad [13]$$

It is seen that z = 2 is still close to the van der Waals limit.

In Fig. 3 we have plotted N_a for z = 2 for $\beta\varepsilon/z = 0,1$, and 2. There is adsorption even when there are no attractive forces ($\beta\varepsilon = 0$) because of the hard core

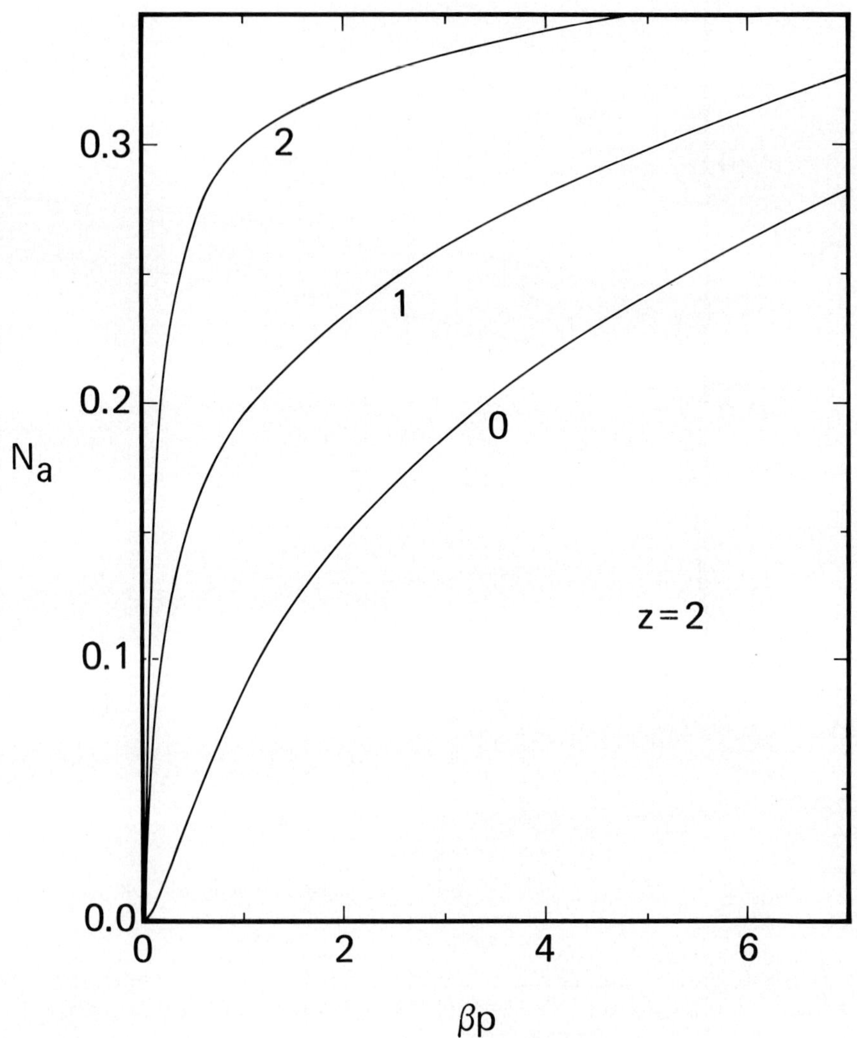

Fig. 3. Adsorption of gas for z = 2

repulsions of the gas. Similar results have been obtained

by Perram and Smith (unpublished work) using a less realistic potential.

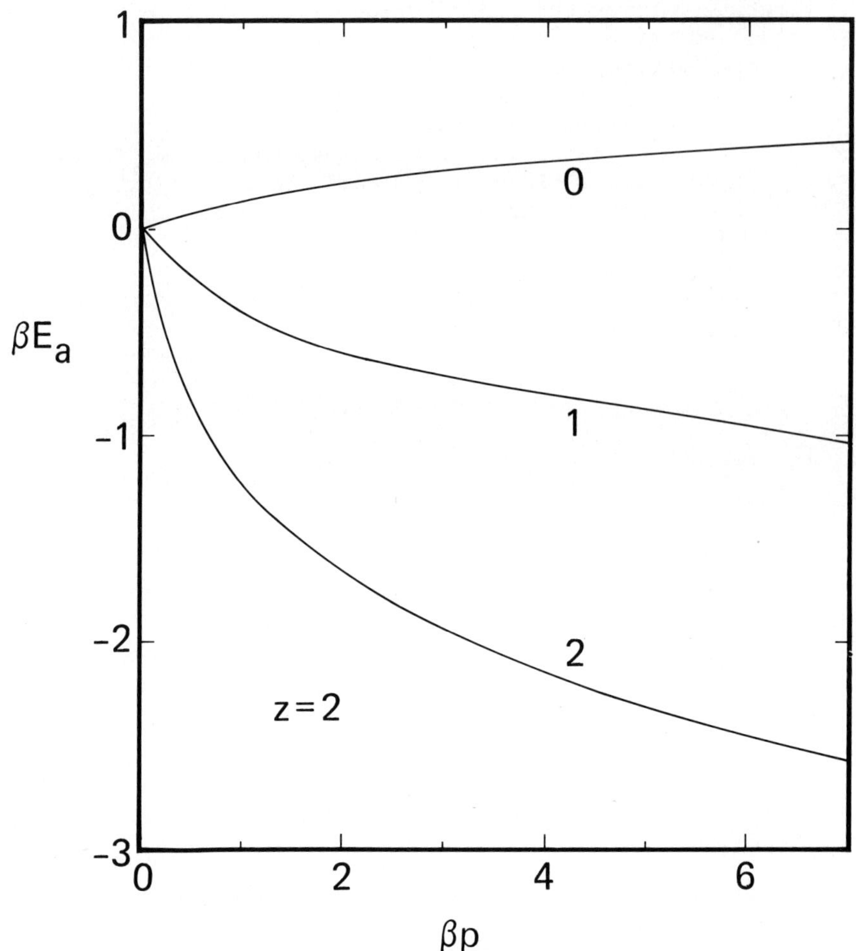

Fig. 4. Excess energy of adsorption for z = 2

The excess energy of adsorption/unit area is given by

$$\beta E_a = \frac{3}{2} N_a + \beta \int_0^\infty w_{12}(x) \rho_{12}(x) dx \ . \qquad [14]$$

This integral may also be obtained analytically. The result is

$$\beta E_a = \frac{3}{2} N_a - \left[\frac{\beta\varepsilon}{z} C(1+2\eta) - \frac{(\beta\varepsilon)^2}{2z} C^2 (1-\eta)^4 \right] \rho \ . \qquad [15]$$

The first term gives the contribution of the kinetic energy. The excess energy of adsorption is plotted, for z = 2, for three temperatures in Fig. 4.

A comparison with experiment is not appropriate here as the case of liquid adsorption is excluded because $w_{11}(r) = 0$. However, these calculations do indicate the utility of applying theories which have proven useful in the theory of bulk fluids to fluids near a wall.

Blum and Stell (9) have also solved [1] when attractive forces are present.

IV. ACKNOWLEDGEMENT

The stimulating remarks of J. A. Barker are gratefully acknowledged.

V. REFERENCES

1. Henderson, D., Abraham, F. F., and Barker, J. A., Mol. Phys. 31, 1291 (1976).

2. Lebowitz, J. L., and Percus, J. K., Phys. Rev. 144, 251 (1966).

3. Percus, J. K., and Yevick, G. J., Phys, Rev. 110, 1 (1958).

4. Lebowitz, J. L., Phys. Rev. 133, A895 (1964).

5. Percus, J. K., J. Stat Phys. (in press).

6. Waisman, E., Henderson, D., and Lebowitz, J. L., Mol. Phys. (in press).

7. Liu, K. S., Kalos, M. H., and Chester, G. V., Phys. Rev. A10, 303 (1975).

8. Barker, J. A., and Henderson, D., J. Chem. Phys. 47, 2856, 4714 (1967).

9. Blum, L., and Stell, G., J. Stat. Phys. (in press).

THE REACTIONS OF ATMOSPHERIC VAPORS WITH LUNAR SOIL (14003)*

E. L. Fuller, Jr.
Oak Ridge National Laboratory

ABSTRACT

Detailed experimental data have been acquired for the hydration of the surfaces of lunar fines. Inert vapor adsorption has been employed to measure the surface properties (surface energy, surface area, porosity, etc.) and changes wrought in the hydration-dehydration processes. Plausible mechanisms have been considered and the predominant process involves hydration of the metamict metallosilicate surfaces to form a hydrated laminar structure akin to terrestrial clays. Additional credence for this interpretation is obtained by comparison to existing geochemical literature concerning terrestrial weathering of primary metallosilicates. The surface properties of the hydrated lunar fines is compared favorably to those of terrestrial clay minerals. The results show that ion track etching and/or grain boundary attack are minor contributions in the weathering of lunar fines in the realm of our microgravimetric experimental conditions.

A. Introduction

Lunar materials react with the water and carbon dioxide of the terrestrial atmosphere (1,2). The mechanism is somewhat ill-defined at present. This study was undertaken to obtain more details concerning the nature of the hydration processes which occur as the metamict surfaces are attacked. Specifically, we hoped to delineate between the three plausible mechanisms of a) ion tract etching (1), b) metallosilicate hydration (2) and/or c) grain boundary attack (3).

---

*Research sponsored by NASA under Union Carbide contract with the Energy Research and Development Administration.

Mechanism a) has been postulated as induced by water migration along the small channels of ion tracks, with subsequent countercurrent etching of larger chambers in the inner reaches of the bulk material, forming an "ink bottle" shaped pore. Presumably the "leached" material migrates to the surface and is rather uniformly distributed thereupon. The entrance channels are presumably blocked by bound water which can exclude inert vapors from the inner chambers. This bound water can be removed by stringent outgassing to permit access to inert vapors.

Mechanism b) has been forwarded as being akin to the hydration of primary terrestrial minerals (metallosilicates) to form laminar (clay like) hydrated metallosilicates from the damaged, amorphous outer surfaces of the particles. The pores are envisioned as the voids between the laminae which are formed as the hydration water is removed at elevated temperatures.

Mechanism c) was presented as involving the grain boundaries inherent in the particles. Water presumably migrates down these boundaries allowing the primary subunits (crystallites) to separate with a simultaneous release of strain energy. It is difficult to associate this mechanism with the observation that the bound water reversibly excludes inert vapors from the inner reaches of the sample as noted previously (1,2) and in this work. Pores formed by this process should be "open" (orifices greater than inner dimensions).

We chose an Apollo 14 sample (14003-60) as being generally characteristic of lunar fine samples in terms of its chemical composition, mineralogy, glass sphere content, color, etc. Due to the paucity of material we have chosen to perform extensive, controlled microgravimetric analyses. The goal was to study the interrelationship of hydration and carbonation reactions and surface structural analyses (inert vapor adsorption). A detailed analyses of surface area, porosity, surface energy, etc., and changes therein should help measure the relative contributions of the three aforementioned mechanisms.

B. Experimental Approach

An alequot of lunar fines was placed on a microbalance in an adsorption chamber described previously (4,5) and subjected to the conditions described in the subsequent section. It was used as received, the fraction of a soil sample which passed through a 1 mm seive. Optical microscopy showed that it was predominantly composed of angular dark fragments 1 to 10 μm in size. A few translucent spheres of ca. 5 to 10 μm were dispersed in the sample. Throughout the experimental sequence we have verified steady state data by allowing the sample to equilibrate at given temperatures and pressures for days to weeks when necessary. In the cases where slow processes were in play, we were able to determine the kinetics of the reactions. The sample was subjected to varying degrees of hydration via multiple exposures to water vapor. The gravimetric technique has proven to be quite informative in simultaneously monitoring several pertinent surface properties as anticipated by previous evaluations (5).

C. Results and Discussion

1. Reaction with Sorbed Water

Representative water vapor isotherms are given in Figure 1. Initial adsorption at pressures up to 0.4 P_o are characteristically (1,2) somewhat limited with respect to apparent monolayer capacity and multilayer formation. Each desorption boundary curve shows hysteresis in the capillary range (0.3 to 1.0 P_o) and marked low pressure and vacuum retention. Each of these isotherms are presented with respect to the vacuum weight of the sample upon completion of the preceding desorption cycle. These incremental vacuum retentions are cumulatively defined as "bound water", Γ (B, H_2O) in subsequent discussions. Steady state weights were achieved in ca. 5 minutes for pressures less than ca. 0.85 P_o. Above this pressure slow, first order processes become predominant and relatively large amounts were adsorbed for each pressure increment.

Excursions to these higher pressures resulted in increasingly greater enhancement of the bound component, Γ (B, H_2O) and apparent monolayer capacity, Γ (m, H_2O). The nature of the irreversible retention for each of these larger excursions is seen more clearly in Figure 2. Here we have deducted the amount adsorbed, Γ (A), from the amount remaining on desorption Γ (D) to evaluate the retention Δ (P) at each pressure. From this we can evaluate the hysteretic concentration, Δ (h) and retention increment Δ (B), [Γ (B) = $\Sigma\ \Delta$ B for successive cycles].

A direct correlation between Γ (m, H_2O) and Γ (B, H_2O) is noted in Figure 3. The initial adsorption (1 a) is greater than the desorption Γ (m) due to the increment of Γ (B) involved. Thereafter each desorption Γ (m) and

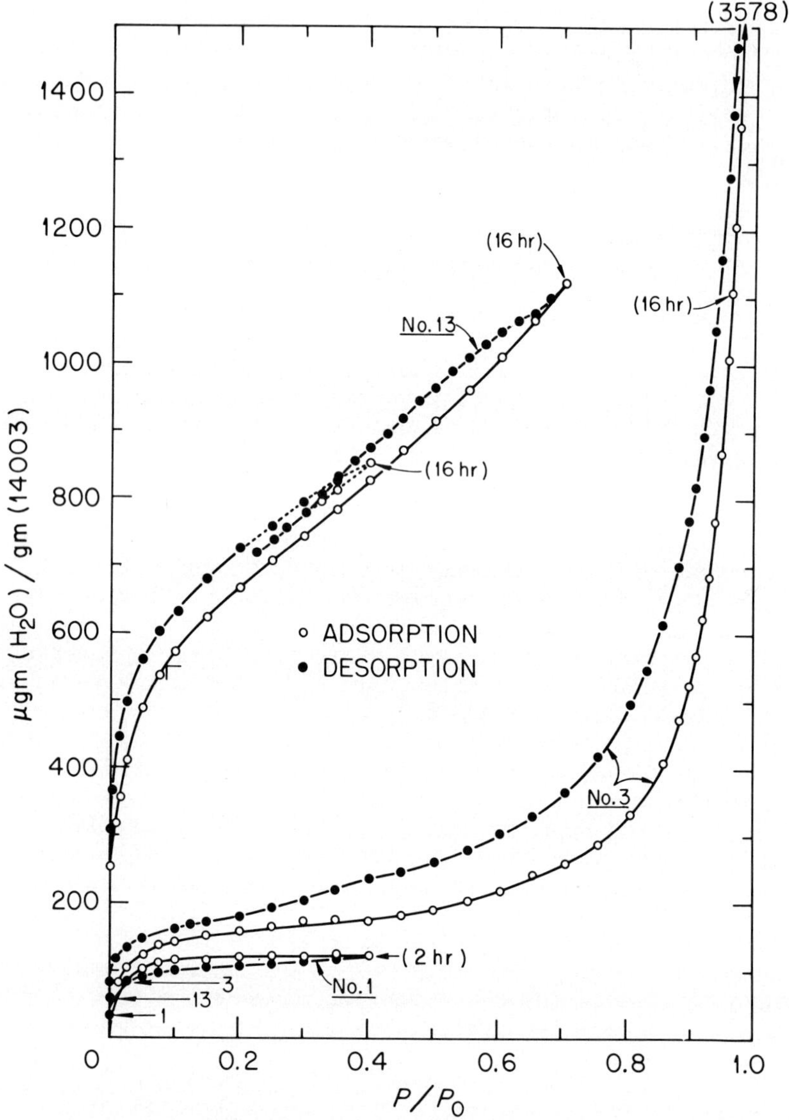

Fig. 1. Water sorption by 14003 at 22.00°C.

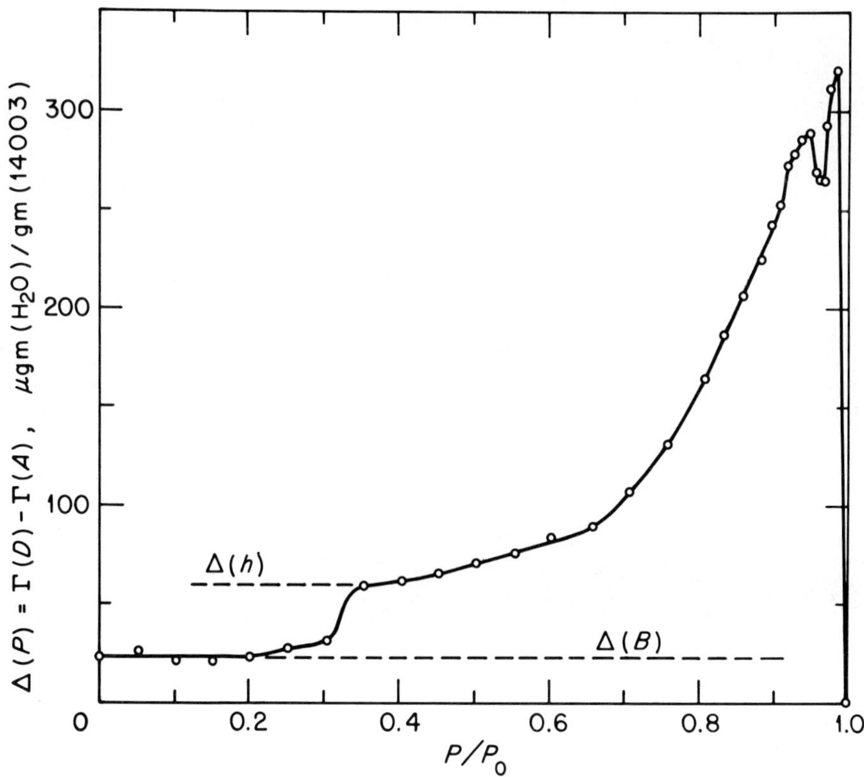

Fig. 2. Retention of H_2O on 14003 at 22.00°C.

subsequent adsorption Γ (m) were identical, for each is referred to the same vacuum weight [sample weight plus the respective Γ (B, H_2O)]. This trend continued for 5 cycles where data was acquired in the closed system mode (5) (balance chamber connected to a liquid water reservoir whose vapor pressure (temperature) was precisely controlled). In view of the displacement phenomena noted previously (2) we subsequently changed to the open mode of isotherm construction. Here we introduced and extracted (discarded) alequots of water vapor as needed to produce the desired pressure (cycles 6 thru 13). Cycle 14 was obtained in the closed mode of operation. The original trend [Γ (m) increasing with increasing Γ (B)] is enhanced initially and returns to the original relationship (cycles 10-14). Apparently the

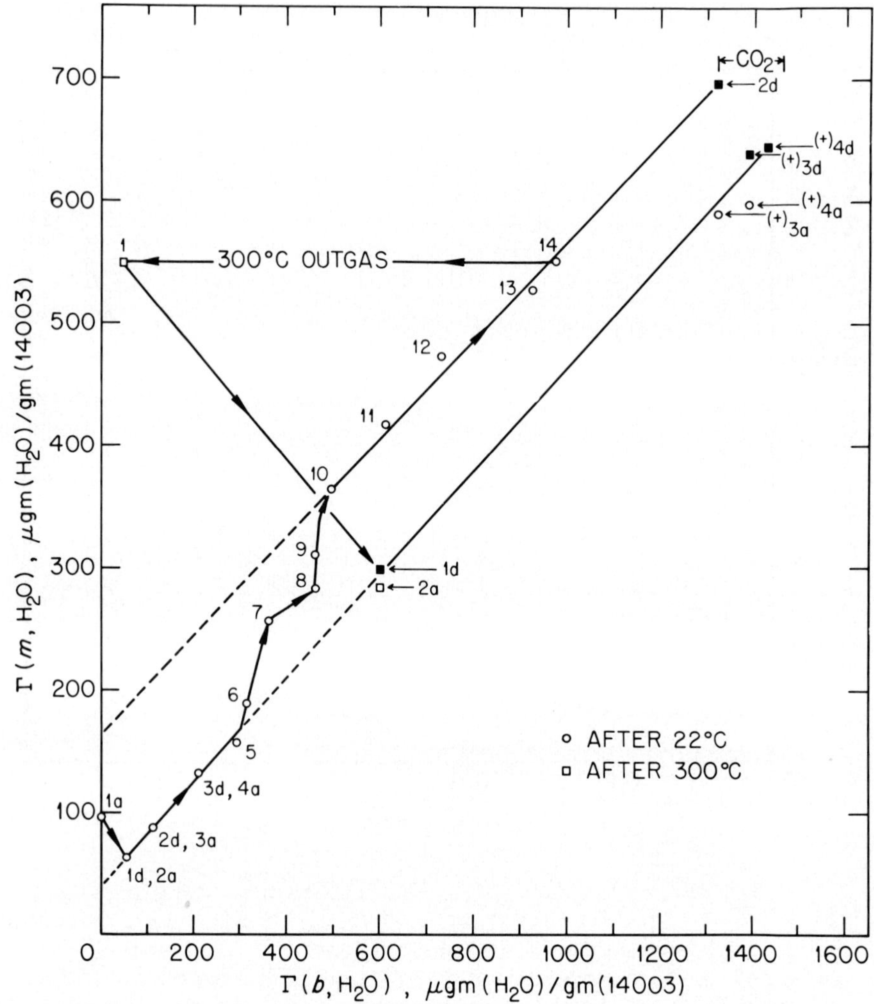

Fig. 3. H_2O sorption on 14003 at 22.00°C.

displaced species (CO_2?) is removed (cycles 6-10) and decreases the increment of apparent Γ (B, H_2O) proportionally until the original (concentration of bound Γ (B, CO_2) is removed. This conclusion is based on the CO_2 readsorption experiments described below.

Noting a continued increase in both Γ (B, H_2O) and Γ (m, H_2O) without any signs of approaching an upper limit we then outgassed the sample at 300°C with a nearly complete loss of the Γ (B, H_2O) complement. Subsequently, readsorption of water vapor (1 a) appears to occur on an area equivalent to the preceding maximal value. Considerable retention follows with the desorption (1 d) data commensurate with the original behavior. Excursion to 1.0 P_o in the next cycle reverted to the behavior (2 d) noted for the decarbonated sample. The sample was then exposed to CO_2 (700 torr) with a concommittal retention noted by the double arrow. Subsequent H_2O data is readily related to the original carbonated sample when Γ (B, H_2O) is corrected for Γ (B, CO_2) as noted in Figure 3.

2. Nitrogen Adsorption

During the course of the preceding experiment we constructed nitrogen isotherms as a measure of surface properties of the sample. Sample results are shown in Figure 4. We see that the initial water sorption (cycles 1 thru 14) enhanced Γ (m, N_2) and induced some porosity, which in turn gave rise to the hysteresis loop (curve B). Removal of the bound water by 300°C outgassing markedly increased both area and porosity (curve C). A small amount of retention was noted for the 300°C outgassed sample as noted in Figure 4. Rehydration and reoutgassing did not remove this disparity (0.0 to 0.5 P_o) at -196°C. However, the original vacuum weight was obtained in both cases by warming the sample to 22.00°C. We suspect that this phenomena is due to the existence of "hot spots" on the sample [defects or reduced states (i.e., metallic iron)]. We were able to oxidize the sample at 300°C and 100 torr (O_2) and deactivate the sites responsible for the nitrogen retention. Nitrogen adsorption may well be an excellent way of measuring the concentration of such active sites on lunar samples.

The data of Figure 5 suggest a linear relationship between Γ (h, N_2) and Γ (m, N_2). Pursuant to this topic we present a more comprehensive picture for the above sequence in Figure 6. In each case, dehydration or rehydration, the data extrapolate to Γ (m, N_2) equal to <u>ca</u>. 114.5 μgm (N_2)/gm (14003) when Γ (h, N_2) is zero. This seems to be a measure of the external surface, Γ (E, N_2), (0.399 m^2/gm) which is virtually unaltered during these cycles. The 700 and 800°C data also indicate that the sintering process is primarily intraparticulate. Furthermore, the linear relationships between Γ (h) and Γ (m) is significant. If the internal dimensions of these pores is much greater than the orifices, the volume capacity [Γ (h)] will be considerably greater than the surface capacity [Γ (m)].

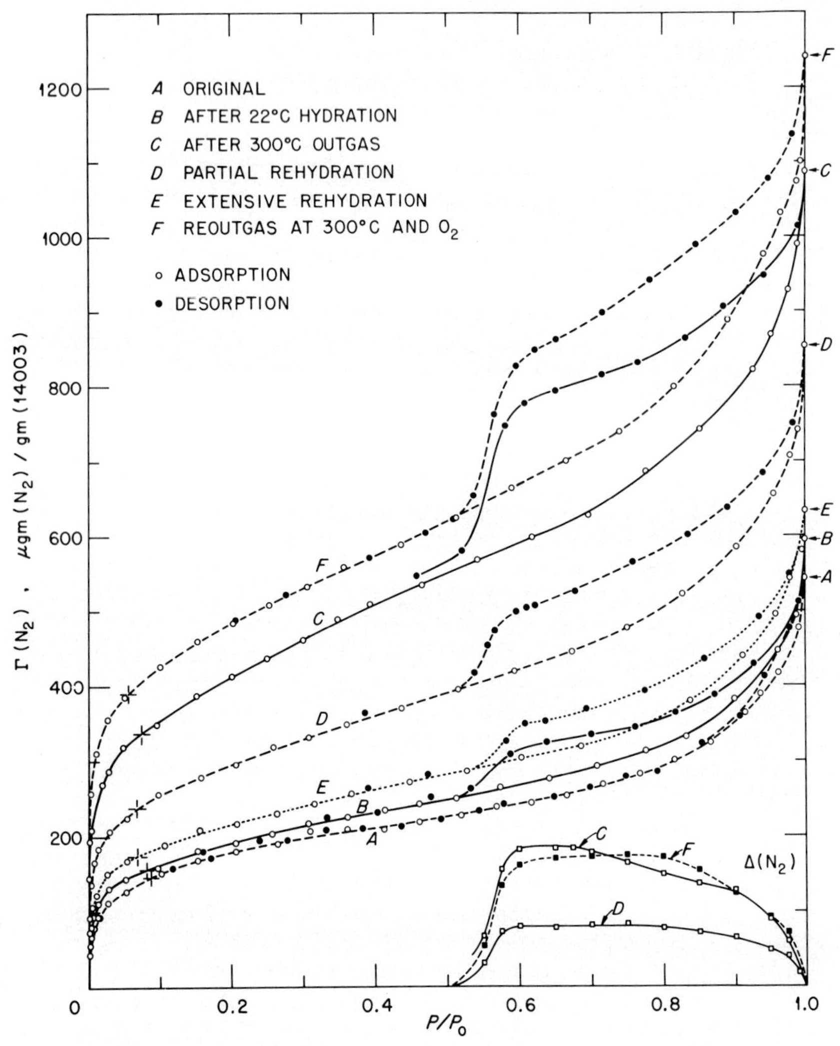

Fig. 4. Nitrogen sorption by 14003 at -196°C.

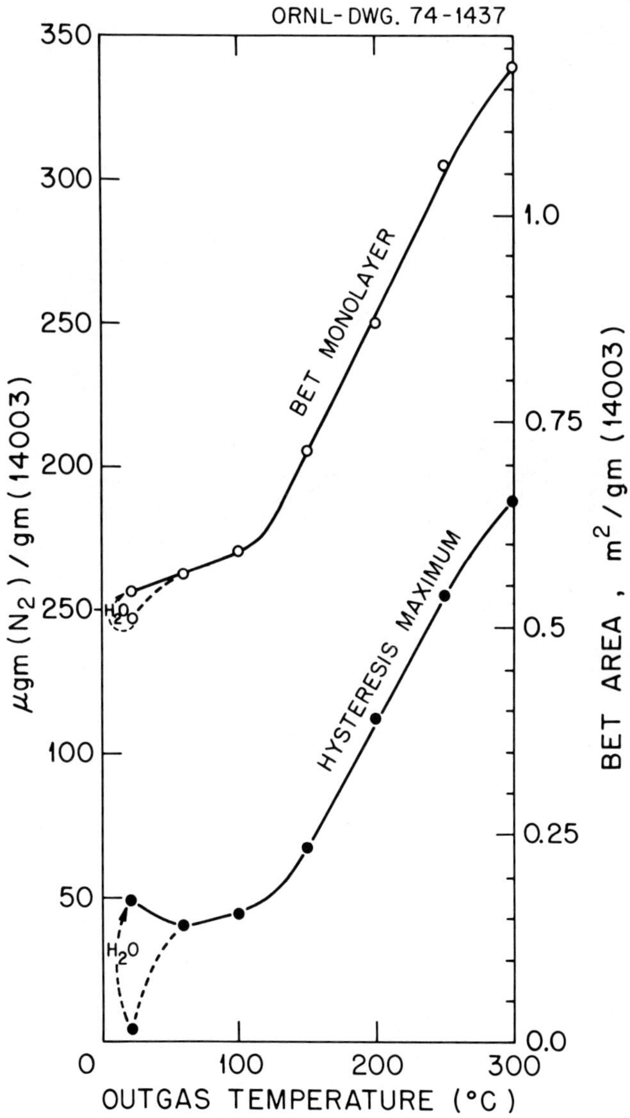

Fig. 5. Nitrogen sorption on 14003 at -196°C.

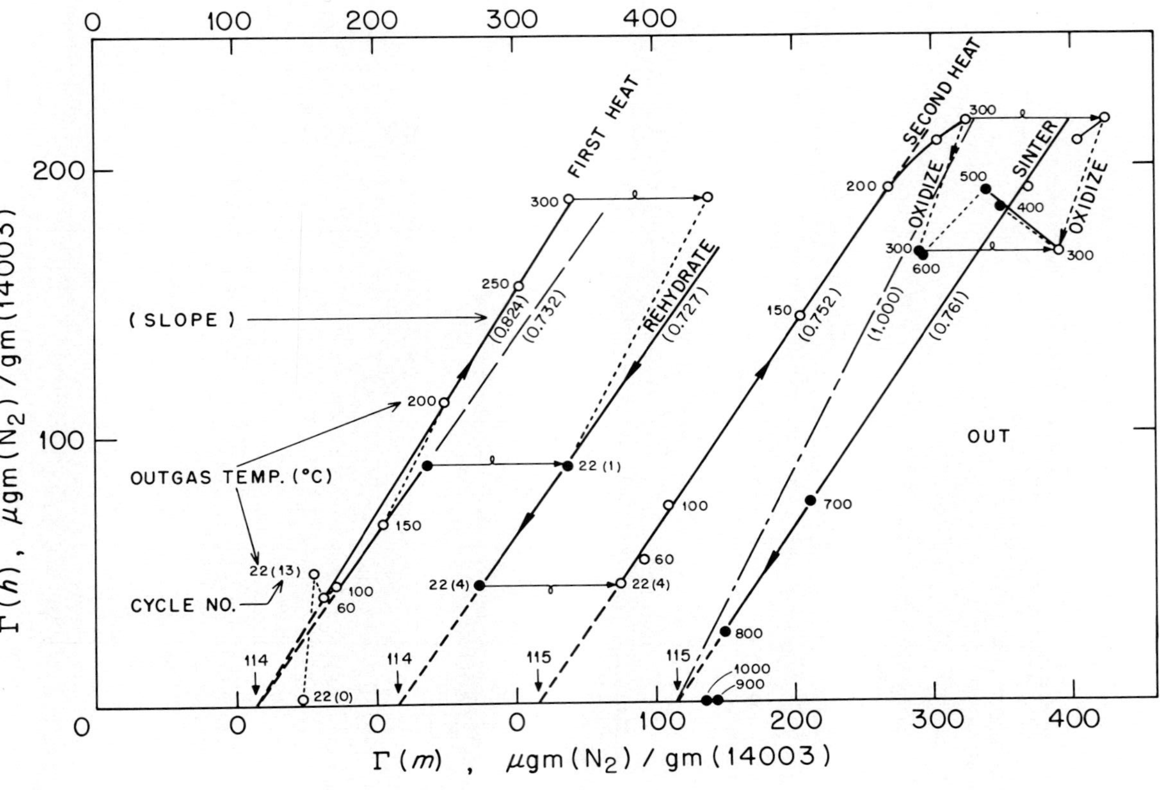

Fig. 6. Nitrogen sorption on 14003 at -196°C.

D. Conclusions

This sample has reacted in a very similar manner as have the majority of other lunar samples (2). Water vapor has attacked the material to form and fill a porous structure. The new surfaces are accessible to inert vapors only after the bound water is removed in vacuum at elevated temperatures. Rehydration reversibly fills the inner reaches to the exclusion of inert vapors. The hydration process requires multiple exposures to relatively high humidity. The cycling to and from vacuum is required to facilitate the hydration reaction whereas prolonged exposures at high humidity does not lead to appreciable further reaction. Water vapor physical adsorption is enhanced in proportion to the degree of hydration even in the presence of the bound water (in contrast to inert vapor (N_2) physical adsorption inhibited by the bound H_2O component).

The ion track etching mechanism can be discredited in terms of the relationship between Γ (h, N_2) and Γ (m, N_2) for the dehydration experiments of Figure 6. If indeed the inner reaches of the induced "ink bottles" are excluded by bound water filling the narrow necks we would expect the increments of volume filling terms Γ (h, N_2) and Γ (S, N_2) to be much greater than the surface term Γ (m, N_2). Furthermore, a limited saturation capacity for water Γ (S, H_2O) should be apparent and is not, as noted in Figure 1. Such behavior should be sensed if the internal voids are equal to or less than ca. 500 Å and are encased in a rigid framework. Conceptually it is difficult to envisage migration of the bulk entities through small orifices (< ca. 30 Å) in either anhydrous molecular form (i.e., Al_2O_3, SiO_2, MgO, NaO) or hydrated form. If appreciable differential migration and deposition on the surface had occurred we would expect an appreciable change in the chemical and physical nature of the surface. Granted one has difficulty differentiating slit or ink bottle pores based solely on the nature of nitrogen hysteresis (Figure 4) (6). However, one would expect water hysteresis (Figures 1 and 2) to be closely related to the nitrogen results for ink bottles in a rigid substrate, both in magnitude and partial pressure dependence. This is definitely not the case for this sample.

Considering the grain boundary mechanism, we have some difficulty in explaining the experimental results. If water indeed penetrates these boundaries with release of the strain energy one would anticipate a folding open with the primary elements spreading to leave an open "V" shaped pore which would not give rise to hystereses in the water and nitrogen isotherms. The enhanced area should be accessible to both vapors in near equal amounts regardless of degree of hydration. However if the surface grains were to remain in contact in a very unique manner (i.e., cubes arranged in very ordered states such that all edges were in contact via a hydration layer) one would sense increased area and volume capacities as the bound water is removed. However, it is unlikely that such an array of microscopic entities are present for they should be discernible in electron microscopic analyses and have not been detected to our knowledge.

The chemical hydration model is far superior in terms of explaining the experimental results. The hydration of anhydrous metallosilicates is a very thermodynamically favored reaction (9) albeit generally a very slow process (10). However the strain energy and disorder (2) in the surfaces of these materials is quite large and probably tends to accelerate the reaction. We can readily explain the enhanced water adsorption Γ (S, H_2O) with respect to Γ (S, N_2) for various degrees of hydration/dehydration. Apparently the metallosilicate laminae are free to expand under the capillary forces of condensed water virtually without limit. Conversely, nitrogen sorption at -196°C is limited to the external surface and only to the interlaminar channels that are voided by high temperature dehydroxylation. Apparently the leaflets or laminae are expanded to a very limited extent, by nitrogen, as observed by others (11,12).

A further interesting comparison can be made, utilizing the data of Figure 7, with respect to what one would anticipate for the ink bottle or laminar surface configuration. In both cases the most labile surface species are probably the hydroxyls and hydrates on the external surface. Outgassing at lower temperatures will remove this complement initially (25 to 100°C) with a concurrent minor enhancement of Γ (m, N_2) due to increased roughness and/or energy of adsorption. Subsequently an ink bottle model would predict a very marked increase in Γ (m, N_2) and Γ (h, N_2) for a small increment of Γ (B, H_2O) (dehydration of restricted orifices) followed by a nearly constant, larger surface and volume capacity as the surfaces of the internal voids are dehydrated. Conversely the laminar model is more consistent with the experimental data. The surface capacity and volume capacity are increased in direct proportion to the amount of internal dehydration in the temperature realm of 100 to 300°C. This is the trend one would anticipate for interlaminar dehydration where accessibility is afforded by removal of the interlaminar water and/or hydroxyls. Their presence tenaciously binds the laminae to the substrate (especially at -196°C) by

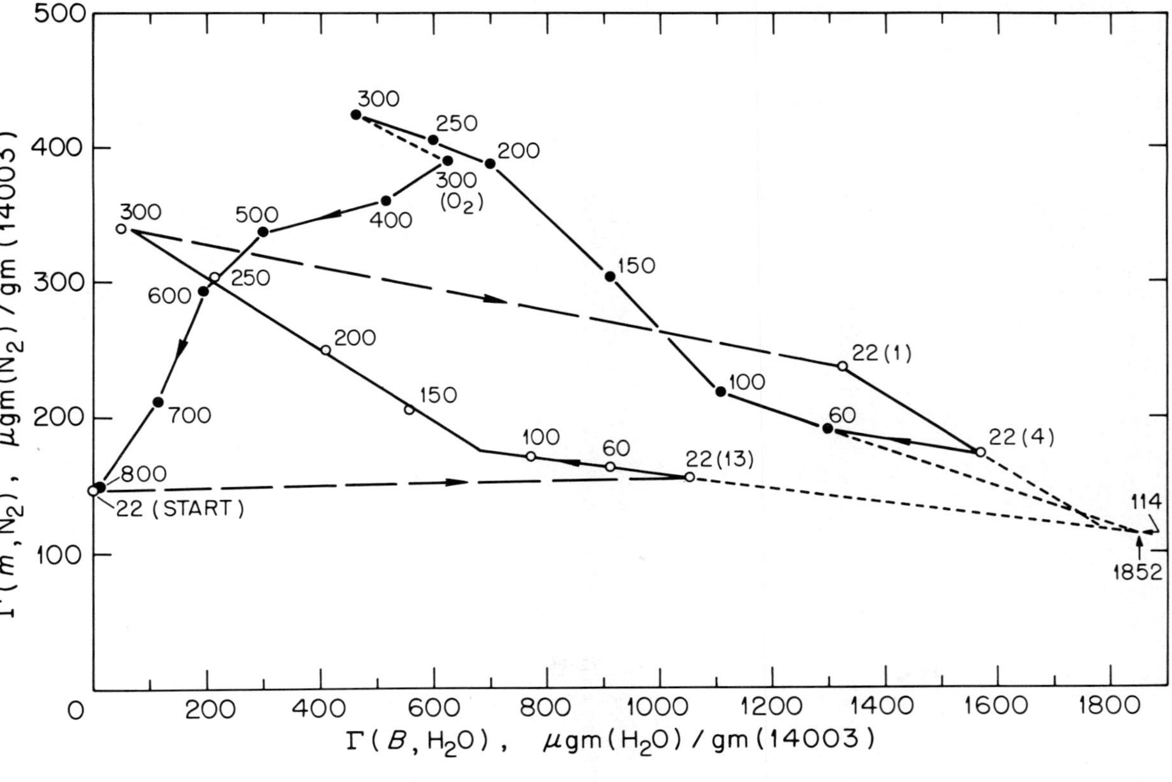

Fig. 7. Surface properties of 14003.

means of cross-linking hydrogen bonds.

Further credit for the laminar model is obtained in comparing our results to that of others. The total surface area enhancement due to dehydration can be calculated as the ratio of Γ (m, N_2) for 300°C outgassing to Γ (E, N_2) as 2.97, which is also semi-quantitatively the 400°C value of 3.05. These values are virtually identical to the 3.0 fold increase in Γ (m, N_2) noted for dehydration of montmorillonite (13). In our case we can see that the hydration mechanism has not proceeded into the bulk appreciably. Apparently we are limited to the formation of a single laminar metallosilicate layer over the external surface of the particles. At this juncture we have encountered the prohibitively slow kinetics that involve geologic time scales of weathering (14). Our limiting case of hydration involves only 0.18 weight percent of the sample. Even then, it is doubtful that we have formed a contiguous laminate over the surface. Rather, the hydration layer is probably comprised of small leaflets which yield even to the rather weak forces involved in the interlaminar condensation of nitrogen.

Additional credence for the metallosilicate model is obtained by the similar analysis of volcanic sand which had been subjected to extensive percussive grinding (5). This material behaved semi-quantitatively identical in virtually every respect to this and other lunar samples. Apparently the grinding process quite adequately simulates the meteoritic bombardment to produce fresh disordered surfaces. Temper this conclusion with the knowledge that ionic bombardment (15) of annealed lunar samples does appear to qualitatively regenerate some of the surface activity. However sample processing for the radiation experiment did involve some abrasion of the particles which assuredly disturbed (activated) surface layers. Moreover, heavy ion tracks are long in bulk material and the majority of lattice disruption occurs far below the crystal surface. In our experiment (15) the small particles (1 to 10 μm) were used. Thus the single particle sizes were less than the length (16) required to slow the ions to allow momentum transfer (and simultaneous lattice disruption). We can be assured that some (if not most) of the incident ions dissipated their energy near the surface of the particles that made up the heterogeneous target. Likewise, a large fraction of the macroscopically homogeneous ion beam (both in experiment and in the cosmic flux on the moon) will be slowed differentially in the bed of small particles (more in particle center and less near edges). This condition soon leads to a high degree of energy dispersion for lower layers of particles in a bed. All in all, it is difficult to envision ionic bombardment of small particles resulting in energy dissipation exclusively in classical ion tracks within the particles. If indeed this is the case we can readily explain the very small amount of reactivity (15) reinduced by experimental ion bombardment.

E. Summary

Water vapor reaction with lunar soil is closely related to the geochemical weathering of terrestrial primary metallosilicates. The product is a hydrated laminar metallosilicate comprised of a two-dimensional array much akin to clay minerals. The hydration process involves only the outermost complement of the particles which have been disordered by the eons of meteoritic impact and radiation damage in the lunar environ. Microgravimetric analyses prove to be very valuable in studying this weathering process in simulated arid climates (humidity fluctuations without solution transport). Alternate proposed mechanisms have been discredited as being minor contributions in terms of experimental results.

REFERENCES

1. Fuller, E. L., Jr., Holmes, H. F., Gammage, R. B., and Becker, K., Proceedings of the Second Lunar Science Conference (Suppl. 2) Geochim. Cosmochim. Acta Vol. 3, The MIT Press, p. 2009 (1971).
2. Fuller, E. L., Jr., J. Colloid and Interface Science (in press).
3. Cadenhead, D. A., Stetter, J. R., and Buergel, W. G., 47th National Colloid Conf., Ottawa (1973), J. Colloid Interface Science 47, 322 (1974).
4. Fuller, E. L., Jr., Holmes, H. F., and Secoy, C. H., Vacuum Microbalance Techniques 4, 109 (1964).
5. Fuller, E. L., Jr. and Agron, P. A. in Progress in Vacuum Microbalance Techniques Vol. 3, p. 71, C. Eyraud and M. Escoubes, eds., Heydon and Sons, London (1976).
6. deBoer, J. H., "The Structure and Properties of Porous Materials" p. 68, Butterworth (1958).
7. Gammage, R. B., Holmes, H. F., Fuller, E. L., Jr., and Glasson, D. R., J. Colloid and Interface Sci. 47, 350 (1974).
8. Borg, J., Maurette, M., Durrieu, L., and Jouret, C., Proc. 2nd Lunar Sci. Conf., Geochim. Cosomochim. Acta

(Suppl. 2) 3, 2027 (1971).
9. Ramberg, H., "Chemical Thermodynamics in Mineral Studies" in Physics and Chemistry of the Earth (Vol. 5) p. 226, Pergamon Press (1964).
10. Loughnan, F. C., "Chemical Weathering of the Silicate Minerals" American Elsevier (1969).
11. Keller, W. D. in Soil Clay Mineralogy, University of North Carolina Press, edited by C. I. Rich and G. W. Kronze (1964).
12. Rouquerol, F., Thesis, Centre d'Etude Nucléares de Saclay (1966).
13. deD.Lopez-Gonzalez, Jr. and Dietz, V., J. Res. NBS 48, 325 (1952).
14. Garrels, R. M., "Rates of Geochemical Reactions at Low Temperatures and Pressures" in Researches in Geochemistry (P. H. Abelson, ed.) John Wiley (1959).
15. Holmes, H. F., Agron, P. A., Eichler, E., and O'Kelley, G. D., Earth and Planetary Science Letters 28, 33 (1975).
16. Northcliffe, L. C. and Schilling, R. F., "Range and Stopping Power Tables for Heavy Ions" in Nuclear Data Tables (K. Way, ed.) Sec. A, Vol. 7, Nos. 3 and 4, Academic Press (1970).

Controlled Pore Silica Bodies Gelled From Silica Sol-Alkali Silicate Mixtures

Robert D. Shoup
Corning Glass Works

Abstract: Mixtures of potassium silicate and colloidal silica sol have been shown to possess reasonable resistance to spontaneous gelation. When caused to gel by addition of a latent acid reagent such as formamide, controlled pore silica structures were obtained. These structures ranged in pore size from 100Å to about 3600Å, and each specified composition had a relatively narrow pore size distribution. This technique is useful in forming high purity silica bodies with intricate geometric shapes.

I. INTRODUCTION

Materials commonly described as silica gels are generally formed by some destabilization technique involving aqueous alkali silicate solutions. Hydrogels can be formed by ion-exchange removal of alkali ions or, more commonly, by acid neutralization[1] of alkali silicates. These three dimension polymerized silica structures are characterized as having very small pore diameters, seldom greater than about 300Å.[2] Because these pores are so small, capillary forces[3] encountered in dehydration make it difficult to form coherent silica structures of practical size or with geometric complexity.

The purpose of this paper is to describe a technique for producing controlled pore silica structures which were formed by gelation of alkali silicate compositions with organic reagents such as formamide. Silica bodies were formed with narrow pore size distributions in the range from 100Å to about 3600Å. Bodies with pore diameters greater than about 600Å possess sufficient bond strength to resist fracture during processing. As a result, coherent intricate configurations were gel-molded.

II. EXPERIMENTAL

A. Materials

Pore size control in the gels was obtained by varying the ratio of potassium silicate to colloidal silica sol in the precursor mixtures. The two silicate solutions used in these studies were purchased from Philadelphia Quartz Company. The potassium silicate (Kasil 1) contained 20.8 weight percent SiO_2 and 8.3 weight percent K_2O. The sodium silicate (S-35) contained 25.3 weight percent SiO_2 and 6.75 percent Na_2O. DuPont's Ludox HS-40 which contained about 40 percent SiO_2 by weight was the colloidal silica source.

The organic gel reagent used in this study was formamide as received in reagent grade. Other reagents which were used successfully include formaldehyde and the solid polymer form, paraformaldehyde.

B. Procedures

A typical procedure used to produce a porous silica body which contains pores with average diameters of about 1800Å is as follows. Eighty grams of Kasil 1 and 20 grams of Ludox HS-40 are mixed to produce a homogeneous solution. To this is added about 10 grams of formamide with stirring. The solution is placed in a plastic container and allowed to gel. At room temperature gelation occurs in several hours, but by increasing the temperature gelation time can be reduced to minutes. The resulting silica body has the shape of the container in which it was gelled. This body can then be leached in weakly acidic solutions to produce a silica structure which contains less than 0.02 weight percent total alkali.

C. Analyses

Porosity data was obtained by the mercury intrusion method. Surface area was determined by nitrogen adsorption using B.E.T. techniques. Microstructures were obtained by standard Transmission Electron Microscopy methods.

III. RESULTS AND DISCUSSION

The study of the interactions of latent acid organic

reagents, such as formamide, with alkali silicate solutions has been reported by many workers. The ability to form coherent silica structures is primarily due to the controlled rate of gelation of these solutions. Polymerization of the silicate is thought to be induced by a decrease in pH of the solution as formamide undergoes hydrolysis. This technique of producing ammonium and formate ions in solution is much

$$HCONH_2 + NaOH + H_2O \rightarrow HCOONa + NH_4OH$$

more desirable than addition of this salt directly to the solution. The latter results in high localized concentrations which produce undesirable gelation conditions. Table 1 shows a series of our alkali silicate (and colloidal silica) compositions which decrease in pH as gelation occurs in the presence of formamide. In each case the supernatant solution has a pH below 11. This corresponds to a pH range described by Iler[(4)] wherein disilicate ions readily polymerize, especially in solutions of high silicate concentration. A dehydration mechanism may also be operative in these solutions,

Table 1
Change in pH During Gelation of Silicates with Formamide

Composition[a] [Kasil 1/Ludox HS-40]	pH Mixed Solution	pH Supernatant Solution
100% Kasil 1	11.90	10.77
9/1	11.70	10.76
7/3	11.64	10.75
5/5	11.58	10.75

a. ratio of weight plus 10% Formamide addition

but this is probably a minor effect. Regardless of the mechanism there definitely is phase separation into silica rich and alkali rich phases. The resultant silica body is easily leached in weakly acidic solutions to produce coherent silica structures retaining their mold shape and possessing less than 200ppm alkali.

Thus far, no differentiation between sodium silicate or potassium silicate has been made. Either silicate can be gelled with formamide to form coherent silica bodies with average pore diameters greater than 2000Å. The average pore diameter does vary somewhat depending on the body thickness and the concentration of silica in solution. A concentration effect is observed in sodium silicate gels which produce pore diameters as large as 6000Å. On the other extreme, the so-called silica gels are formed from very small particles which produce pore diameters seldom larger than 300Å. In an effort to bridge the gap between these extremes in pore structure, studies were undertaken to gel colloidal silica-alkali silicate mixtures with formamide.

As predicted, colloidal silica sols (Ludox HS-40) were found to gel spontaneously after addition of only 20 weight percent sodium silicate. Gels formed from mixtures with less than 20 percent sodium silicate had pore sizes less than 200Å.

On the other hand, homogeneous solutions of potassium silicate and Ludox HS-40 have reasonable stability over the entire compositional range. Upon gelation with formamide (10 weight percent), strong controlled pore silica structures were obtained. As Ludox HS-40 was increased in the mixture (Fig. 1), the average pore diameter of dealkalized silica bodies decreased from 3600Å to about 100Å. In addition to the

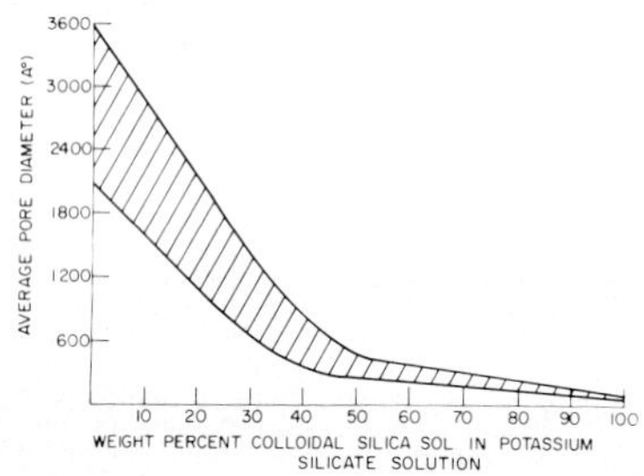

Fig. 1 Controlled pore silica bodies gelled from potassium silicate, Ludox HS-40 and formamide mixtures.

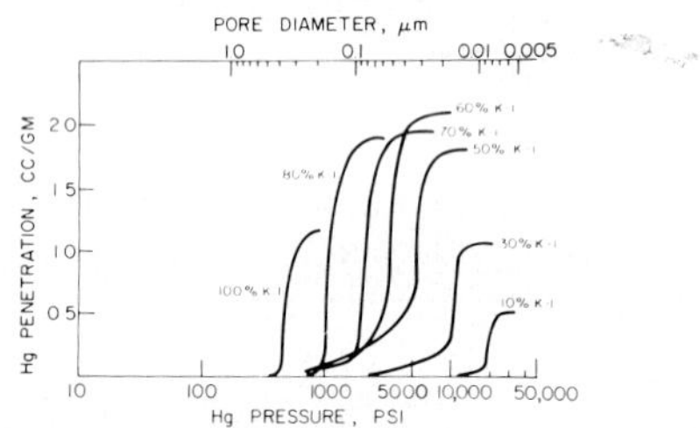

Fig. 2 Pore distribution curves for controlled pore dealkalized silica bodies as gelled in Fig. 1.

broad range of pore sizes that can be prepared, each range has a relatively narrow pore size distribution (Fig. 2). At least 80% of the pores are no greater than ± 30% and, commonly, as little as ± 10% from the average.

One theory which serves to explain the greater compatibility of potassium silicate (versus sodium silicate) in colloidal silica sols is based on the difference in effective size and charge density of the two hydrated alkali ions. The smaller size and higher charge density of the hydrated sodium ion permits it to more easily approach the colloidal silica particle and thus break down the energy barrier of the electrical double layer. Apparently, the lower charge density of the potassium ion is more effectively shielded by its large hydration sphere so as to prevent neutralization of charge and thus slow down the destabilization process. A similar response was observed when 0.1M NaOH or 0.1M KOH was added to Ludox HS-40. The NaOH induced spontaneous gelation whereas KOH was compatible with the silica sol for at least 24 hours. Allen and Matijevic[5] have observed higher critical coagulation concentrations for KCl as compared to NaCl at a pH greater than 10 in diluted Ludox AM sols. However, this did not occur in the Ludox AS system[6] where only CsCl had a higher c.c.c. at high pH values. Other works, such as Brady, et al.[7] have observed more rapid polymerization of silicate solutions by sodium salts than by potassium salts.

The shape of the silica body is determined by the container in which the precursor solution is gelled. The final dimensions of the body, however, vary somewhat with composition of the solution. Figure 3 shows that there is about

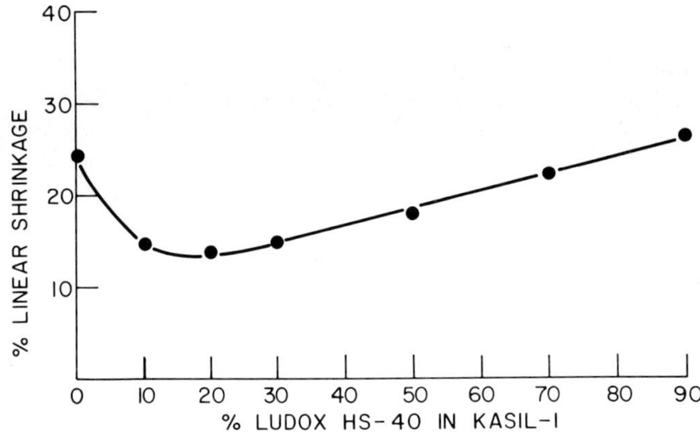

Fig. 3 Dimensional Change Accompanying Syneresis of Gelled Silicate Compositions.

13% to 25% linear shrinkage during gelation and syneresis of the 3-dimensional structure. The minimum in the curve may reflect a concentration effect in that Ludox HS-40 (40 weight percent silica) is being added to Kasil 1 which has a lower solids content (21% silica). Of course, it may also be related to establishing a critical concentration of colloidal nuclei in the mixture.

The most plausible explanation for the formation of controlled pore bodies with narrow pore size distributions is one involving nucleation. It appears that the colloidal silica remains as a stable dispersed phase and serves as nucleation or growth sites for polymerization of the molecular silicate. At high silica concentrations a small number of nuclei grow extensively to produce large pore dimensions. As the concentration of silicate in the mixture decreases, less growth occurs on a larger number of nuclei. Thus, pore size becomes ever increasingly the function of the small colloidal silica particle, 100 to 150Å. Transmission Electron Micrographs (Fig. 4) support this mechanism of particle growth.

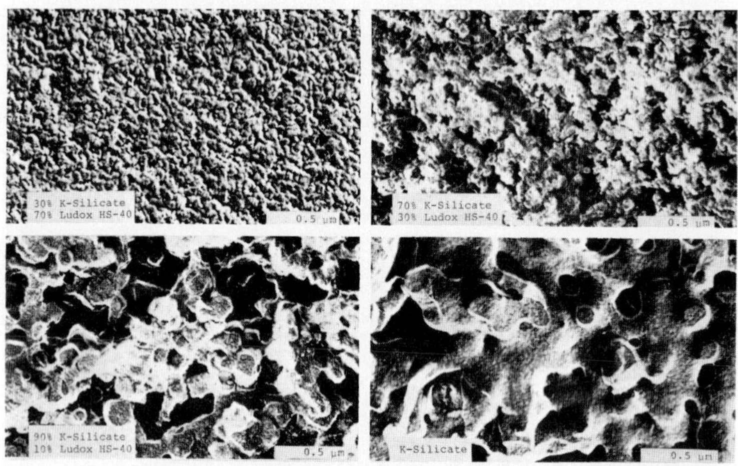

Fig. 4 TEM's of controlled pore silica structures gelled from different silicate mixtures.

The particle sizes estimated from these photomicrographs increase from about 300Å to about 1800Å as the potassium silicate increases in the precursor mixture. The number of discernible spherical particles also decreases as the silicate increases. Pore sizes estimated from these micrographs agree very closely with those obtained by the mercury intrusion technique in Fig. 2.

Finally, the nucleation-growth process which is induced in colloid-silicate mixtures is primarily responsible for developing increased body strengths. The bond strengths obtained in silica bodies with greater than about 600Å pore diameters are able to withstand shrinkage that accompanies syneresis. These larger pore structures are also able to withstand capillary forces involved in drying processes. The overall effect of this technique is to give one the capability of forming relatively large porous silica bodies in complex shapes that are not common to gelled silica materials.

REFERENCES

1. Vail, J. G., Soluble Silicates, Volume 2, Chapter 9, 511, Reinhold Pub. Co., N. Y., 1952.
2. Mober, P. K., Kirk-Othmer Encyclopedia of Chemical Technology, 2nd Ed., Volume 18, Amorphous Silica, 61, Interscience, 1969.
3. Alexander, A. E. and Johnson, P., Colloid Science II, 607, Oxford: Clarendon Press, 1946.
4. Iler, R. K., The Colloid Chemistry of Silica and Silicates, Chapter II, Cornell University Press, 1955.
5. Allen, L. H. and Matijevic, E., J. Colloid and Interface Science, 31, 287-296, 1969.
6. Allen, L. H. and Matijevic, E., J. Colloid and Interface Science, 33, 420-429, 1970.
7. Brady, A. P., Brown, A. G. and Huff, H., J. Colloid Science, 8, 252-276, 1953.

SURFACE AND BULK PROPERTIES OF AMORPHOUS CALCIUM PHOSPHATE

Stuart S. Barton and Brian H. Harrison
Dept. of Chemistry and Chemical Engineering
Royal Military College of Canada

Amorphous calcium phosphate (ACP) samples have been prepared by separation of the precipitate from the mother liquor after varying ageing times and subsequent freeze drying. Changes in Ca^{2+}/PO_4^{3-} ratio, water content and heat of solution in 1M HCl have been measured. Quantitative differential thermal analysis and heat of solution measurements indicate that the water content of ACP is essentially bulk water of hydration and that the process of ageing in the mother liquor is in part one of dehydration. Thermogravimetric and temperature programmed desorption experiments support the chemical analysis data and suggest that bulk water of hydration is removed completely at 773 K. B.E.T. surface area measurements show that surface area increases with time of ageing in the mother liquor to a constant value. For a single ACP sample, prepared by five minute ageing, it is shown that the surface activity as measured by heat of immersion in 2-propanol is not greatly influenced by the presence of weakly bound hydrate water except that a large increase in heat of immersion is observed when a small amount of water is initially removed from the surface. A further increase in heat of immersion is observed when the final small amount of water is removed.

I. INTRODUCTION

Hydroxyapatite (HA) is the final product of the reaction between high concentrations of calcium and phosphate ions in basic solution.[1] The initial precipitate from a solution having a Ca^{2+}/PO_4^{3-} ratio of 1.66 is not, however, hydroxyapatite ($Ca_{10}(PO_4)_6(OH)_2$) but an amorphous product with a Ca^{2+}/PO_4^{3-} ratio of about 1.5, corresponding to tricalcium phosphate. The amorphous calcium phosphate (ACP) is converted into HA on remaining in contact with the basic mother liquor for several hours. During this change the Ca^{2+}/PO_4^{3-} ratio of the solid phase increases and the Ca^{2+} content of the mother liquor decreases. It has been shown,[1], that the conversion process follows a sigmoid type time curve and at high pH is strongly influenced by the amount of HA already present. Increased pH of the mother liquor causes an increase in the half-life of ACP. Indeed at pH = 10, an induction period of some 100 minutes is observed so that it is convenient to isolate and study ACP by precipitation at pH>10.

In this study, ACP was prepared at pH = 10.5 from solutions having an initial Ca^{2+}/PO_4^{3-} ratio of 1.7. Samples were separated and stabilised after increasing periods of ageing time and their surface and bulk properties investigated.

II. EXPERIMENTAL

A. Preparation of ACP

ACP precipitates were prepared, at room temperature, by pouring simultaneously 500 ml of 0.75M calcium nitrate (reagent grade) and 0.44M dibasic ammonium phosphate (reagent grade) into one liter plastic centrifuge bottles. Both solutions were prepared using carbon dioxide free water and were adjusted to pH = 10.5 by the addition of concentrated ammonia. Immediately after mixing the centrifuge bottles were tightly stoppered, excluding air except for the small air space left in the bottles. The white precipitate was allowed to age in contact with the mother liquor for the prescribed periods of time with gentle agitation of the centrifuge bottle on a mechanical roller.

At the end of the ageing period the precipitate and mother liquor were separated in a refrigerated centrifuge (IEC, PR 6000) at 0° and washed and centrifuged four times with pH 10.5 ammonia water which had been pre-cooled to 0°C. Following the final washing the wet product was frozen in a glass vessel and pumped on a trapped vacuum line until a free-flowing powder was obtained. In this manner five ACP samples were prepared with ageing times of 5, 30, 60, 90 and 120 minutes. The ageing time is the time from solution mixing to the start of the centrifuge separation.

B. Apparatus and Procedures

B.E.T. nitrogen areas were measured as previously described assuming that the cross-sectional area of the nitrogen molecule is 16.2Å^2.(2)

Enthalpies of solution in 1M HCl and enthalpies of immersion in dry 2-propanol were determined in thermistor type calorimeters.(3) Quantitative differential thermal analyses were carried out on a Series 300 Fisher Thermalyzer and thermogravimetric experiments were done on a Cahn Microbalance system(4) using an Edwards Pirani Gauge 8/2 to monitor pressure between 10^{-4} and $5x10^{-3}$ torr.

Calcium and phosphate determinations were made following the procedure outlined earlier[3].

III. RESULTS AND DISCUSSION

Analytical, weight loss and nitrogen surface area data are given in TABLE 1.

TABLE 1 Analytical, Weigyt Loss and Surface Area Data of ACP

ACP	Ca^{2+}/PO_4^{3-}	Area m^2g^{-1}	% Weight Loss: Calc. from analysis	% Weight Loss: $273K^a$	% Weight Loss: $773K^b$	% Weight Loss: $1073K^b$
5 min	1.42±.01	41.5	16.9	8.1	17.0	18.1
30 min	1.50±.01	59.3	15.8	6.3	15.0	15.9
60 min	1.50±.01	92.5	13.6	5.9	13.9	14.3
90 min	1.52±.01	89.0	11.8	4.3	12.0	12.5
120 min	1.54±.01	90.0	9.92	1.65	9.2	9.8
HA 4 day[c]	1.62±.01	198	9.0	4.0	7.6	-

a. *After pumping for 24 hours.*
b. *Heating rate 1/2° min^{-1}.*
c. *See reference 3.*

The analytical data is in accord with the findings of other workers[1],[5] and indicates that the calcium-phosphate stoichiometry is close to that of tricalcium phosphate. The analytical data were also used to calculate, on the basis of hydrated sample weight, the percentage of ACP which was not calcium and phosphate. There is reasonable agreement between these calculated figures and the weight loss (at a heating rate of 1/2°-min^{-1}) up to 773 K. The smaller weight loss which occurred between 773 K and 1073 K may represent either the desorption of very strongly bound water or the removal of surface OH groups with the formation of pyrophosphate structures.

The specific surface area data are of the same order as those found by Holmes and Beebe[5] for 5 min. ACP. Ageing in the mother liquor results initially in an increase in surface area but a levelling off occurs after 60 minutes.

The weight loss which occurred above 300 K was investigated more closely in the following manner. After pumping at room temperature for 24 hours on the Cahn Microbalance, to remove the weakly bound water, the sample temperature was increased at 5°-min^{-1} and the pressure over the sample monitored at constant pumping speed.(6) The results are shown in Fig. 1.

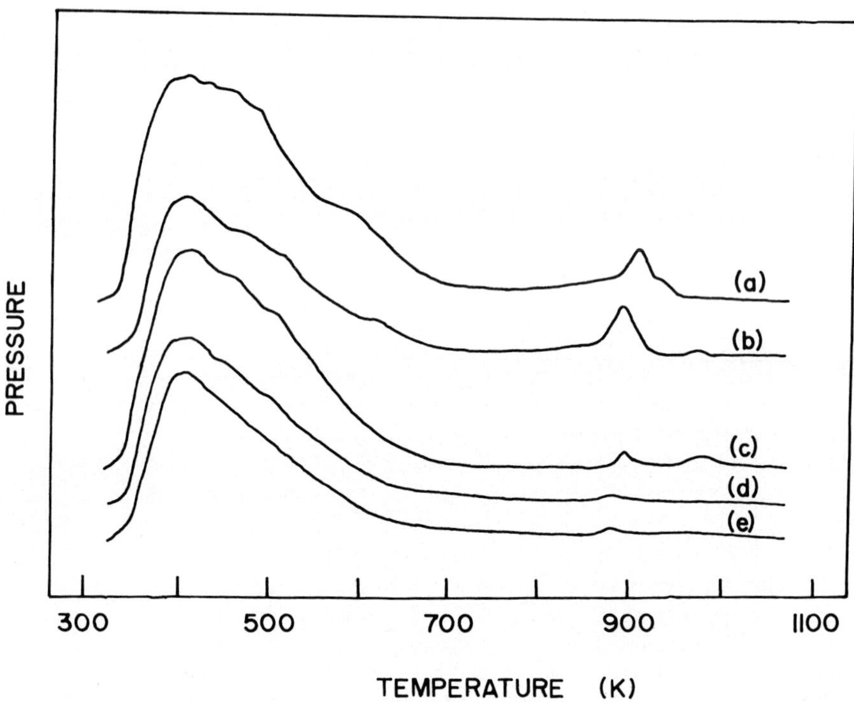

Fig. 1. Temperature programmed desorption spectrum of ACP.

a. 5 min, b. 30 min, c. 60 min, d. 90 min, e. 120 min.

With increasing ageing time the peak representing desorption around 400 K decreases slightly and becomes narrower indicating that the binding energy of the 400 K water becomes more homogeneous with ageing time. The higher energy shoulder which is very apparent in the 5 min. ACP diminishes with increasing ageing time is absent for the 120 min ACP. The 900 K peak, which presumably represents either the desorption of very strongly held water or the formation of pyrophosphate structures from hydroxyl groups, becomes smaller with increasing ageing time and has almost disappeared for the 120 min ACP sample

The binding energy of the weakly bound water which can be pumped off at room temperature was determined for the 5 min ACP by quantitative differential thermal analysis. Two types of run were made. In the first a 5 min ACP sample which had not been previously evacuated was heated at $10°\text{-min}^{-1}$ under vacuum. A clearly defined endothermic peak at about 360 K was observed. In the second type of experiment the 5 min ACP sample was evacuated at room temperature for 24 hours. The thermogram after room temperature evacuation showed no peak at 360 K. From the difference in the two thermograms, a calibration of the instrument with calorimetric standard indium and the assumption that the endothermic peak is due to water lost on room temperature evacuation, a value of $15.4 \pm .9$ k J-mole^{-1} for the binding energy of this water is obtained. This hydrate water is probably similar to that discussed by Sedlak and Beebe(7).

The question of the location of the bound water in ACP is difficult to answer. It is reasonable to assume that water is present both on the surface and in the bulk of the solid.

The observed B.E.T. surface areas can be used to calculate the number of layers of water on the surface if all the water present is confined to the surface. For the 5 min ACP a monolayer of water should be achieved at 1.2% water content (based on $41.5\ m^2\text{-}g^{-1}$ and the cross-sectional area of the water molecule being $10.5 Å^2$). To account for the total water content about 14 adsorbed layers would be needed. If, on the other hand, only a monolayer is present then 7% of the water is surface water and 93% is located in the bulk of the solid.

A further difficulty, that the water content of ACP is variable and depends strongly on the history of the particular sample under investigation was pointed out by Holmes and Beebe(5). These authors considered that 298 K degassing should be used to establish a reference point and that only further weight losses should be considered in making comparisons between samples. Since, except for the ageing period, all of our samples were prepared in the same manner we have not had recourse to this procedure but rather have focussed our attention initially on this weakly bound water. To check the result of the quantitative differential thermal analysis it was thought to be desirable to make a direct measurement of the enthalpy change when ACP dehydrated at room temperature is immersed in water. This procedure proved not to be feasible since, in addition to the expected rapid heat evolution, a large but slow exothermic heat was also observed. This large, slow heat evolution is probably due to the ACP to HA conversion which occurs at a reasonable rate at pH 7.[1]

A value for the binding energy of the weakly bound water was obtained by comparing the heat of solution of ACP, before and after room temperature pumping, in 1M HCl. Typical results for 5 min ACP are shown in Fig. 2.

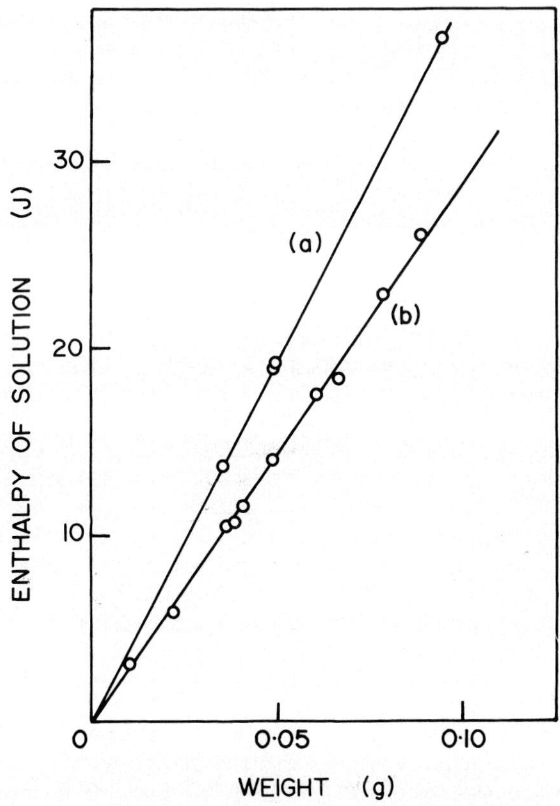

Fig. 2. Enthalpy of solution of 5 min ACP in 1M HCl (55 ml).

a. pumped at room temperature for 24 hours, 384 J-g^{-1}.
b. not evacuated. 287 J-g^{-1}.

If the heat of dilution of 1M HCl by the small amount of water present is ignored, then the difference in the slopes of the two lines divided by the molar water weight loss on pumping should give the enthalpy of hydration for this type of weakly bound water. The results obtained are shown in TABLE 2.

TABLE 2 Enthalpy of Hydration of ACP

ACP	Wt. Loss at 293 K mol g^{-1}	Enthalpy of Solution in 55 ml 1M HCl: Degassed J g^{-1}	Not Degassed J g^{-1}	ΔH kJ mol^{-1}
5 min	.0045	384	287	-22
30 min	.0035	402	334	-19
60 min	.0033	388	354	-10
90 min	.0024	392	382	-4.2
120 min	.00092	393	391	-2.2

These results indicate that the weakly bound water of ACP becomes increasingly less strongly bound as ageing time increases and that the small amount of pumpable water on the 120 min sample is probably simply occluded in the solid mass.

The question of how much of the weakly bound water is associated with the ACP surface remains. ACP is stable in ethanol and acetone(1) and no exothermic heat due to ACP to HA conversion should be expected on immersion in 2-propanol

Enthalpy of immersion measurements were carried out on the 5 min ACP. Small amounts of water were removed by rough pumping of calorimeter samples for periods of from 5 minutes to 2 weeks at room temperature. Weight losses were obtained by weighing the thin walled sample bulb, attached stem and sample before pumping and after careful sealing-off of the thin-walled sample bulb. Weight losses of from 2 to 10% were realized in this way. Further weight loss was achieved by pumping at 373, 473, 573 and 773 K for 24 hours. The enthalpy of immersion data are shown in Fig. 3.

A sharp rise in the enthalpy of immersion is observed from about 2% to 5% weight loss. Thereafter, the enthalpy value remains constant to about 17% weight loss where a slower rise occurs over the range 17% to 22%.

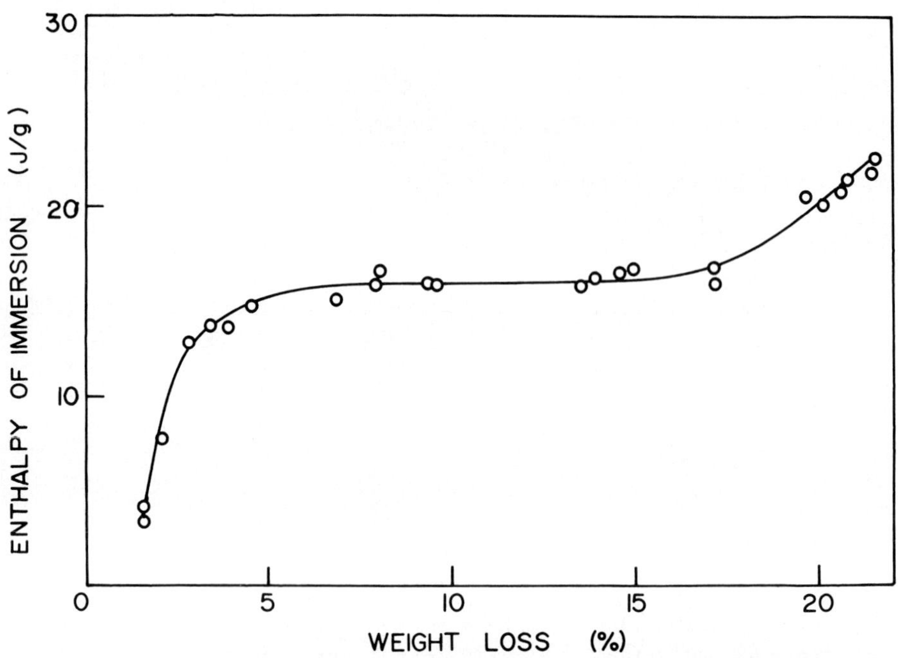

Fig. 3. Enthalpy of immersion of 5 min ACP in 2-propanol.

If it is assumed that wetting in 2-propanol affects only the surface of the ACP then the weight loss of 3% (from 2% to 5%) completely uncovers the surface and further dehydration takes place from the interior of the solid which is not accessible to the 2-propanol. The final rise in enthalpy value between 17% and 22% may be ascribed to the removal of very strongly bound surface water or to an increase in surface activity due to loss of surface OH groups and the formation of pyrophosphate structures. This increase in enthalpy of immersion occurs over the same temperature range where the shoulder on the 400 K peak of the desorption spectrum of 5 min ACP is observed (see Fig. 1).

It is interesting to observe that if the room temperature removable water of the 120 min ACP (1.65%) is very weakly bound, occluded water which is present to the same extent on all samples, then its removal should not affect the surface properties. This amount of water can then be subtracted from the 3% weight loss, which, according to the 2-propanol enthalpy of immersion data may be associated with the removal of the surface water from the 5 min ACP, arriving at a weight loss of 1.4% as representing the surface bound water on the 5 min ACP. This value agrees well with the calculated monolayer coverage of this sample.

It follows from this argument that there should be no surface water on the 120 min ACP sample. This point is being checked at the present time.

IV. REFERENCES

1. Boskey, A.L. and Posner, A.S., J. Phys. Chem. 77, 19 (1973), J. Phys. Chem. 80, 40 (1976).
2. Barton, S.S., Beswick, P.G. and Harrison, B.H., J. Chem. Soc. Far. Trans. 68, 1647 (1972).
3. Barton, S.S. and Harrison, B.H., J. Colloid Interface Sci. in press (1976).
4. Brown, J.G., Dollimore, J., Freedman, C.M. and Harrison, B.H., in Vacuum Microbalance Techniques, Vol. 8, p. 17, Plenum Press, 1971.
5. Holmes, J.M., and Beebe, R.A., Calc. Tiss. Res. 7, 163 (1971).
6. Cvetanovic, R.J., and Amenomiya, Y., Catal. Rev. 6, 21 (1972).
7. Sedlak, J.M., and Beebe, R.A., J. Colloid Interface Sci. 47, 483 (1974).

THE CHEMISTRY AND SURFACE CHEMISTRY OF OXIDES IN THE SILICA-ALUMINA RANGE

P.G. Rouxhet, P.O. Scokart, P. Canesson, C. Defossé, L. Rodrique, F.D. Declerck, A.J. Léonard, B. Delmon,
Groupe de Physico-Chimie Minérale et de Catalyse, Université Catholique de Louvain,

and J.P. Damon
Laboratoire de Chimie Générale, Université Scientifique et Médicale de Grenoble

ABSTRACT

Correlations are presented between the surface properties (surface area, acidity), the catalytic properties (cracking, isomerization) and the chemical composition of silica-aluminas prepared by coprecipitation and covering the complete range from pure silica to pure alumina. Attention is focused to the solid phases present, to the protonic acidity of the surface and to their dependance on chemical aspects of sample preparation and activation.

I. INTRODUCTION

Non crystalline silica-aluminas are an important type of catalysts used in petroleum processing (1,2,3). They are also relevant to soil science, due to the natural occurence of allophane, which presents similarities with the product obtained by coprecipitation of SiO_2 and Al_2O_3 (4,5,6). These important fields of application have motivated extensive research on silica-aluminas, either fresh or heat treated. However there is still much progress to be done in characterizing the surface properties and the structural organization of these compounds and in understanding the relationship between both aspects.

The aim of this contribution is to present a synthesis of relationships observed between the surface acidity, the catalytic properties and the chemical composition of silica-aluminas. It deals with a collection of samples covering the complete composition range,from pure silica to pure alumina, synthesized according to the same procedure and calcined at 500-550°C.

II. EXPERIMENTAL

A. Materials

The silica-alumina samples have been prepared from aluminum isopropylate and ethylsilicate, both purified by distillation, following a procedure similar to that used in reference (7). The two reagents were mixed together, while heated gently to lower viscosity, and introduced progressively into a 0.1 F acetic acid solution (10 ml per g oxide). The mixture was then refluxed at 80°C for 6 h, kept overnight at room temperature and, after addition of distilled water (1 ml per g oxide), refluxed again for 8 h; vigorous stirring was applied during the refluxing treatments and prolonged for 1 h. After one night the pH was raised progressively to 7 by an ammonia solution and kept at that value during 1.5 h; the material became then viscous but good stirring was maintained. It was then refluxed for 2 h, centrifuged and washed three times (3 ml water per g oxide). The recovered solid was heated up to 500°C with a heating rate of 30°C/min and maintained 24 h at this temperature.

The samples are identified by the sign SA- followed by their approximate SiO_2 weight percent and listed in table 1. Their chemical composition corresponds to that expected on the basis of the amounts of reagents used.

B. Methods

The surface area has been measured by adsorption of nitrogen at liquid nitrogen temperature (BET method). The cation exchange capacity is the amount of NH_4^+ retained after treatment by an ammonium acetate solution and washing by methanol (8).

The infrared spectra of the solids have been recorded in absorbance by a spectrograph Perkin Elmer 180; the powder was homogeneously dispersed in a KBr pellet, with a concentration corresponding to 1-2 mg oxide/cm^2. The X-ray diffraction patterns have all been obtained under the same conditions, using a Philips diffractometer and a Cu K_α radiation.

The other methods of characterization have been described in other publications. Those concerning the surface properties were preceded by a treatment at 500-550°C under vacuum.

TABLE 1

Surface area of the silica-aluminas and data on their chemical composition (Al_2O_3 content determined by conventional analysis and ratio of EMMA signals I_{Al}/I_{Si})

Sample SA- [a]	Al_2O_3 content (weight %)	I_{Al}/I_{Si} EMMA	Surface area (m^2/g)
100	0.0	0	452
95	4.5	0.047 ± 0.016	408
90	10.8	0.118 ± 0.027	295
85	14.8	0.216 ± 0.049	185
80	19.2	0.225 ± 0.050	213
75	26.1	0.278 ± 0.045	185
70	29.4	0.363 ± 0.037	189
65	35.9	0.431 ± 0.043	249
60	40.7	0.515 ± 0.053	266
55	46.3	0.711 ± 0.074	388
50	50.6	0.995 ± 0.107	229
45	54.9	1.17 ± 0.12	299
40	62.2	1.68 ± 0.27	369
35	(64.8)[b]	1.94 ± 0.27	348
30	70.5	2.55 ± 0.80	447
25	74.6	3.24 ± 0.59	357
20	80.5	4.13 ± 0.82	362
15	86.8	6.88 ± 1.38	373
10	(88.3)[b]	8.67 ± 2.16	345
05	93.6	19.51 ± 3.41	302
00	98.4	–	198

a *The number following SA- indicates the approximative SiO_2 content.*
b *Determined from EMMA measurements.*

III. RESULTS

A. Surface Properties

The surface acidity of the oxides has been characterized by non aqueous titrations using benzene as medium, butylamine as titrating agent, and carbonium ions forming indicators (9, 10). The series of indicators chosen allowed to titrate acidities covering the range from $H_R \leqslant -13.3$ to $H_R \leqslant +4.75$.

The shape of the acidity spectrum (density of sites vs H_R) is the same for all the samples in the range SA-95 to SA-30, and is similar to that reported for commercial silica-aluminas

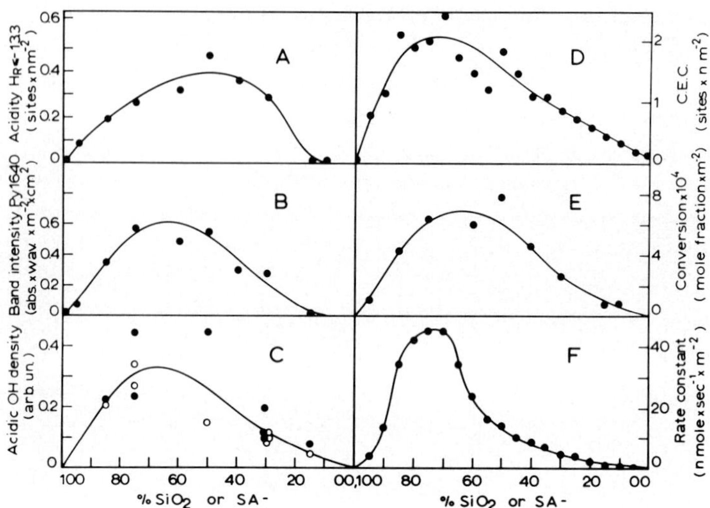

Fig. 1. Relationship between surface properties and chemical composition of silica-aluminas :

A. Strong acidity determined by titration ($H_R \leqslant -13.3$).

B. Maximum intensity reached by the 1640 cm^{-1} band of pyridinium formed by adsorption of pyridine.

C. Density of strongly acidic hydroxyls estimated through adsorption of benzene (●) and acetonitrile (○).

D. Cation exchange capacity.

E. Catalytic activity for dehydration of 4-methyl-2-pentanol with skeletal isomerization.

F. Catalytic activity for cracking of cumene.

(11,12). In particular it reveals the presence of very strong acid sites ($H_R \leqslant -13.3$), while moderately strong sites ($-13.3 < H_R \leqslant -4.04$) are practically absent. The surface density of these strong acid sites varies according to the composition of the silica-alumina as illustrated by Fig. 1A.

The infrared study of the adsorption of pyridine has been used to reveal the presence of protonic acid sites (13). Fig. 1B shows the variation, as a function of chemical composition, of the maximum intensity reached by the band at 1640 cm^{-1} characteristic of pyridinium ion. This intensity has been normalized with respect to the surface area and is thus proportional to the surface density of protonic acid sites.

The total quantity of strongly adsorbed pyridine, held on protonic and non protonic acid sites ,has been determined from correlations between the intensity of infrared bands and the adsorbed amount. It provided an estimate of the total density of acid sites, which is plotted as a function of chemical composition of the adsorbents in Fig. 5A. X-ray photoelectron spectroscopy (XPS) has also been used to study the adsorption of pyridine (unpublished data); the deduced density of acid sites is in good agreement with the infrared data, as illustrated by Fig. 5A.

The infrared study of the adsorption of H bond accepting compounds (benzene, acetonitrile) has been used to characterize the properties of the OH groups of silica-alumina surfaces (14,15). The presence of strongly acidic OH is revealed by a stretching band of bonded hydroxyls which shows a very high wavenumber shift with respect to the free OH band.The maximum intensity of that band, normalized with respect to the surface area of the adsorbent, gives the surface density of strongly acidic hydroxyls in arbitrary units. Its variation as a function of chemical composition is illustrated by Fig. 1C; for drawing this graph the data obtained from the adsorption of acetonitrile and benzene were superposed by considering that the absorption coefficient is about 3.1 times higher for OH interacting with acetonitrile than for OH interacting with benzene (16).

Fig. 1D gives a plot of the cation exchange capacity as a function of chemical composition.

The catalytic activity of the samples for dehydration of 4-methyl-2-pentanol has been determined at 360°C in a flow system (10). Some olefins produced result from a skeletal isomerization which requires strongly acidic sites (9). Fig. 1E gives a plot of the yield of these olefins, given by the degree of conversion of the alcohol divided by the surface area of the catalyst.

The catalytic activity for cracking of cumene into propylene and benzene has been determined at 420°C in a recycle flow reactor, allowing to eliminate diffusion limitation (unpublished data). Fig. 1F illustrates the variation of the rate constant, measured when the system has reached a stationary reaction rate and divided by the catalyst surface area.

B. Bulk Characterization

Fig. 2 gives representative infrared spectra of the solids. The absorption coefficients of the two more intense bands of pure silica have been measured by dividing their area (absorbance unit x wavenumber) by the effective sample thickness (mg/cm^2) and are plotted as a function of chemical composition in Fig. 4.

The evolution of the absorption coefficient of the Si-O stretching bands (1100-1200 cm^{-1}) is not different from what would be expected for a mechanical mixture of silica and

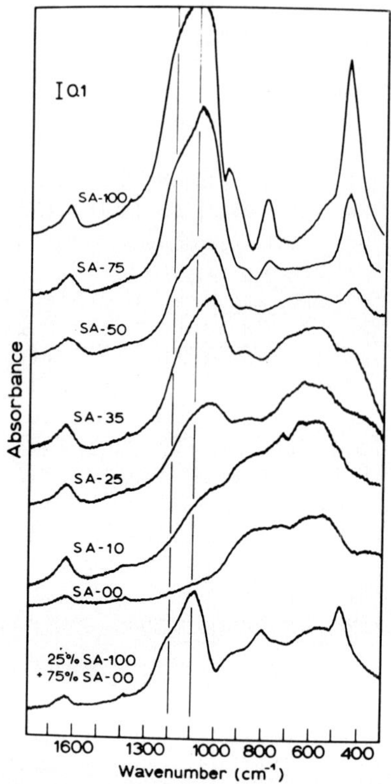

Fig. 2. Infrared spectra of representative silica-aluminas and of a mixture of the pure oxides.

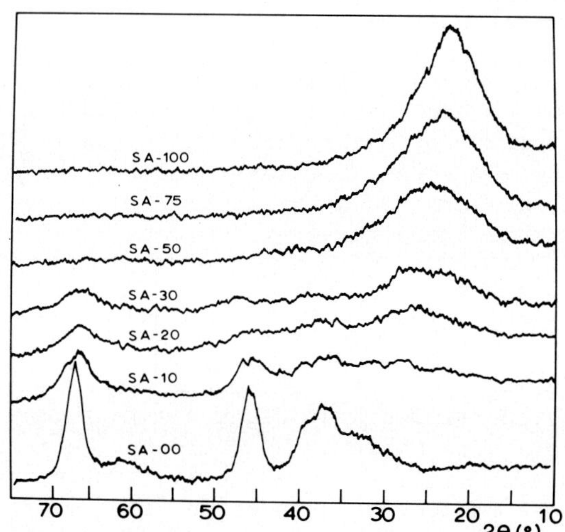

Fig. 3. X-ray diffraction patterns of representative silica-aluminas.

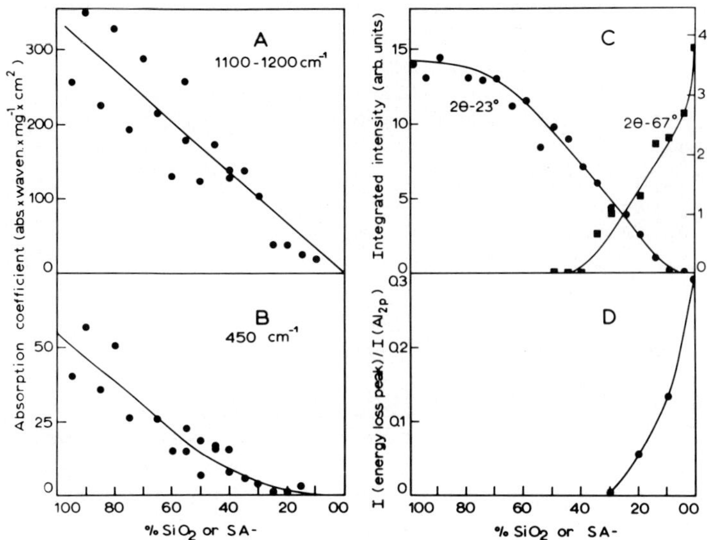

Fig. 4. Relationship between bulk characteristics and chemical composition of silica-aluminas :

A. *Absorption coefficient of the infrared bands around 1100 cm^{-1}.*
B. *Absorption coefficient of the infrared band at 450 cm^{-1}.*
C. *Area of the X-ray diffraction bands at 2θ around 23° (●) and 67° (■).*
D. *Intensity ratio of the energy loss peak over the parent peak for Al_{2p} XPS signals.*

alumina. However the bands broaden toward low wavenumber: the main band shifts from 1090 cm^{-1} for SA-100 to about 1050 cm^{-1} for SA-50; the shoulder observed at 1200 cm^{-1} for SA-100 shifts in a similar way and does no longer appear distinctly below 40% SiO_2.

The spectrum of SA-100 shows a shoulder at 970 cm^{-1} which is attributed to Si-O stretching of Si-OH groups (17). This absorption feature is very weak for SA-95 and is no longer observed for SA-90. For solids with a lower SiO_2 content, the main Si-O stretching band is extended to that spectral range. The band at 450-460 cm^{-1}, which is due to O-Si-O bending, drops quickly as the SiO_2 content decreases and is hardly observed below 30% SiO_2 (Fig. 4B). A similar trend is followed by the band at 800 cm^{-1} .

The evolution of the bands characteristic of the silicic component is also illustrated by a comparaison of spectra of SA-25 and of a mechanical mixture of 25% SA-100 and 75% SA-00

(Fig. 2). In addition, the incorporation of Al_2O_3 gives rise to the progressive development of broad bands between 500 and 1000 cm^{-1}. The profile of these bands seems to vary as a function of Al_2O_3 content; however the complexity of the spectra does not allow to refine these observations.

The X-ray diffraction patterns of SiO_2 rich samples show a broad band around $2\theta = 23°$ (Fig. 3). As the SiO_2 content decreases, the band becomes broader and shifts towards higher 2θ values; its area does not vary much between 100 and 75% SiO_2 and decreases regularly below 75% SiO_2 (Fig. 4C).

The Al_2O_3 rich samples present broad diffraction features which appear clearly for SA-00 and may be attributed to a η like alumina. A plot of the area of the peak around 67° as a function of chemical composition (Fig. 4C) shows that this alumina phase appears only below 40% SiO_2 and develops progressively as the SiO_2 content decreases.

The XPS spectra of the Al_2O_3 rich compounds present unexpected peaks which are due to a discrete energy loss of the electrons. The explanation of these energy loss peaks is not known; however they seem to be associated with the alumina phase (unpublished data). Fig. 4D gives a plot of the intensity ratio of the energy loss peak over the parent Al_{2p} peak; the variation obtained is indeed very similar to that of the X-ray diffraction peak at $2\theta = 67°$, which is characteristic of a transition alumina phase.

If account is taken of the energy loss peaks, the intensity ratio of the Al_{2p} and Si_{2p} photoelectron peaks is in very good agreement with the chemical composition (unpublished data); this shows that the chemical composition of the external surface of the oxide grains is the same as the bulk composition.

The chemical composition of the silica-alumina grains (1 to 10 μm diameter) obtained by grinding the calcined product has also been examined by the microanalysis attachment of an electron microscope (EMMA), following a method described previously (18). The intensity ratio of the Al over Si signals has been determined for about 30 or 60 grains of each silica-alumina and the results are listed in table 1. The average value gives a good relationship with the atomic ratio deduced from conventional chemical analysis:

$$I_{Al}/I_{Si} = 0.88 \ Al/Si$$

The relative standard deviation is about 20% of the average value and does not vary significantly as a function of chemical composition. As this deviation includes the analytical errors, it may be concluded that, contrarily to what has been suggested on the basis of the investigation of a limited number of samples (18), the chemical variability is smaller than 20% at the scale of a few μm and does not change appreciably with composition.

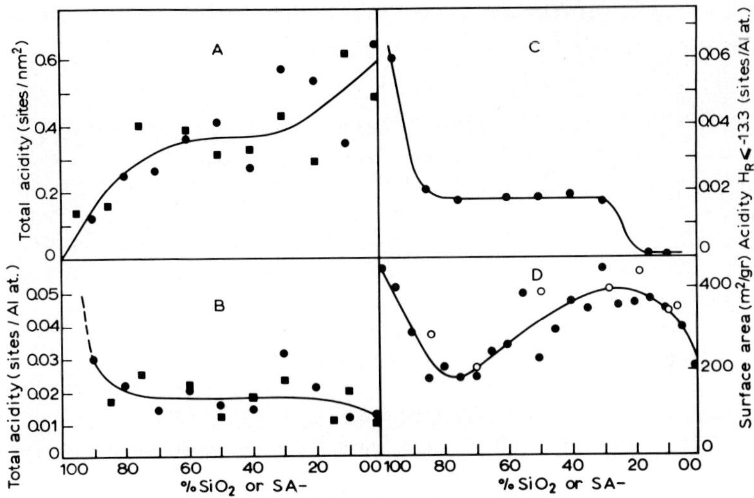

Fig. 5. Relationship between surface properties and chemical composition of silica-aluminas :

A. *Total acidity measured by pyridine adsorption, normalized with respect to unit area (■, Infrared; ●, XPS).*

B. *Same as A, normalized with respect to Al content*

C. *Strong protonic acidity determined by titration (cfr Fig. 1A), normalized with respect to Al content.*

D. *Surface area, (●, samples of the present work; O, samples of reference 7).*

IV. DISCUSSION AND CONCLUSION

A. Acid Sites

The strong acidity measured by titration corresponds essentially to strongly acidic hydroxyls detected by formation of pyridinium ions or by adsorption of H bond accepting compounds (13); this explains the close similarity of graphs A,B and C in Fig. 1.

The density of these protonic sites in the range of 85 to 50% SiO_2 (0.2 to 0.4 sites/nm^2 or 0.3-0.6 μmoles/m^2) is similar to that reported for silica-aluminas of various origins (11,12,19,20). The total density of acid sites is also in remarkable agreement with data reported for samples precipitated under quite different conditions (21). It seems thus that the surface properties of silica-alumina, related to unit area, do not vary dramatically as the preparation procedure changes.

It may be noted in Fig. 1 (graphs A,B,C) that the protonic acidity per unit area does not vary appreciably in the composition range of 75 to 40% SiO_2. If the protonic acidity is normalized with respect to the aluminum content instead of surface area (Fig. 5C), there is also no variation observed between 80 and 30% SiO_2. The constancy of the protonic acidity developed per unit area as well as per Al atom is explained by the evolution of the surface area, illustrated by Fig. 5D; it appears that the aluminum introduced in the products as the SiO_2 content decreases from 75 to 40% does not change appreciably the intrinsic properties of the surface but helps in increasing surface area.

The evolution of the surface area below 25% SiO_2 is such that the total acidity, mainly non protonic, developed per Al atom does not change appreciably (Fig. 5B) while the density of sites on the surface increases (Fig. 5A). At present it is not possible to assess the significance of that observation.

B. The Mixed Phase

The protonic acidity is due to a phase in which SiO_2 and Al_2O_3 are mixed at the atomic scale and therefore called mixed phase. The latter is responsible for catalytic activity in reactions such as polymerization, isomerization and cracking (Fig. 1,E and F; 1-3, 22-24). Careful measurements performed on the same silica-aluminas show however that the plot of the activity for cumene cracking (Fig. 1F) as a function of composition is appreciably narrower than the plot of protonic acidity. The relationship between the activity and the nature of surface sites is thus not completely clarified yet.

The mixed phase contains presumably a tetrahedral network of silica, with partial substitution of silicon by aluminum. Incorporation of aluminum in a silica network explains the shift and the shape modification of the Si-O stretching bands (25-27) as well as the change of the broad X-ray diffraction band of silica.

It has been found before (3) that silica-aluminas coprecipitated in presence of cations such as Na^+ or NH_4^+ have cation exchange properties, which are attributed to negative charges arising from substitution of Si^{4+} by Al^{3+}. According to the preparation procedure used in the present work it is thus expected that the fresh products in the range of 100 to 75% SiO_2 consisted in a tetrahedral network, with NH_4^+ as exchangeable cation. The first calcination may have produced a high density of acid sites like $\equiv Al^{IV}-\overset{H}{O}-Si\equiv$, similar to those occuring on a decationated zeolite (1,28) or H^+ exchanged layer silicates such as vermiculite or beidelite.

However such acidic surfaces are not stable in presence of water (29,30) and hydrolysis is expected to occur, either during calcination or upon subsequent exposure to humidity. This must alter a fraction of the solid and lead to saturation of the negative charge due to residual tetrahedral aluminum by aluminum containing cations. Recent measurements of the phosphorescence spectra of Fe^{3+} doped silica-aluminas (31) reveal indeed that extraction of Fe^{3+} ions from tetrahedral coordination sites results from calcination of NH_4^+ saturated samples.

Below 70% SiO_2, all the aluminum cannot be accomodated in a tetrahedral network and the amount in excess must contribute to balance the charges of the latter after calcination and possibly in the fresh gels.

When a silica-alumina is prepared by hydrolysis in absence of cation other than Si^{4+} or Al^{3+}, the negative charges of the tetrahedral network are already balanced by hydroxyaluminum cations in the fresh product (32). After calcination there may thus be a great similarity between silica-aluminas originating from a NH_4^+ saturated sample or prepared by simple hydrolysis; in both cases it has indeed been found that the ratio of aluminum in fourfold coordination to aluminum in sixfold coordination is fairly close to 1 (32,7). The situation should not be very different in the case of silica-alumina obtained by acid treatment of a Na^+ saturated gel. This explains that the surface properties of silica-alumina, related to unit area, do not depend strongly on the preparation procedure.

The protonic acidity of calcined silica-alumina may be attributed to $\equiv Al^{IV}-\overset{H}{O}-Si\equiv$ sites (15), which are formed accidentally upon dehydration of the entity constituted by the tetrahedral network and the hydroxyaluminum cations. Indeed it is seen in Fig. 5 that the acidity preserved is only about 1 site for 50 Al atoms present in the solid or 0.2 to 0.4 sites/nm^2. The much higher value of the cation exchange capacity may be due either to partial displacement of Al containing cations by NH_4^+ or to simultaneous adsorption of NH_4^+ and acetate ions, when the silica-alumina is contacted with a solution of ammonium acetate. For having a clear meaning the cation exchange capacity measurement should be accompanied by the determination of acetate adsorption and aluminum complexes desorption.

C. Organization of the Solid

As surface properties per unit area are concerned, the

mixed phase appears to build up as the SiO_2 content decreases from 100 to 75%. However the surface area developed by the mixed phase rises as the SiO_2 content decreases from 75 to 30% (Fig. 5D); this observation is reproducible, as shown by data obtained for another collection of samples prepared in the same way (7). According to the preparation procedure used here, the increase of surface area may be due to the presence in the fresh gel of a weakly polymerized alumina which affects the coagulation of the colloidal particles. This behavior must be very sensitive to experimental conditions; for instance substitution of aluminum isopropylate by aluminum nitrate gives a quite different evolution of the surface area (7).

The segregation of an alumina phase is only observed below 40% SiO_2; its development is revealed by X-ray diffraction and affects the XPS spectra. This phase results presumably from the decomposition of aluminum hydroxides which have crystallized separately (32); some characteristics of its surface hydroxyls and non protonic acid sites have been described previously (13,15).

The evolution of some parameters (Fig. 5, C and D) suggests that, between 40 and 0% SiO_2, the mixed phase is progressively diluted by the alumina phase, as regards surface properties (Fig. 1). However according to other data (Fig. 5, A and B), properties of the mixed phase do no longer show up below 20% SiO_2; the catalytic activity measurements (Fig. 5, E and F) do not provide additional information. In fact this concerns the environment of silicon in a silica-alumina of low SiO_2 content. It has been suggested (7) that in this concentration range, the silicon may occur in a mullite like phase, which is characterized by a high Al^{IV} over Al^{VI} ratio and by isolated silica tetrahedra, with a low apparent coordination number for Si. This may be supported by the fact that the intensity of the band at 450 cm^{-1}, due to O-Si-O bending, decreases more quickly than that of the Si-O stretching bands (Fig. 4 A and B). However no firm conclusion can be drawn at present, due to the low accuracy of the experimental data and the complexity of their meaning.

As the silica-aluminas present some heterogeneity in their organization, it is worth to recall that they do not show significant variation of chemical composition at the scale of a few μm.

The authors acknowledge the support of Fonds National de la Recherche Scientifique and of Musée Royal de l'Afrique Centrale (Belgium).

V. REFERENCES

1. Tanabe, K., "Solid Acids and Bases", Academic Press, London, 1969.
2. Tamele, M.W., Discussions Faraday Soc. 8, 270 (1950).
3. Milliken, Jr., T.H., Mills, G.A., and Oblad, A.G., Discussions Faraday Soc. 8, 279 (1950).
4. Kitagawa, Y., Amer. Miner. 56, 465 (1971).
5. Henmi, T., and Wada, K., Clay Miner. 10, 231 (1974).
6. Cloos, P., Herbillon, A., and Echeverria, J., Trans. 9th Intern. Congr. Soil Sci., Vol. II, 733 (1968).
7. Léonard, A.J., Ratnasamy, P., Declerck, F.D., and Fripiat J.J., Discussions Faraday Soc. 52, 98 (1971).
8. Mackenzie, R.C., J. Colloid Sci. 6, 219 (1951).
9. Damon, J.P., Bonnier, J.M., and Delmon, B., J. Colloid Interface Sci. 55, 381 (1976).
10. Damon, J.P., Delmon, B., and Bonnier, J.M., J. Chem. Soc., Faraday Trans. I, to be published.
11. Benesi, H.A., J. Phys. Chem. 61, 970 (1957).
12. Hirschler, A.E., J. Catalysis 2, 428 (1963).
13. Scokart, P.O., Declerck, F.D., Sempels, R.E.,and Rouxhet, P.G., J. Chem.Soc., Faraday Trans. I, to be published.
14. Rouxhet, P.G., and Sempels, R.E., J. Chem. Soc., Faraday Trans. I, 70, 2021 (1974).
15. Sempels, R.E., and Rouxhet, P.G., J. Colloid Interface Sci. 55, 263 (1976).
16. Sempels, R.E., and Rouxhet, P.G., Bull. Soc. Chim. Belg. 84, 361 (1975).
17. Moenke, H.H.W., in "The infrared spectra of minerals" (V.C. Farmer, Ed.) p. 365, Mineralogical Society, London, 1974.
18. Rodrique, L.,and Declerck, F.D., C.R. Acad. Sc. Paris 280, Sér. C, 1061 (1975).
19. Hirschler, A.E., and Schneider, A., J. Chem. Eng. Data 6, 313 (1961).
20. Take, J., Tsuruya, T., Sato, T., and Yoneda, Y., Bull. Chem. Soc. Japan 45, 3409 (1972).
21. Schwarz, J.A., J. Vac. Sci. Technol. 12, 321 (1975).
22. Holm, V.C.F., and Clark, A., J. Catalysis 2, 16 (1963).
23. Ward, J.W. and Hansford, R.C., J. Catalysis 13, 154 (1969).
24. Amenomyia, Y., and Cvetanovic, R.J., J. Catalysis 18, 329 (1970).
25. Léonard, A., Suzuki, S., Fripiat, J.J., and De Kimpe, C., J. Phys. Chem. 68, 2608 (1964).
26. Milkey, R.G., Amer. Miner., 45, 990 (1960).
27. Flanigen, E.M., Khatami, H., and Szymanski, H.A., in "Molecular Sieve Zeolite I" (R.F. Gould, Ed.), p. 201, Adv. Chem. Ser. 101, Amer. Chem.Soc., 1971.

28. Uytterhoeven, J.B., Christner, L.G., and Hall, W.K., J. Phys. Chem. 69, 2117 (1965).
29. Grim, R.E., "Clay Mineralogy", Mc Graw Hill, 1968.
30. Kerr, G.T., J. Phys. Chem. 73, 2780 (1969).
31. Pott, G.T., 6th Intern. Congress Catalysis, London 1976, B3.
32. Cloos, P., Léonard, A.J., Moreau, J.P., Herbillon, A., and Fripiat, J.J., Clays Clay Miner. 17, 279 (1969).

INFRARED SPECTRA AND SURFACE DECOMPOSITION OF NH_3 AND ND_3 ADSORBED ON GERMANIUM FILMS

Z. Calahorra, A. Lubezky and M. Folman
Technion - Israel Institute of Technology

ABSTRACT

Infrared spectra of NH_3 and ND_3 adsorbed on high surface area evaporated germanium films were investigated over a range of temperatures utilizing a U.H.V. system.

On admission of NH_3 to freshly prepared films maintained at 195°K adsorption bands due to physisorbed molecules appeared in the spectrum at frequencies 3350 cm^{-1}, 3275 cm^{-1}, 1620 cm^{-1}, and 1060 cm^{-1}. The first two bands are assigned to ν_3 and ν_1 modes and are shifted to lower frequencies by 64 cm^{-1} and 61 cm^{-1}. The two remaining absorptions are assigned to ν_4 and ν_2 modes and are shifted to lower and higher frequencies by 7 cm^{-1} and 110 cm^{-1} respectively. In addition a band at 1260 cm^{-1} is observed with a peak frequency strongly coverage dependent. It is believed that this band is also due to the ν_2 mode and may be assigned to molecules coordinatively bound to the surface.

On heating the system to room temperature decomposition of the adsorbate takes place. An absorption due to formation of Ge-H bonds appears at 1900-1950 cm^{-1} and a second band, which is assigned to Ge-N bonds, is present at 550 cm^{-1}. On increasing the temperature, the absorptions due to physisorbed NH_3 disappear. The decomposition process is noticed already at 195°K.

A series of measurements were done with ND_3. For physically adsorbed molecules the four absorptions were obtained at frequencies expected from isotope shifts. An additional band at 950 cm^{-1} was assigned, as in the previous case, to ND_3 molecules bound coordinatively to the surface. The Ge-D vibration appeared at 1380 cm^{-1} whereas the Ge-N band appeared at 550 cm^{-1} similarly to the band obtained with NH_3.

INTRODUCTION

There exist a large number of investigations concerning

adsorption of gases and vapours on germanium.[1] These studies were undertaken mainly to investigate changes in the electric properties of this element on adsorption. Also some i-r spectroscopic measurements in which Ge served as adsorbent were made in the past where different i-r techniques were employed.[2]

The present investigation was undertaken to study the nature of NH_3 adsorption on this substrate. Spectra of NH_3 and ND_3 adsorbed at different temperatures on high surface area Ge films were recorded. At low temperatures, absorption bands due to physisorbed NH_3 and Ge-H, Ge-N(weak) bonds were obtained. At higher temperatures the absorptions due to physisorbed NH_3 disappeared and the absorptions due to Ge-H and Ge-N strengthened. Spectra recorded on adsorption of ND_3 showed features resulting from isotopic substitution.

EXPERIMENTAL

Due to the extremely high reactivity of Ge towards oxygen and H_2O, an i-r adsorption cell was designed for work under U.H.V. conditions. The cell in general resembled the low temperature i-r cell used and described previously.[3] To enable evacuation at high temperatures (up to 550°K) an electric heating element was inserted into the metal holder and frame of the central window. The main gas outlet was made of thin stainless steel and could be raised to the mentioned temperature by direct electric heating. The two electric elements enabled heating of the whole adsorption cell, including the two CsBr windows. The transparent windows were glued to thin copper foils by means of Eccoband 104 adhesive and were coated with a layer of silicon vacseal (G.E.). The copper foils were tightened to the body of the cell by means of gold o-rings. All the seals were made of stainless steel flanges and gold o-rings *(Fig. 1.)*. The cell was attached to a U.H. vacuum bakable system using a Granville-Phillips membrane valve.

The all-metal U.H.V. pumping system included a 3" oil diffusion pump (150 l/sec), and a liquid nitrogen trap backed by a mechanical oil pump. The low pressure in the system was measured by means of a Bayard-Alpert ionization gauge, and higher gas pressures by means of two thermocouple vacuum gauges (Hastings-Raydist) for pressure ranges 0-0.1 torr and 0-20 torr.

The gases used in the adsorption measurements were stored in glass bulbs attached through metal valves to a side line of the system.

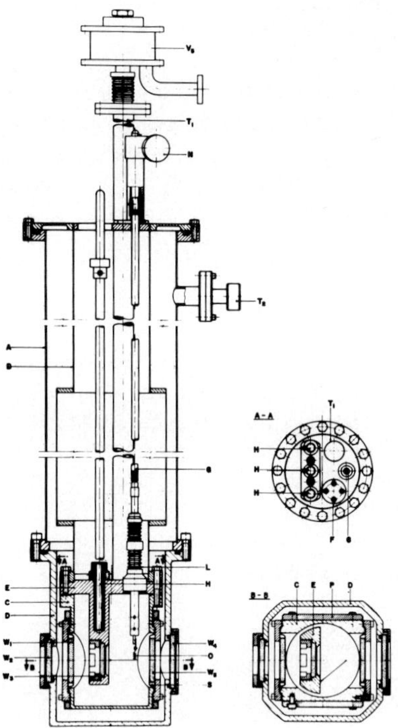

Fig. 1. The adsorption cell.

After evacuation and baking of the system for 24 hours, pressures as low as 10^{-8} torr could be obtained. It is believed that the pressure inside the dewar cell maintained at 77°K was still lower, which may be inferred from the fact that the Ge films could be kept for several hours without noticeable contamination, as judged from the i-r spectrum.

The germanium films were prepared by evaporation of small crystals and deposition of the vapours on the cold CsBr window (77°K). The crystals were placed inside a basket-shaped tungsten filament and outgassed <u>in vacuo</u> for 2 hours at 1100-1150°K. Then the temperature of the filament was raised to ~1500°K and the electric current was passed until the evaporation of the germanium was complete. To increase the surface area of the films, the evaporation was performed in an inert gas atmosphere. Helium of spectroscopic grade at a pressure of 0.15 torr appeared most adequate for this purpose. No accurate B.E.T. measurements could be made due to the large dead space of the cell, but it is believed that under the specified working conditions the surface area of the films was of the order of tens of square meters per gram.

The films prepared in the He atmosphere had a mat non-reflecting surface as distinguished from the films prepared in vacuo which had a mirror-like appearance, showed much better adherence to the window but were of much lower specific surface area. The films were usually annealed at 195°K which was also the temperature for NH_3 adsorption.

The spectra were recorded by means of a 521 Perkin-Elmer grating spectrophotometer with spectral slit width of the order of 1.5 cm^{-1} (at 3000 cm^{-1}) - 2.2 cm^{-1} (at 500 cm^{-1}).

RESULTS

The transparency of the evaporated germanium films was good over the spectral region studied (4000 cm^{-1} - 350 cm^{-1}). After evaporation, the i-r radiation was attenuated by about 15%. NH_3 and ND_3 were adsorbed on the Ge films at 195°K at different equilibrium pressures (0.1-0.3 torr). On introduction of NH_3 to the cell, absorptions appeared at 3350 cm^{-1}, 3275 cm^{-1}, 1880 cm^{-1}, 1620 cm^{-1}, 1260 cm^{-1} (low coverage, 1290 cm^{-1} high coverage), 1060 cm^{-1} and 550 cm^{-1} *(Figs. 2. and 3.)*.

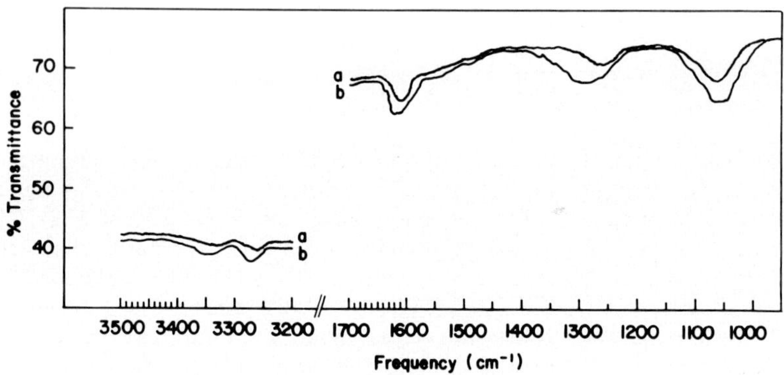

Fig. 2. I.R. spectrum of NH_3 adsorbed on a Ge film at 195°K. Equilibrium pressures a. 0.1 torr, b. 0.3 torr.

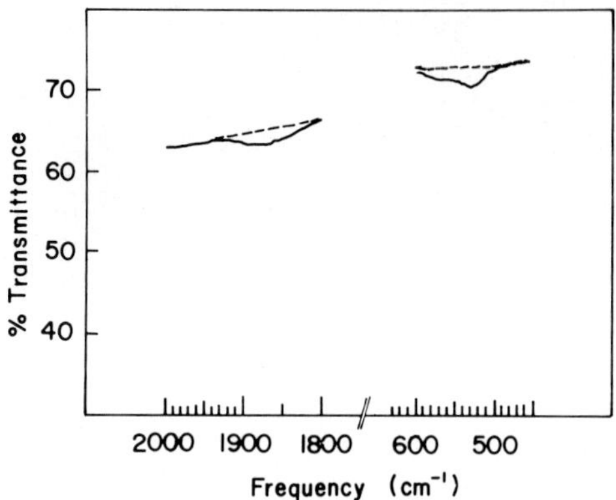

Fig. 3. I.R. spectrum of NH_3 adsorbed on a Ge film at 195°K in the 1900-1800 cm^{-1} and 600-500 cm^{-1} regions. ------- background ——————— at 0.3 torr equilibrium pressure.

On pumping the system, the bands at 3350 cm^{-1}, 3275 cm^{-1}, 1620 cm^{-1} and 1060 cm^{-1} became much weaker. Whereas no intensity changes were observed in the bands at 1880 cm^{-1} and 550 cm^{-1}, the absorption at 1260 cm^{-1} changed only slightly *(Fig. 4.)*.

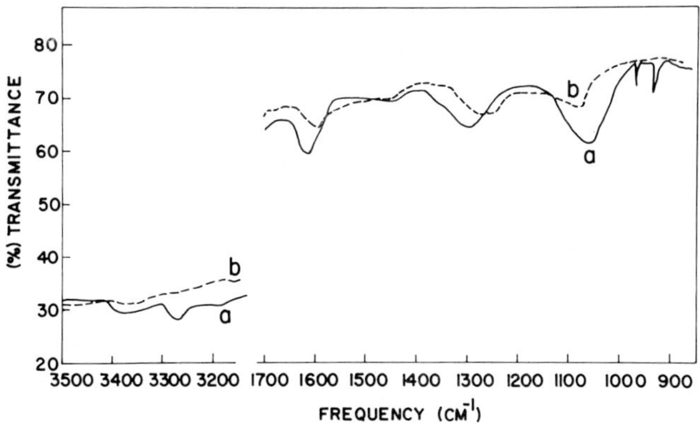

Fig. 4. I.R. spectrum of NH_3 adsorbed on a Ge film on desorption. Equilibrium pressures a. 3 torr, b. 3 x 10^{-3} torr.

These intensity changes were quite reproducible. When the cell temperature was raised to 300°K in the presence of NH_3, only two absorptions were seen in the spectrum. The one at

1880 cm^{-1} became much more intense and shifted to 1950 cm^{-1}, whereas the absorption at 550 cm^{-1} remained unchanged in frequency and its intensity increased *(Fig. 5.)*.

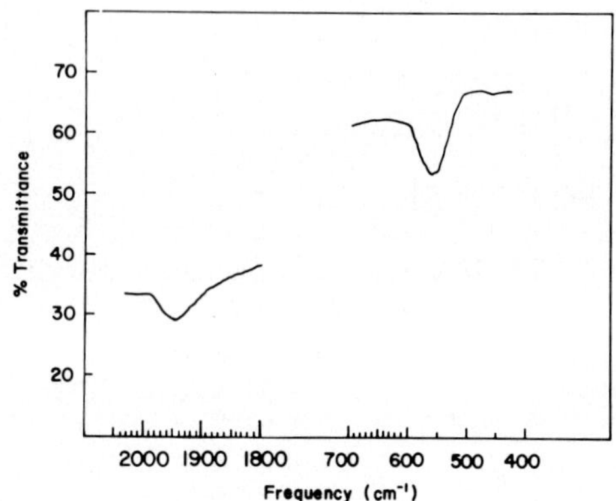

Fig. 5. I.R. spectrum of NH_3 in the 2000-1900 cm^{-1} and 600-500 cm^{-1} regions at 300°K.

When ND_3 was adsorbed at 195°K, the spectrum shown in *Fig. 6.* was obtained.

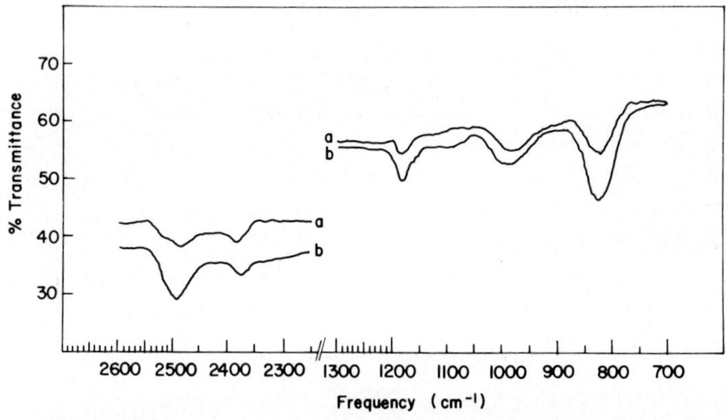

Fig. 6. I.R. spectrum of ND_3 adsorbed on a Ge film at 195°K. Equilibrium pressures a. 5 x 10^{-2} torr, b. 1 torr.

Absorptions at 2490 cm^{-1}, 2380 cm^{-1}, 1180 cm^{-1}, 950 cm^{-1} (low coverage) - at high coverage 990 cm^{-1} and 820 cm^{-1} were seen in the spectrum. On desorption, all the absorptions became much weaker except for the absorption at 950 cm^{-1} which showed only moderate changes *(Fig. 7.)*.

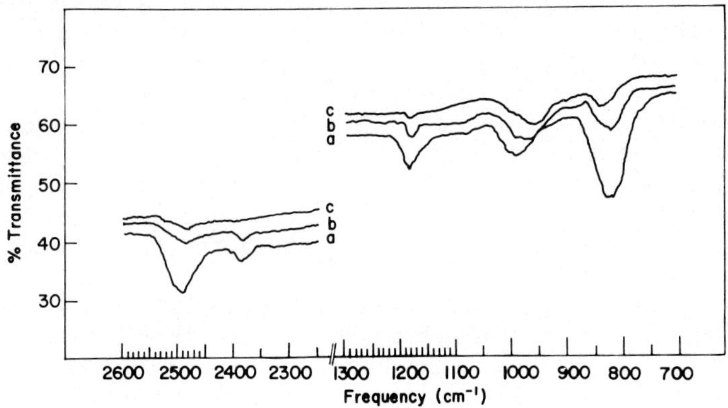

Fig. 7. I.R. spectrum of ND_3 on a Ge film on desorption. Equilibrium pressures a. 1.7 torr, b. 2.5 x 10^{-2} torr, c. 2 x 10^{-3} torr.

On introducing ND_3 to the cell at 195°K and raising the temperature to 300°K, absorptions at 1380 cm^{-1} and 550 cm^{-1} and a weak and broad band at 1910 cm^{-1} were obtained *(Fig. 8.)*.

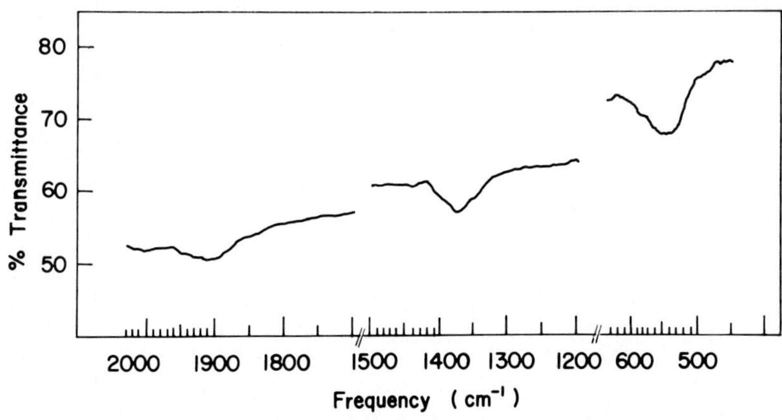

Fig. 8. I.R. spectrum of ND_3 adsorbed on a Ge film at 300°K.

DISCUSSION

The gas phase spectrum of NH_3 and ND_3 is well-known and the assignments of the absorptions are given in the literature.[4] The i-r spectrum of NH_3 adsorbed on different substrates was also studied.[5] To our knowledge no spectra of NH_3 adsorbed on Ge were investigated until now.

The results described in the previous section indicate that the adsorption of NH_3 on Ge is a rather complex process.

All the absorptions obtained at 195°K (except the bands at 1880 cm^{-1} and 550 cm^{-1}) are assigned to NH_3 adsorbed physically, since on evacuation of the film they disappeared gradually even at low temperatures.

The absorption obtained at 3350 cm^{-1} is ascribed to the ν_3 vibration. It is shifted to lower frequencies by 64 cm^{-1} with respect to the gas phase band. The band at 3275 cm^{-1} is assigned to the ν_1. The band at 1620 cm^{-1} corresponds to ν_4 whereas the absorption at 1060 cm^{-1} is assigned to ν_2.

The ν_1 and ν_4 bands are shifted to lower frequencies with respect to the gas phase by 61 cm^{-1} and 7 cm^{-1} whereas the ν_2 is shifted to higher frequencies by 110 cm^{-1}.

The absorption at 1260 cm^{-1} is assigned to the ν_2 vibration mode of NH_3 molecule bound coordinatively to the surface and is shifted by as much as 310 cm^{-1}, as compared to the gas phase. The existence of two absorptions assigned to the ν_2 mode indicates that NH_3 is adsorbed in two different ways. The absorption at 1060 cm^{-1} which disappears easily on pumping, is connected with molecules bonded weakly to the surface, whereas the band at 1260 cm^{-1} belongs to molecules attached to the surface in a much stronger way. This bonding is probably of coordinative nature. There exist examples of coordinative bonding of NH_3 to adsorbent surfaces. NH_3 on γ-alumina gives a band at 1280 cm^{-1} which is ascribed to the ν_2 mode of NH_3 coordinated to Al atoms.[6] In NH_3 complexes with transition metals the ν_2 absorptions are obtained in the region 1100 cm^{-1} - 1290 cm^{-1}.[7] The possibility of coordinative bonding, in which the nitrogen of the NH_3 molecule acts as electron donor, may be explained quite plausibly on the grounds of known properties of germanium surfaces. It is known that in evaporated germanium films, the equilibrium crystal face is mainly the (111) plane. In that plane a surface germanium atom forms three bonds with three neighbouring germanium atoms leaving one free bond (dangling bond). This bond may accommodate one electron from the valence band, leaving a hole (space charge layer) which gives the p-type nature to the surface region.[8] Negatively-charged sites such as these may give rise to certain chemical reactions, like reduction of NO to N_2O.[9] The second possibility may be a transfer of the free electron from the dangling bond to the conduction band or to an existing hole in the valence band.[10] The remaining Ge^+ atom may accept a pair of electrons from a donor molecule leading to a coordinative type of bonding. Of all the vibration modes of adsorbed NH_3, only the ν_2 showed noticeable difference in frequency for physically and coordinatively bound species. This may be due to the fact that this absorption is the most intense in the NH_3 spectrum.

The two absorptions at 1950 cm^{-1}(at 300°K) and at 550 cm^{-1} which are stable on pumping are assigned to Ge-H and Ge-N surface bonds. These bonds are obtained as a result of decomposition of NH_3 at that temperature. This decomposition commences already at 195°K (at that temperature the Ge-H gives a band at 1880 cm^{-1}) and is much more pronounced at the higher temperature. In the case of ND_3, the Ge-D band is obtained at 1380 cm^{-1}(at 300°K) a frequency expected from the isotope shift, while the Ge-N frequency remains unchanged as expected.

It is known that germanium hydrogen compounds show absorptions due to the Ge-H stretching modes in the 1900 - 2000 cm^{-1} region. For example, in GeH_4 two absorptions appear at 1988 cm^{-1} and 2110 cm^{-1} which belong to the symmetric and anti-symmetric stretching modes.[11]

As was already mentioned, in the initial stages of NH_3 decomposition, the Ge-H vibration is obtained at 1880 cm^{-1} and on further decomposition as a result of heating, the frequency shifts to 1950 cm^{-1}. This shift may be connected with incorporation of the H-atoms into the Ge lattice, similarly to the known phenomenon of incorporation of oxygen which is accompanied by a frequency shift to higher energies of the Ge-O stretching vibration.[12]

The absorption due to the stretching of the Ge-N bond in germanium-nitrogen compounds appears in the region 450-500 cm^{-1}. In $Ge(N_3)(CH_3)_3$ this frequency is 456 cm^{-1}.[13]

The absorptions in the ND_3 spectrum obtained at frequencies 2490 cm^{-1}, 2380 cm^{-1}, 1180 cm^{-1}, 950 cm^{-1} and 820 cm^{-1} parallel those obtained with NH_3 adsorbed physically. The first two correspond to ν_3 and ν_1 modes, the third to the ν_4 mode, whereas the last one to the ν_2 mode. The 950 cm^{-1} band is assigned as in the NH_3 case to molecules coordinatively bound to the surface. All these frequencies correspond to values expected from the isotope effect.

It was mentioned in the previous section that at 300°K a weak and broad band at 1910 cm^{-1} appeared in the spectrum. This is ascribed to decomposition of small traces of H_2O on the Ge surface and formation of Ge-H bonds. These traces of H_2O are most probably obtained as a result of desorption from the walls of the cell.

ACKNOWLEDGEMENT

This research was supported by a grant from the United States - Israel Binational Science Foundation (BSF), Jerusalem, Israel.

REFERENCES

1. Boonstra, A.H., Philips Res. Rep. Suppl. nr. 3 (1968).
2. (a) Sharpe, L.H., Proc. Chem. Soc., 461 (1961).
   (b) Harrick, N.Y., "Surface Phenomena in Semiconductors", 101, 463 (1963).
   (c) Lubezky, A. and Folman, M., Israel J. Chem. 9, 469 (1971).
3. Kozirovski, Y. and Folman, M., Trans. Faraday Soc., 62, 808 (1966).
4. Herzberg, Z., "Infra-red and Raman Spectra of Polyatomic Molecules", Van-Nostrand, New York, 1945.
5. Little, L.H., "Infra-red Spectra of Adsorbed Molecules", Academic Press, London, 1966.
6. Eishens, R.P. and Pliskin, W.A., "Advances in Catalysis", Vol. 10, p. 1, Academic Press, New York, 1958.
7. Wilmshurst, J.K., Can. J. Chem., 38, 467 (1960).
8. Handler, P., J. Phys. Chem. Solids, 14, 1 (1960).
9. Calahorra, Z. and Folman, M., Jap. J. Appl. Physics, Suppl. nr. 2, Pt. 2, 307 (1974).
10. Gatos, H.C., "The Surface Chemistry of Metals and Semiconductors", p. 384, John Wiley, 1959.
11. Straley, Y.W., Tindel, C. and Nielsen, H.H., Phys. Rev., 62, 161 (1942).
12. (a) Lubezky, A., D.Sc. Thesis, Technion, Haifa, Israel, October 1971.
   (b) Howe, R.F., Liddy, J.P. and Metcalfe, A., J. Chem. Soc., Faraday Trans. I, 68 (1972).
13. Cradok, S. and Ebsworth, E.A., J. Chem. Soc. (A), 1422 (1968).

AES STUDY OF CHEMISORPTION OF WATER TO GOLD AT ROOM TEMPERATURE

Malcolm E. Schrader

David W. Taylor Naval Ship
Research and Development Center
Annapolis, Maryland 21402

ABSTRACT

Polycrystalline gold surfaces were exposed to water vapor at room temperature and 5×10^{-7} torr and the adsorption determined by following the appearance of oxygen with AES. All clean ion-bombarded surfaces reached the same maximum which was estimated at 5% of a monolayer. This adsorbate was strongly chemisorbed, remaining stable to heating in ultrahigh vacuum at temperatures as high as 500° C. Surfaces which contained detectable amounts of calcium that had segregated to the surface as a result of heating following ion bombardment adsorbed additional water estimated as high as 10% of a monolayer. The amount increased, but not linearly, with increased calcium segregated. The number of water molecules adsorbed were substantially in excess of the number of calcium atoms catalyzing the adsorption. This component of the adsorbate was fairly stable in ultrahigh vacuum for long periods at room temperature but desorbed in a few hours at 200° C.

I. INTRODUCTION

It is quite natural to consider gold as the prime candidate for use as a material or coating in applications where a pure metallic passive surface is required. However, although it is the most noble of all metals and would not be expected to form bulk compounds with any of the common atmospheric ambients, its high surface free energy provides the possibility that these ambients might chemisorb to form stable surface species (1). We have previously investigated the chemisorption of molecular oxygen to the surface of gold at various temperatures (2). The present work is concerned with the possible chemisorption of water vapor to the gold surface.

This investigation is also of interest to the theoretically important problem of the wettability of the gold surface. A long controversy on the experimental aspects of this problem was settled by two independent experimental approaches (1, 3) which found zero contact angles for water on gold under conditions which precluded surface segregation of bulk impurities. While it had once been supposed that an approach which attempts to isolate the contribution of dispersion forces (4) to the attraction between gold and water could not predict this wettability, newly calculated experimental (5) and theoretical (6, 7) values for the gold Hamaker constant indicate that the spreading of water on clean gold is not inconsistent with this approach (8). A complete theory of the process would have to involve an understanding of the entire adsorption isotherm of water vapor on gold from zero to saturation pressure. The present investigation yields information concerning the initial stage of such an isotherm.

Our previous work on the chemisorption of oxygen to (111) oriented gold indicated that chemisorption on annealed surfaces proceeded via an initial rapid chemisorption followed by a slow chemisorption (2). The mechanism proposed involved an initial rapid dissociative chemisorption of gas phase molecular oxygen to a small number of active sites, followed by diffusion of oxygen atoms to the less active sites constituting the main portion of the annealed surface. The active sites catalyzing the chemisorption consist either of surface defects or of traces of calcium segregating at the surface during annealing. When these active sites are not present, chemisorption of oxygen, when it occurs, must take place via a different mechanism.

In the present work the adsorption of water vapor at 5×10^{-7} torr to the surface of gold was studied at room temperature. The adsorption was studied as a function of time and nature of surface preparation. Stability of the chemisorbed species was estimated relative to exposure to ultrahigh vacuum at room and elevated temperatures.

II. EXPERIMENTAL SECTION

The adsorption studies were performed in an ultrahigh vacuum system equipped with AES as described previously (9). The water vapor was leaked in from a distilled water reservoir thoroughly degassed as described previously (9). No evidence of molecular oxygen in the water vapor was found within the limits of detection of the quadrupole residual gas analyzer.

All experiments, unless designated otherwise, were performed on a polycrystalline gold sample with a purity of 99.999%. The surface referred to as (111) oriented was on a single crystal also of 99.999% purity. Adsorption of water to the gold surfaces was followed by means of the oxygen Auger peak. All surfaces prepared on the polycrystalline sample were free of detectable carbon.

III. RESULTS

A. Adsorption of H_2O to Ion-Bombarded Gold Surfaces

Ion bombardment of this polycrystalline gold sample produced surfaces which remained free of any detectable carbon. Exposure to H_2O at 5×10^{-7} torr and room temperature resulted in a time dependent adsorption which reached a maximum after approximately 30 minutes. There was no observed pressure dependence for the value of this maximum. It remained stable to ultrahigh vacuum pumping at room temperature for days and did not desorb substantially even on heating in vacuum as high as 400°C for a few hours. The amount adsorbed was estimated as roughly 5% of a monolayer. This result was also obtained on a (111) oriented surface.

B. Adsorption of H_2O to Polycrystalline Gold Heated after Ion Bombardment

When the surface was heated to 500°C or higher following ion bombardment, then cooled to room temperature and exposed to H_2O at 5×10^{-7} torr, there was a dramatic increase in both the rate and final amount adsorbed. The maximum amount adsorbed was approximately three times that on the unheated surface. It remained unchanged after a few hours of pumping at room temperature but decreased 25% after a few days in ultrahigh vacuum. Vacuum heating at 200°C desorbed most of the remainder in a few hours. A residue approximately equal to the maximum amount adsorbed on an unheated surface remained, however, and was stable to further vacuum heating even at 400°C.

C. Auger Spectrum of Polycrystalline Gold Heated to 500°C or Higher Following Ion Bombardment

The appearance of a calcium signal was observed subsequent to heating the clean gold surface at 500°C or higher. The amount varied with the history of the sample as well as the time and temperature of heating, but was generally estimated as less than 15% of the total amount of H_2O subsequently adsorbed.

D. Adsorption of H_2O to Surfaces with Varying Amounts of Calcium

The amount of calcium present in the bulk was reduced through a continuous process of surface segregation and cleaning by ion bombarding the sample at temperatures of 600°C to 800°C. This generated the possibility that the effect of calcium segregation could be separated from that of surface structural changes occurring from annealing during heating. Hopefully a given set of anneal conditions would segregate less calcium. Fig. 1 shows adsorption of H_2O versus time for a series of surfaces with decreasing amounts of calcium segregated, where the amounts of calcium were not varied by systematically changing the conditions of anneal. It can be seen from examination of H_2O adsorption at 1200 seconds that the order of increasing amounts of calcium segregated is exactly the order of increasing H_2O adsorbed.

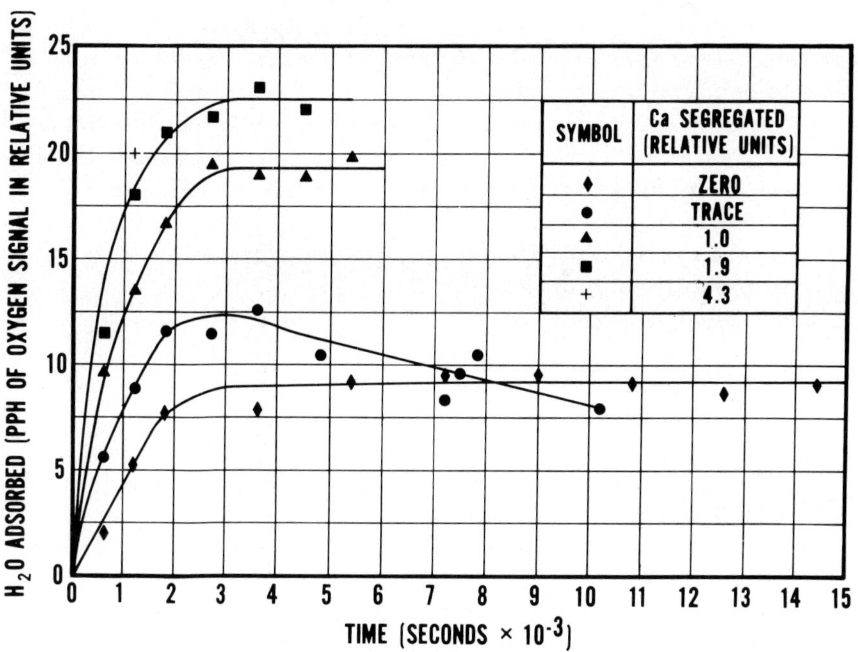

Fig. 1. Chemisorption of water vapor at 5×10^{-7} torr to polycrystalline gold at room temperature

IV. DISCUSSION

A. Adsorption to Clean Ion-Bombarded Surfaces

The adsorption of H_2O taking place at 5×10^{-7} torr on clean ion-bombarded gold surfaces at room temperature must be labeled as strong chemisorption. As expected, variations in the H_2O pressure ranging from an increase to the 10^{-6} torr decade to a decrease to the 10^{-10} torr decade merely change the rate of chemisorption without changing the final amount. This adsorption which is roughly estimated at 5% of a monolayer probably takes place at a specific type of surface defect. The high stability of this species to vacuum heating resembles that of chemisorbed oxygen, which suggests the possibility of dissociative chemisorption of the water molecule. This dissociation, if it takes place, could occur on adsorption at room temperature or upon subsequent heating in vacuum. Regardless of its nature, the chemisorption is too irreversible to allow measurements to be made to incorporate it into an adsorption isotherm. Any room temperature adsorption isotherm of water on gold would therefore have to treat this species as an unavoidable contaminant. Once exposed to water vapor in the atmosphere, gold surfaces in outer space would retain this chemisorbed species under all conditions except those of heating to very high temperatures such as 800°C.

B. Room Temperature Adsorption to Surfaces Heated after Ion-Bombardment: Effect of Calcium

The surface segregation of calcium upon heating gold in the presence of oxygen has been reported by Chesters and Somorjai (10). In the present work it is found that this segregation also takes place upon vacuum heating in the absence of oxygen (11), although to a lesser extent. The enhanced adsorption of H_2O taking place on these heated surfaces is due either to structural smoothing of the surface or to segregation of calcium. Experience would lead one to expect the opposite effect for the case of structural smoothing. A role for Ca in an increase of H_2O adsorption seems quite plausible, on the other hand. Since the annealing temperature is approximately the same as the threshold temperature for surface segregation (approximately 500°C), it is rather difficult to separate the two effects experimentally, unless the gold is further purified of Ca to eliminate or reduce the amount segregated at a given temperature. We have succeeded in partially accomplishing this by ion bombarding the sample at temperatures of 600°C or higher, thus continuously segregating and removing the Ca. The resulting surfaces were then heated without ion bombardment, the amount of Ca segregated after cooling was determined from the AES spectrum, and the adsorption of H_2O was measured at room temperature. In the experiments performed thus far the exact annealing procedures, with respect to both time and temperature, have not been duplicated both before and after a purification, so the results demonstrating the enhanced adsorption as due to Ca rather than structure are not completely conclusive. However, there is no correlation between the types of anneal and H_2O adsorption, whereas the order of amounts of Ca segregated and amounts of H_2O adsorbed are the same for all five experiments. This evidence combined with the greater theoretical plausibility of the Ca hypothesis leads to the conclusion that the enhanced adsorption of H_2O is in all probability due to the surface segregation of Ca.

C. Mechanism of Promotion by Ca of H_2O Adsorption

Adsorption of H_2O to surface segregated Ca on the basis of one or two molecules per atom cannot account for the effect. For example, if the maximum of curve ◇ in the figure is subtracted from that of curve □, the amount of H_2O adsorbed as a result of the presence of Ca is obtained. After correcting for the different sensitivities of the Auger Ca and O peaks, the ratio of H_2O molecules adsorbed to Ca atoms segregated is estimated as 6.7. A mechanism in which H_2O molecules initially adsorb to Ca atoms and then diffuse freely over the rest of the surface is not quite satisfactory, since H_2O adsorption does not reach the same maximum level on surfaces containing varying amounts of calcium. A modified mechanism of this sort in which H_2O molecules initially adsorbed to Ca transfer to neighboring sites where they become immobile may provide the answer. The fact that the H_2O adsorption is not linear with respect to Ca surface concentration may be due to a tendency for segregated atoms to nucleate, so that as the surface Ca concentration increases atoms are added to existing islands which supply H_2O to the gold only from the periphery.

Unlike the strong chemisorption to favored sites on clean ion-bombarded surfaces, Ca-catalyzed adsorption to gold may be described as moderately weak chemisorption. Judging from the relative ease or difficulty of desorption, the attachment seems to be somewhat stronger than that found in hydrogen bonding.

V. CONCLUSIONS

Water vapor chemisorbs strongly to the surface of clean ion-bombarded gold at room temperature to yield an oxygen-containing species equivalent to approximately 5% of a monolayer. The adsorbed layer is stable in ultrahigh vacuum to at least 400°C. This adsorption occurs even at very low pressures and is apparently associated with a specific type of surface defect. The presence of small amounts of calcium on the surface as a result of prior heating catalyzes the chemisorption of roughly 10% of a monolayer. These water molecules, which apparently reside on gold sites in the neighborhood of the calcium atoms, are chemisorbed less strongly than those forming the first 5% of a monolayer. They desorb in vacuum in days at room temperature or hours at 200°C.

VI. REFERENCES

1. Schrader, M. E., J. Phys. Chem. 74, 2313 (1970).
2. Schrader, M. E., Abstracts of 171st National Meeting of the American Chemical Society, April 1976. Manuscript in preparation.
3. Bernett, M. K., and Zisman, W. A., J. Phys. Chem. 74, 2309 (1970).
4. Fowkes, F. M., Ind. Eng. Chem., 56 (12), 40 (1964).
5. Derjaguin, B. V., Muller, V. M., and Rabinovich, Ya. I., Kolloid Zh. 31, 304 (1969).
6. Matsunaga, T., and Tamai, Y., Surface Sci. (in press).
7. Parsegian, V. A., Weiss, G. H., and Schrader, M. E., in preparation.
8. Schrader, M. E., J. Phys. Chem. 78, 87 (1974).
9. Schrader, M. E., J. Phys. Chem. 79, 2508 (1975).
10. Chesters, M. A., and Somorjai, G. A., Surface Sci. 52, 21 (1975).

11. Isa, S. A., Joyner, R. W., and Roberts, M. W., J. Chem. Soc. Farad. Trans. (1) 72 (2), 540 (1976).

THE INFLUENCE OF THE CATALYST PRETREATMENT ON THE SELECTIVITY OF ALCOHOL CONVERSION OVER METAL OXIDES.

Burtron H. Davis
Potomac State College of West Virginia University

Most metal oxide catalysts have been found to have a different dehydrogenation-dehydration selectivity for the conversion of 2-octanol when pretreated with hydrogen than when pretreated with oxygen. In general the hydrogen pretreated sample is more selective for dehydration and also more selective in producing 1- + cis-2-alkene with little trans-2-alkene. Metal sulfide catalysts are selective dehydration catalysts but not selective for 1- + cis-2-alkene. It is proposed that there are two types of dehydration occurring: type I is selective and produces a high percentage of 1- + cis-2-alkene and type II which is not selective and produces about equal amounts of 1-, trans-2- and cis-2-alkenes.

I. INTRODUCTION

The conversion of alcohols over metal oxides has been one of the most widely studied catalytic reactions. One facet of the reaction is the selectivity for dehydration vs. dehydrogenation. Many of the early studies of this aspect were undertaken to verify Professor Taylor's concept of active sites (1). Later Professor Schwab (2) advanced the concept that dehydrogenation occurred on the flat portion of the surface while dehydration occurred in the small pores. In a recent review Mars (3) presented data to show that there was a parallel dehydration/dehydrogenation selectivity for the conversion of ethanol and formic acid over a variety of oxide catalysts; in this correlation alumina was almost exclusively a dehydration catalyst while magnesium oxide was exclusively a dehydrogenation catalyst. De Vleesschauwer (4) studied the influence of the preparation of magnesia on the selectivity and activity for isopropyl alcohol decomposition. He found that it took an appreciable time to reach a steady decomposition rate and that this period could be shortened by subjecting the catalyst to an oxygen pretreatment. He found that magnesium oxide treated at 550°C or higher was almost exclusively a dehydrogenation catalyst; however, magnesia heated to 450°C was about 14 times as active for dehydration as for dehydrogenation. On the other hand, Davis (5) found that the selectivity of alumina catalysts could vary widely depending on whether a hydrogen or an oxygen pretreatment was used. For some alumina samples, an oxygen pretreatment provided an alumina catalyst that was more active for dehydrogenation than for dehydration. Thus, alumina and magnesia, rather than having the extremes in selectivity observed by Mars, are found to have a selectivity that depends on the pretreatment. Knozinger (6) in a recent review outlined the problem in determining the influence of the pretreatment on the catalytic selectivity and concluded that it was as yet an unsolved problem. Krylov (7) has attempted to correlate the selectivity for alcohol conversion with some property of the oxide catalyst such as band-width.

II. EXPERIMENTAL

The runs were carried out in a flow apparatus consisting of a syringe pump to provide a constant flow of reactant, a glass reactor, liquid product collector, and a burette to measure gas flow. Approximately 30 ml of silicon carbide chips were placed above the catalyst bed to serve as a preheater. The runs were carried out with the reactant initially at atmospheric pressure. Liquid samples were collected at intervals and analyzed for conversion using g.l.c. with a Carbowax 20M column. The alkene products were analyzed using one of the following columns: Carbowax 20M, Ucon-W or β,β'-oxydipropio-nitrile.

Hydrogen, oxygen, or air was passed over the catalyst at atmospheric pressure with a flow of 10 cc/minute at 500°C (unless otherwise specified) during the 3-4 hour pretreatment.

The catalysts, with the exception of the rare earth oxides, were prepared by precipitation from a 1-3M nitrate solution by the rapid addition of a large excess of concentrated ammonium hydroxide. Exceptions to this preparative procedure will be noted in the Results and Discussion section.

The rare earth oxides were low surface area (< 5 m^2/g) materials of unknown history; they were reported to be at least 99.99% pure.

III. RESULTS AND DISCUSSION

The conversion of alcohols may proceed in two directions: dehydrogenation to ketones and dehydration to alkenes. In this paper we will define this selectivity as (ketone/alkene). Another selectivity is the amount of 1-alkene to 2-alkene formed from 2-alcohols and we will consider this selectivity to be (1-ene/2-ene).

We have previously reported that high surface area aluminas are sensitive to pretreatment for the determination of the dehydrogenation-dehydration selectivity. In Table 1 we compare the high surface area alumina (approximately 200 m^2/g) with some other alumina samples. Alon-C and Alucer MA have been reported to exist as individual single crystals formed during the calcination process and to have a surface area in the range of 100 m^2/g. These two "nonporous" aluminas differ from the porous higher surface area transitional aluminas in that both the hydrogen and oxygen

TABLE 1
The Selectivity for the Conversion of 2-Octanol over Aluminas.

		Selectivity	
Catalyst	*Temp., °C*	*H_2 Pretreat*	*O_2 Pretreat*
Nordstrandinite[a]	*180-*	*ca. 0.0*	---
α-Al_2O_3	*350*	*3.0-2.8*[b]	
	330	*2.5-2.2*	*1.1*
Alucer MA[c]	*200*	*ca 0.0*	*ca 0.0*
Alon-C	*200*	*ca 0.0*	*ca 0.0*
Al_2O_3-A[d]	*180*	*ca 0.0*	*1.0*

TABLE 1 (Cont'd)
The Selectivity for the Conversion of 2-Octanol over Aluminas.

		Selectivity	
Catalyst	*Temp., °C*	H_2 *Pretreat*	O_2 *Pretreat*
Al_2O_3-Cl[d]	180	0.1	0.25
Al_2O_3-K[d]	250	0.15	0.5-0.15[b]

a. *Prepared according to U.S. Patent 3,328,122.*
b. *Range of selectivities, first to last sample.*
c. *Sample provided by Professor W. H. Wade; see Ref. 8.*
d. *Values from Ref. 5.*

pretreated samples are active for the dehydration reaction but are inactive for the dehydrogenation reaction. However, the lower surface area, nonporous α-alumina is more selective for dehydrogenation than for dehydration. In addition, the influence of the pretreatment for the α-alumina was opposite to that of the transitional alumina in that the hydrogen pretreated sample was more active for dehydrogenation that was the oxygen pretreated sample.

The selectivity of alumina for olefin formation by dehydration is presented in Table 2. We note that α-Al_2O_3 has a much different olefin selectivity than the other aluminas. This is also evident in the conversion of 3-pentanol where the only primary alkene products allowed are cis-2- and trans-2- pentene; approximately 80% of the 2-pentene was the cis-2- isomer with Al_2O_3-A but with α-Al_2O_3 the cis-2- isomer was only 56% of the total 2-pentenes. Note that with all of the aluminas except the α-Al_2O_3, the cis-2-octene is 80-90% of the total 2-octenes; this value is in good agreement with the 80% cis isomer obtained from 3-pentanol conversion over these catalysts. Likewise, with 2-octanol over α-Al_2O_3 the cis-2-octene comprises 55% of the total 2-octenes and is in good agreement with that for 3-pentanol conversion over this catalyst. Thus, it appears that α-alumina also differs from the other aluminas in olefin selectivity as well as the dehydration-dehydrogenation selectivity.

TABLE 2
Olefin Distribution for the Conversion of 2-Octanol over Various Aluminas Pretreated with Hydrogen.

	Alkenes		
Catalyst	*1-*	*trans-2-*	*cis-2-*
Al_2O_3-A (acidic)	45	5	50
Al_2O_3-K (nonacidic)	40	10	50
Alon-C	45	5	50
Alucer MA	40	8	52
α-Al_2O_3	40	27	33

It is illustrative to compare alumina to other oxides of the IIIA family. Gallium oxide, when pretreated with oxygen, was a very selective dehydration catalyst. (Gallium flouride was dissolved in nitric acid, the hydroxide precipitated with ammonia, redissolved and reprecipitated three times, washed ten times with distilled water and dried at 120°C.) The alkene distribution was similar to that of high surface area alumina except for a smaller amount of 1-octene and a corresponding increase in the amount of trans-2-octene; the distribution at 200°C was: 1-octene, 30%; trans-2-octene, 20%; and cis-2-octene, 50%. Due to the limited amount of the gallia sample and the ease of reduction of the oxide with hydrogen we did not attempt a hydrogen pretreatment. Indium oxide was easily reduced in hydrogen. As can be seen in Fig. 1, a hydrogen pretreatment at 300°C did produce a catalyst that was less active for dehydrogenation than an oxygen pretreated catalyst. However, the oxide was not stable in the presence of the alcohol charge and was reduced to a lower oxidation state during the reaction. Thus, the dehydrogenation-dehydration selectivity of the oxygen pretreated sample approaches that of the hydrogen pretreated sample after some time on stream. The yellow color of the In_2O_3 quickly disappeared when the alcohol was passed over the catalyst; the catalyst first turned grey and during the run became progressively blacker. Thus, we are unable to

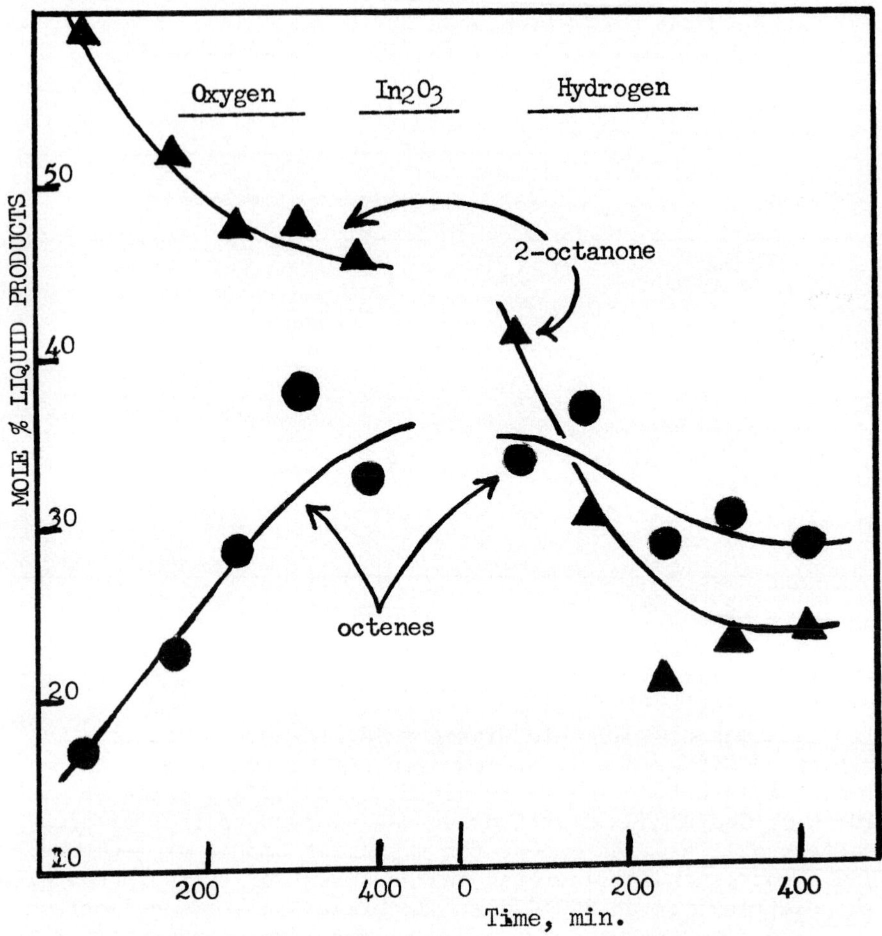

Fig. 1. The influence of pretreatment with hydrogen or oxygen on the selectivity for 2-octanol conversion over In_2O_3.

obtain a true selectivity for the oxidized catalyst. The early high dehydrogenation activity of the oxygen pretreated sample may in fact be due to the alcohol serving as a reducing agent for the oxide. In any event, the reduced indium oxide is as active for dehydrogenation as it is for dehydration. The olefin distribution also is very different from that of alumina and gallia; the olefins from 2-octanol over india were approximately: 1-octene, 90%; trans-2-octene, 6%; and cis-2-octene, 4%. Thus, in going from alumina to the larger cation of india, there is a change in selectivity to favor dehydrogenation and a change in olefin selectivity to favor the formation of 1-octene.

In Table 3 we present the dehydration-dehydrogenation selectivity for some rare earth oxides. These are low area materials ($< 5\ m^2/g$) and we suspect that much different results could be obtained with high surface area materials. In

TABLE 3
Influence of Catalyst Pretreatment on the Dehydrogenation/ Dehydration Selectivity of Rare Earth Oxide Catalysts for the Conversion of 2-Octanol.

	Selectivity, -one/-ene	
Catalyst	*H_2 Pretreat*	*O_2 Pretreat*
Sm_2O_3	4.2-4.6[a] (260°C) 0.8-1.8 (290°C)	64-43[a] (260°C) 1.7-5.3 (290°C)
Nd_2O_3	0.7 (250°C)	16-8 (250°C)
Eu_2O_3	1.7-9.9 (255°C)	90-50 (255°C)
Ce_2O_3	4.6-8.0 (255°C)	b
Pr_2O_3	2.6-1.2 (255°C)	1.9-1.3 (260°C)
Gd_2O_3	0.4-0.15 (255°C)	0.4-0.5 (255°C)
Er_2O_3	0.4-0.5 (270°C)	0.5-0.9 (265°C)
Tm_2O_3	1.9-2.5 (265°C)	0.4-0.8 (263°C)
ThO_2 (nitrate decomposition)[c]	0.01-0.05 (250°C)	1.1-3.3 (250°C)

TABLE 3 (Cont'd.)
Influence of Catalyst Pretreatment on the Dehydrogenation/ Dehydration Selectivity of Rare Earth Oxide Catalysts for the Conversion of 2-Octanol.

	Selectivity, -one/-ene	
Catalyst	*H_2 Pretreat*	*O_2 Pretreat*
ThO_2 (Oxalate decomposition)[c]	*0.1-0.4 (250°C)*	*0.7-3.6 (250°C)*
ThO_2 (carbonate)[c]	*0.03-0.1 (250°C)*	*4.6-5.8 (250°C)*

a. Range of selectivities from first to last sample.
b. The oxygen pretreated sample was too active (oxidation) to run; the temperature rapidly increased when the alcohol contacted the catalyst.
c. See Reference 9.

general, the oxygen pretreated material is more selective for dehydrogenation than for dehydration. The olefin distribution over these catalysts was not very conclusive; in most cases about equal amounts of each of the three isomers were formed. A high area thoria is selective for 1-octene formation. With a few of the other rare earth low area samples there was an indication of a slight selectivity for the 1-octene. However, for most samples the three olefins were produced in nearly equal amounts with a slight increase in percentage in going from 1- to trans-2- to cis-2-octene. It has been reported recently (10) that the thoria most selective for both dehydration and 1-alkene formation results from precipitating thoria onto a low area thorial "support." We plan to prepare some of the rare earth catalysts by a similar procedure and test them for alcohol conversion.

Zirconium oxide presented a complex character in that catalysts prepared by various procedures varied widely. In addition to varying with preparation, the preparation of reproducible catalyst from the same batch of hydroxide was not always possible. The first zirconia sample was precipitated from a zirconium acetate solution with ammonium hydroxide, the hydroxide was redissolved in nitric acid and reprecipitated with concentrated ammonium hydroxide, washed ten times with distilled water and dried at 120°C. Figure 2 shows the influence of pretreatment on this particular catalyst; the

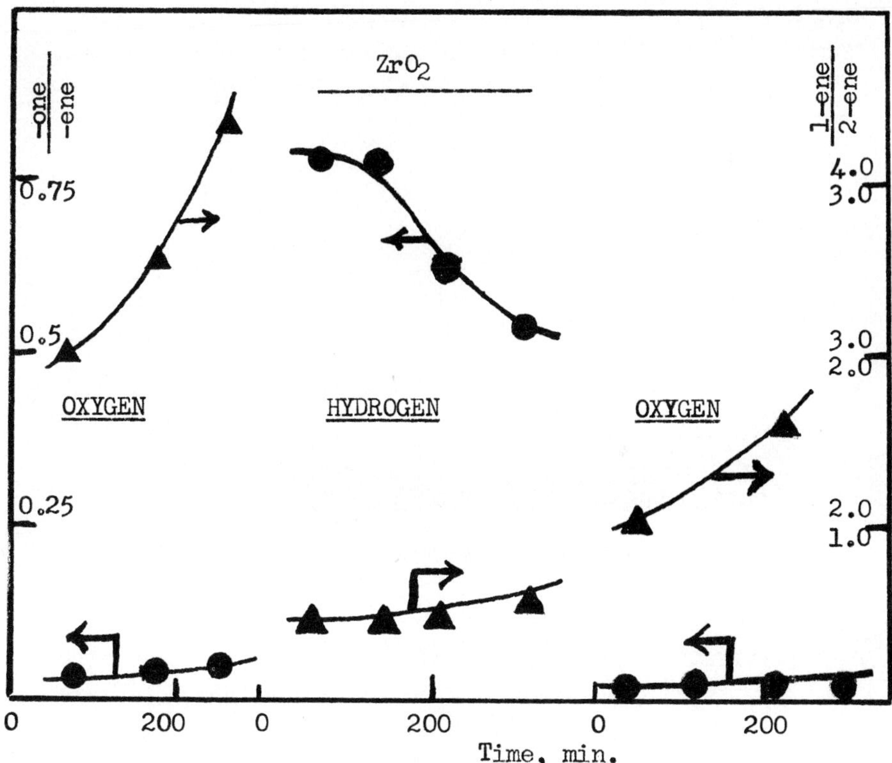

Fig. 2. The influence of the pretreatment with hydrogen or oxygen on the selectivity for dehydrogenation and the formation of 1-octene from the conversion of 2-octanol over zirconia.

oxygen pretreated catalyst was selective for dehydration while the hydrogen pretreated sample was much less selective for dehydration. The 1-octene selectivity for these pretreatments is also shown in the figure. It is apparent that the most selective catalyst for dehydration was the most selective catalyst for 1-octene formation. Another zirconia catalyst was designated Sample G by Dr. Holmes (see Ref. 11) and co-workers who studied water adsorption on the sample and found it to be complicated. For this sample the hydrogen pretreated catalyst was much more active for dehydrogenation than the oxygen pretreated sample (Table 4). However, the hydrogen pretreated sample was more selective for the

TABLE 4
Conversion of 2-Octanol over Zirconia Prepared by Various Methods and with Hydrogen or Oxygen Pretreatment.

	H_2 Pretreat		O_2 Pretreat	
	-one/-ene	1-ene/2-ene	-one/-ene	1-ene/2-ene
ZrO_2 (acetate)[a]	0.8-0.5	0.4-0.5	0.03-0.07	2.0-3.5
ZrO_2 (cl)[b]	0.1-0.2	1.9-2.2	0.7-0.5	2.9-7.3
ZrO_2 (anhyd. Cl)[c]	ca. 0.0	ca 0.5	---	---
ZrO_2 (Samp. G)[d]	1.5-2.3	1-1.3	0.3-0.15	0.4
ZrO_2 (nitrate)[e]	0.3	.30	5.7-0.4	2.9-4.0

a. *Prepared from the acetate; see text for description.*
b. *From $ZrCl_4$ hydrate solution with conc. ammonia.*
c. *Same as b except use anhydrous $ZrCl_4$.*
d. *Sample from Dr. Holmes; see Reference 11.*
e. *From $ZrO(N)_3)_2$ solution with conc. ammonia.*

formation of 1-alkene; the oxygen pretreated sample was not selective and at least a part of the reason was the ability of this catalyst to readsorb and isomerize alkenes after they had initially desorbed to the gas phase. Hydrogen pretreated zirconia, obtained by dissolving anyhdrous $ZrCl_4$ in water, precipitation with concentrated ammonia and washing to a negative Cl^- test with silver nitrate, was almost exclusively a dehydration catalyst; this is just opposite that of the first two zirconia catalysts that we have described above. However, this catalyst was not selective for 1-octene formation from 2-octanol. Another catalyst was prepared by precipitation from a solution prepared from zirconium oxychloride hydrate and this catalyst was more selective for dehydration when pretreated with hydrogen than when pretreated with oxygen. In this case, both pretreatments were selective for 1-octene with the oxygen pretreated one being more selective for 1-octene. Another zirconia was prepared by precipitation with concentrated ammonia from a solution prepared from $ZrO(NO_3)_2$ and, at later time on stream, the same selectivity for dehydration was obtained for the hydrogen or the oxygen pretreated sample. This catalyst was the most selective for 1-octene formation and the hydrogen pretreated sample was more selective than the oxygen pretreated sample. It appears that it would require a major research effort to learn how to prepare reproducible zirconia catalysts, let alone solve the problem of the mechanism of alcohol conversion over these catalysts.

In Table 5 are presented dehydrogenation-dehydration selectivity for metal oxides from group A and B of the periodic table. For some of the oxides the oxygen pretreated sample is more active for dehydrogenation than for dehydration; with other oxides the opposite is true. The olefin distribution for some of these catalysts are presented in Table 6. For the catalysts shown a large amount of the trans-2 alkene is formed. The large amount of trans isomer formed over the hydrogen pretreated WO_3 is probably due to the more acidic nature of this oxide; other acidic catalysts such as silica-alumina and acid ion exchange resins produce a similar high, or higher, trans isomer. This is true even though added heptene is not isomerized in the presence of the 2-octanol conversion; hence, the high trans isomer results during the olefin formation or by isomerization of the olefin before it desorbs to the gas phase. On the other hand, the oxygen pretreated WO_3 is selective for the formation of the 1- and cis-2 isomer. The other oxides produce a similar alkene distribution with nearly equal amounts of the three isomers being formed.

The sulfides of Zn, Cd, In, Cu and Ag were tested for the conversion of 2-octanol at 250-300°C. With the exception of zinc sulfide, they were very selective dehydration catalysts. ZnS reacted with the alcohol charge so that very little liquid product was collected until a volume of alcohol greater than a one to one mole alcohol:ZnS had been pumped into the reactor. A similar olefin distribution was obtained for the catalysts.

TABLE 5
Influence of Pretreatment on the Dehydrogenation Selectivity for Group A and B Metal Oxide Catalysts for 2-Octanol Conversion.

Catalyst	Selectivity, -one/-ene			
	H_2 Pretreat		O_2 Pretreat	
SiO_2 (Cab-O-Sil HS-5)	5.2-3.9	(260°C)	0.46-0.14	(255°C)
CaO (from hydroxide)	6.2-6.4	(290)	0.8-1.0	(290)
CaO (from carbonate)	1.3	(312)	1.1	(320)
			1.8-0.6	(340)
			1.2-0.6	(360)
CaO (from carbonate)[a]	2.1-4.0	(318)	1.3-0.65	(330)
	5.0-4.0	(335)	0.95-0.4	(350)
	5.0-3.3	(368)	0.5-0.3	(360)
MgO (from hydroxide)	60	(250)	60-40	(250)
MgO (from carbonate)	2.4-2.6	(315)	0.2	(305)
	2.8-3.4	(340)	0.6-0.5	(340)
	5.1-6.2	(360)	0.5-0.3	(360)
SrO	----------		0.1-0.2	(280)
Ga_2O_3	----------		ca. 0.0	(200)
In_2O_3	1.2-0.8	(270)	2.2-1.6	(210)
			3.5-1.4	(270)
CdO	11-25	(280)	21-34	(265)
ZnO	7.6-4.0	(295)	2.2-1.3	(300)
Y_2O_3	1.8-1.5	(288)	9.5-8.4	(240)
	0.9-0.7	(314)		
	0.5	(345)		
La_2O_3	200	(210)	15-7	(260)
MoO_3[b]	ca. 0.0	(200)	ca. 0.0	(250)
WO_3[b]	ca. 0.0	(200)	ca. 0.0	(250)

a. *Selectivity for conversion of 4-methyl-2-pentanol.*
b. *Assumed to be the trioxide before H_2 treatment.*

TABLE 6
Olefin Distribution from 2-Octanol over some Group A and B Metal Oxide Catalysts.

	alkene		
Catalyst	*1-*	*trans-2-*	*cis-2-*
CdO (H_2)	36	30	34
CaO (H_2)	37	29	34
ZnO (H_2)	42	25	33
(O_2)	36	26	37
SiO_2 (H_2)	33	40	27
(O_2)	33	41	26
WO_3 (H_2)[a]	23	44	33
(O_2)[a]	32	16	52

a. *The absence of heptene added to the charge indicates that this is the distribution of the primary products.*

TABLE 7
Olefin Distribution from 2-Octanol over Metal Sulfide Catalysts.

	alkene		
Sulfide Catalyst	*1-*	*trans-2-*	*cis-2-*
In	32	33	35
Cd	26	26	48
Cu	25	31	44
Zn	31	32	37

After completing the work on transitional aluminas and high surface area thoria, we felt that the selectivity was determined by the surface oxygen content. A catalyst treated with hydrogen would be expected to be more deficient in oxygen ions (and have more metal ions exposed) so this would provide the dehydration site. Likewise, the oxygen pretreated sample would be expected to present a surface with few, if any, exposed metal ions and this would provide the dehydrogenation catalyst. It was anticipated that the different metal ion size, ion charge, and crystal location would provide a variety of dehydration sites and that much could be learned about the dehydration site from the alkene selectivity. However, the results from additional oxide catalysts indicate that this simple idea will not be sufficient.

A number of workers have advanced an explanation for the reason for the selectivity of alcohol conversion. For example, Vol'kenshtein (12) proposed that lowering the Fermi level of the catalyst would poison dehydrogenation and stimulate dehydration and that displacing the Fermi level upward would have the opposite effect. On the other hand, Hauffe's theory (13) indicated that lowering the Fermi level would enhance dehydrogenation and reduce dehydration. Khoobiar et al. (14) have examined the electrical conductivity of η-alumina and concluded that the oxygen pretreated (500°C) sample was a p-type and the hydrogen pretreated sample was an n-type conductor. On this basis, it would appear that Hauffe's theory was correct for high area transitional aluminas (5).

Likewise there are contrasting viewpoints about the chemisorbed species leading to the two products. Wicke (15) proposed two structures to account for dehydrogenation (a) and dehydration (b):

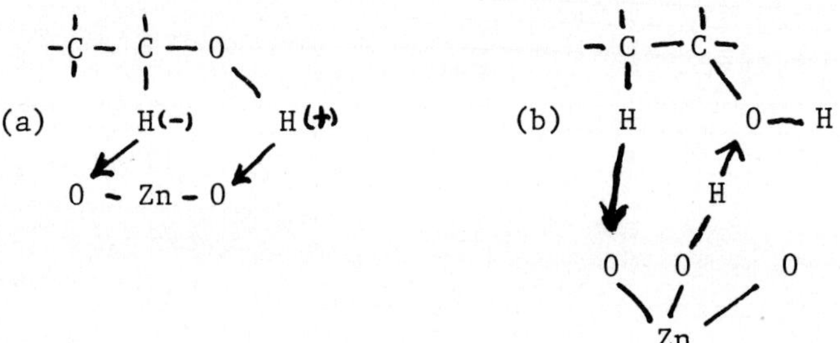

Numerous other workers have accepted this or very similar intermediates to explain their results. However, Hauffe (13)

did not believe that the position of the metal and oxygen ions had an appreciable effect on the selectivity; only the chemical potential of the electrons and the height of the potential barrier of the electric field in the boundary layer were important.

From the work to date, it appears that there are two types of alcohol dehydration. Type I is selective in that it leads to a high ratio of (1-ene + cis-2-ene/trans-2-ene) and occurs on surfaces with exposed metal ions. Type II is not selective since it leads to approximately equal amounts of each of the three olefins and occurs on a surface without exposed metal ions. Type I dehydration would occur on hydrogen pretreated samples, those with small anions, small particle size, and/or a "rough surface." Type II would occur on oxygen pretreated samples, those with large anions, and/or large particle size. It is inviting to attribute the intermediate (b) above as responsible for type I dehydration and intermediate (a) as being responsible for both type I dehydration and intermediate (a) as being responsible for both type II dehydration and dehydrogenation.

ACKNOWLEDGMENT: The financial support of this research by a West Virginia University Senate Research Grant is appreciated. We also thank Dr. Holmes for the zirconia sample and Dr. Wade for the alumina sample.

IV. REFERENCES

1. Taylor, H. S., Proc. Roy. Soc. A108, 105 (1925).
2. Schwab, G.-M., and Agallidis, E. S-., J. Am. Chem. Soc. 71, 1806 (1949).
3. Mars, P., in "The Mechanism of Heterogeneous Catalysis" (J. H. de Boer, Ed.), p. 49. Elsevier, Amsterdam, 1961.
4. DeVleesschauwer, W. F. N. M., in "Physical and Chemical Aspects of Adsorbents and Catalysts" (B. G. Linsen, Ed.), p. 265. Academic Press, London, 1970.
5. Davis, B. H., J. Catal. 26, 348 (1972)
6. Knozineger, H., in "The Chemistry of the Hydroxyl Group" (S. Patai, Ed.), p. 641. Interscience Pub., New York, 1971.
7. Krylov, O. V., "Catalysis by Nonmetals." Academic Press, New York, 1970.

8. Wade, W. H., and Hackerman, N., J. Phys. Chem. 64, 1196 (1960).
9. Davis, B. H., and Brey, W. S., J. Catal. 25, 81 (1972).
10. Traynard, P., Damon, J. P., and Bonnier, J. M., Bull. Soc. chim. 1972, 2306.
11. Agron, P. A., Fuller, E. L., Jr., and Holmes, H. F., J. Colloid Interface Sci. 52, 553 (1975).
12. Vol'kenshtein, F. F., "The Electronic Theory of Catalysis on Semiconductors." Macmillan Co., New York, 1963.
13. Hauffe, K., in "Advances in Catalysis" (W. G. Frankenburg, V. I. Komarewsky, and E. K. Rideal, Eds.), Vol. 7, p. 250. Academic Press, New York, 1955.
14. Khoobiar, S., Carter, J. L., and Lucchesi, P. J., J. Phys. Chem. 72, 1682 (1968).
15. Wicke, E., Z. Elektrochem. 53, 279 (1949).

THE CARBON MONOXIDE OXIDATION ON A PLATINUM CATALYST

H.D. Cochran, R.G. Donnelly*, M. Modell and R.F. Baddour
Massachusetts Institute of Technology

Oxidation of carbon monoxide on a silica-supported platinum catalyst was studied from 438° to 483°K, from 0.7 to 2.1 Torr carbon monoxide, and from 0.1 to 0.6 Torr oxygen in a differential flow reactor. The amount of chemisorbed CO was monitored during reaction by infrared transmission spectroscopy. Transitions were observed between two kinetic regimes — one characterized by moderate rates and nearly full CO surface coverage; the other, by higher rates and low surface coverage.

With nearly full coverage the kinetics were +1 order in P_{O_2} *and -0.62 order in* P_{CO} *with activation energy 26.2 kcal/gmol. Kinetics and CO surface coverage correlated with a mechanism involving surface reaction between adsorbed CO and adsorbed* O_2 *as the rate limiting step but did not correlate with other rate limiting steps or other proposed mechanisms. In the low coverage regime differential reactor conditions were not maintained, and the rate was limited by CO pore diffusion to the platinum. The transition in kinetics (at* $\theta_{CO} \sim 0.6$*) was postulated to be a result of a phase change in the chemisorbed CO.*

I. INTRODUCTION

The conventional approach in studying solid-catalyzed gas phase reactions has been to measure the dependence of the reaction rate on temperature and on the pressures of reactants and products, then to infer information about the reaction mechanism by assuming some relationship, e.g., a Langmuir isotherm, between the gas phase pressures and concentrations on the surface. There are now available experimental techniques (temperature programmed desorption and infrared spectroscopy, for example) which allow the direct observation of surface concentrations of chemisorbed species.

* *Corresponding author*

The objective of this investigation was to elucidate the mechanism of carbon monoxide oxidation on a platinum catalyst by monitoring the surface concentration of CO using infrared transmission spectroscopy while simultaneously observing steady state kinetics in a differential flow reactor. The experimental apparatus is similar to that of Sills (1). This and the detailed experimental procedure are described elsewhere (2). In order to observe significant changes in θ_{CO} (assumed proportional to infrared band area) the kinetics were studied at relatively high temperatures and at relatively low CO pressures. In order to avoid mass transfer limitations the catalyst contained Pt at 1 wt %, with the particle size of the SiO_2 support at about 500 Å (Cabosil L-5).

Kinetics with simultaneous infrared spectroscopy were observed at five temperatures — 438°K, 450°K, 460°K, 469°K, and 483°K. Oxygen partial pressure in the helium diluent was varied from 0.15 to 0.58 Torr in one run, but was held constant at about 0.6 Torr during most runs. Carbon monoxide partial pressure was varied at each temperature from the upper limit imposed by gas chromatograph detectability of CO_2 to the lower limit imposed by kinetics. At 483°K, P_{CO} was varied between 0.07 Torr and 2.2 Torr. The range of variation of P_{CO} was narrower at the lower temperatures.

II. EXPERIMENTAL RESULTS

Two regimes of oxidation kinetics were observed: the normal regime was characterized by a strong IR absorption band at about 2070 cm^{-1} and moderate reaction rate; the low coverage regime was characterized by the absence of the IR absorption band and very high reaction rate. Transitions between regimes were abrupt and reversible and qualitatively similar to the published results of Hugo and Jakubith (3) for CO oxidation on a Pt wire mesh. They observed abrupt, reversible changes in kinetics as a result of changing P_{CO}. In shifting from the high P_{CO} regime (characterized by negative order in P_{CO}) they observed increases in the rate of four to five fold. In the present investigation transitions were not observed as a result of changing P_{O_2}. Figure 1 shows kinetic data in the two regimes.

Multiple linear regression analysis of the kinetic data in the normal regime leads to the following rate expression:

$$R = K \frac{P_{O_2}^{1.0}}{P_{CO}^{0.62}} \exp-(26200/RT) \tag{1}$$

Figures 2,3, and 4 show the effect on rate of varying P_{O_2},

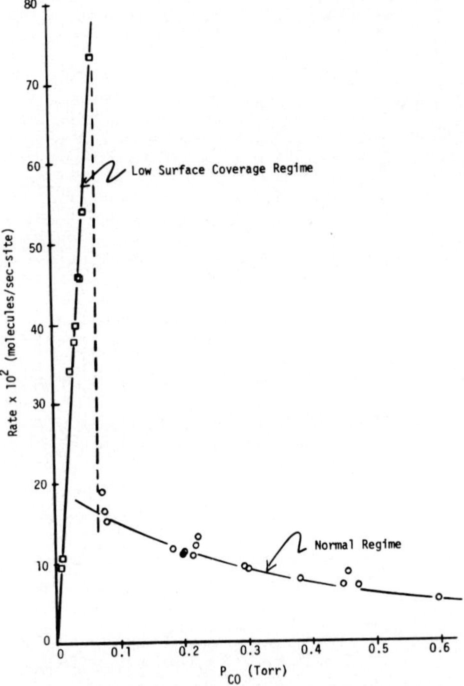

Figure 1. Transition between kinetic regimes

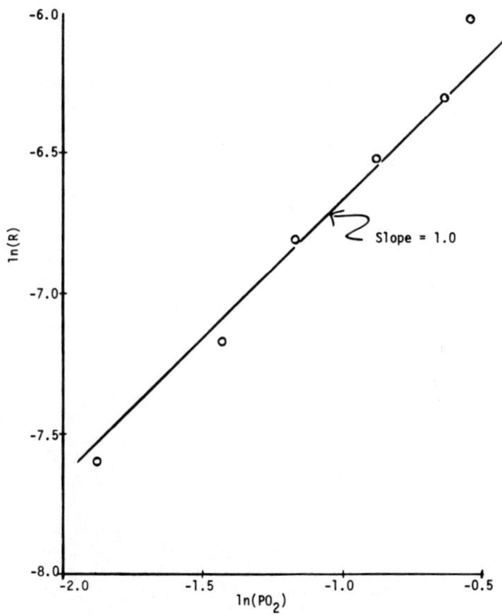

Figure 2. Rate dependence on oxygen pressure

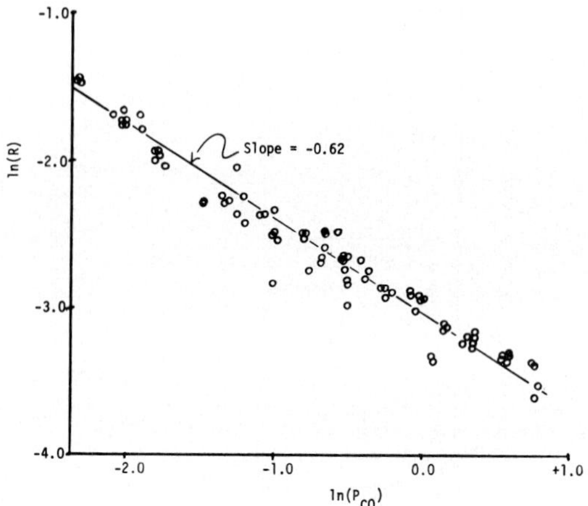

Figure 3. Rate dependence on carbon monoxide pressure

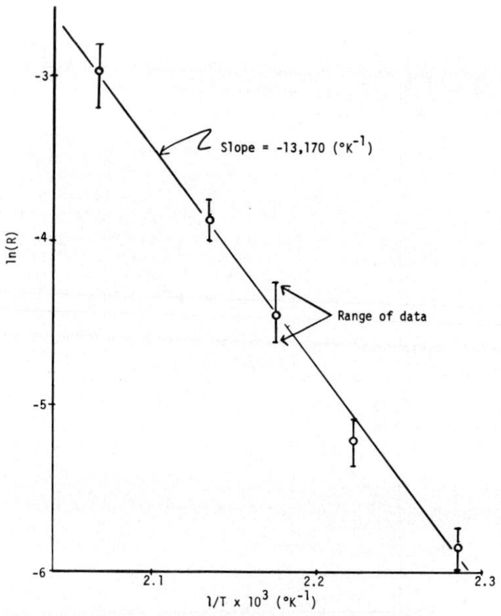

Figure 4. Arrhenius plot of kinetic data

P_{CO}, and T, respectively, the other two variables held constant in each case. The reaction rate has been corrected using equation (1) for small variations in the constant variables.

Typical infrared absorption spectra obtained during a steady state kinetic run are shown in Fig. 5 for 483°K, $P_{O_2} \simeq$ 0.6 Torr and three CO pressures in the range 0.07 to 0.60 Torr. The observed variation in band area is shown on the figure. The presence of oxygen resulted in only minor variations in this band (2). Spectra were recorded digitally and coadded to suppress noise.

Typical simultaneous infrared and kinetics results are presented in Figures 6 and 7 for reaction at 483°K. In Figure 6 the infrared band area, A, is plotted versus rate, R, where changes in rate are a result of varying P_{O_2} at 483°K and P_{CO} = 0.2 Torr. In Figure 7 A is plotted versus R for varying P_{CO} at 483°K temperatures with P_{O_2} = 0.6 Torr.

In the low coverage regime no IR band was observed. Because of the high rate, differential reactor conditions were not maintained. Assuming perfect mixing, the rate is independent of P_{O_2}, is first order in P_{CO}, and has a low activation energy of about 2 kcal/gmol.

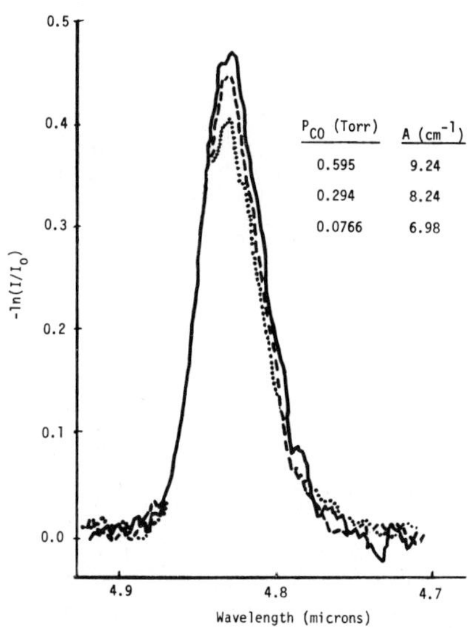

Figure 5. Effect of CO pressure on spectrum

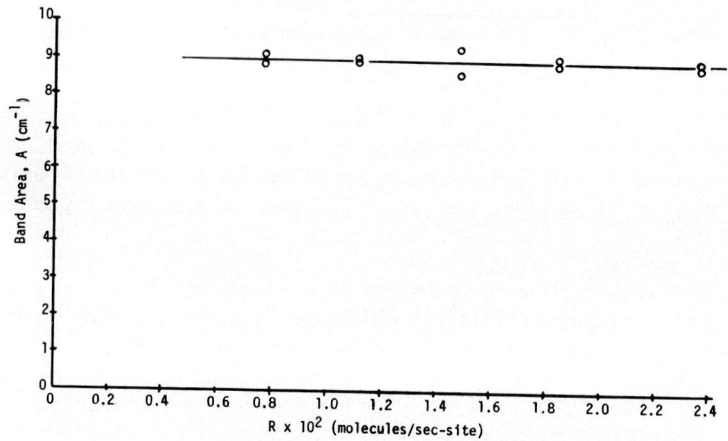

Figure 6. Band area vs rate; variation due to oxygen pressure

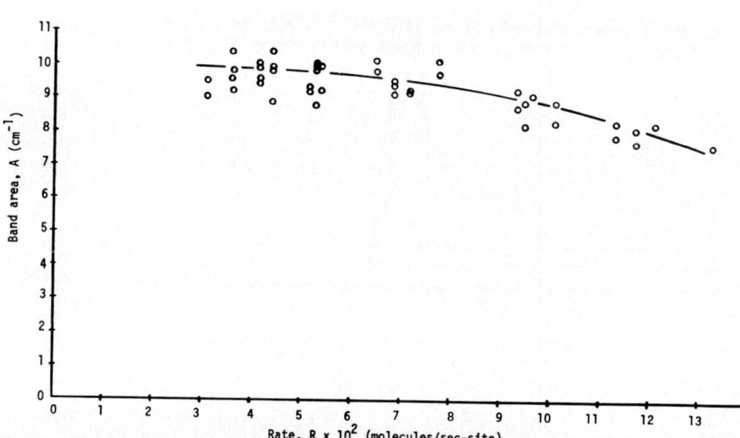

Figure 7. Band area vs rate; variation due to carbon monoxide pressure at 483°K

III. DISCUSSION

A. Mechanism of Carbon Monoxide Oxidation on Platinum

Langmuir (4) first considered a mechanism of CO oxidation on platinum, postulating that chemisorbed atomic oxygen reacted with chemisorbed CO, chemisorbed CO serving also to block the surface and hinder oxygen adsorption. Heyne and Tompkins (5) proposed a mechanism of CO oxidation on platinum in which chemisorbed molecular oxygen and chemisorbed CO react to produce an intermediate product, chemisorbed CO_3, which reacts rapidly with chemisorbed CO to produce two molecules of CO_2. Sills (1) analyzed his results within the framework of a reaction mechanism between chemisorbed CO and chemisorbed oxygen. His analysis indicated that reaction with chemisorbed atomic oxygen was rapid, but that only reaction with chemisorbed molecular oxygen was consistent with his overall rate expression. Wood, *et al*. (6) showed conclusively that chemisorbed CO can react with oxygen, establishing that the Hinschelwood-type mechanisms considered by Langmuir, Heyne and Tompkins, and Sills are, at least, possible. Wood, *et al*. postulated a reaction with atomic oxygen.

Sklyarov, *et al*. (7) analyzed their results in terms of a reaction mechanism between chemisorbed molecular oxygen and gas phase CO. Reaction of chemisorbed atomic oxygen was considered to be fast, and the formation of chemisorbed atomic oxygen from the gas phase was considered to be slow in comparison with the rate of oxidation by chemisorbed molecular oxygen. Oki and Kaneko (8) demonstrated that their results were not consistent with the Hinshelwood-type mechanisms proposed by Langmuir, Heyne and Tompkins, and Sills. Oki and Kaneko concluded that reaction between chemisorbed oxygen and gas phase CO was an appropriate mechanism of CO oxidation in their experiments. Bonzel and Ku (9) analyzed their results in terms of Sklyarov's mechanism and showed qualitative agreement. At constant reactant pressures the rate showed a maximum with respect to temperature as expected from Sklyarov's mechanism. At constant temperature and oxygen pressure the rate showed a maximum with respect to carbon monoxide pressure, also expected from Sklyarov's mechanism. At high temperature, $T > 400°C$, the observed decrease in activation energy is also expected.

Hugo and Jakubith (3) showed that their dynamic results were qualitatively consistent with a reaction mechanism in which CO adsorbs in an unreactive form which converts on the surface to a reactive form.

Two main themes characterize the mechanisms considered in the literature: Hinschelwood-type reactions in which chemisorbed CO and chemisorbed oxygen react; and Rideal-Eley mechanisms in which chemisorbed oxygen reacts with gas phase CO. All data which appear to support the Hinschelwood-type mechanism were obtained at relatively low temperatures. All data which appear to support the Rideal-Eley type mechanism were obtained at relatively high temperatures.

In none of the existing investigations of CO oxication on platinum is there any direct examination of the participation of chemisorbed species in the reaction. Conclusions have, of necessity, been based on observations of gas phase concentrations and extremely simple models of the adsorption process. Direct observation of the participation of chemisorbed species allows more confident postulation of reaction mechanism, as will be shown.

B. Mechanism Analysis in the Normal Regime

The Hinshelwood-type mechanistic model based on a surface reaction between chemisorbed CO and chemisorbed O_2 can be written in typical form as follows:

$$O_2(g) + * \underset{K_{-1}}{\overset{K_1}{\rightleftharpoons}} O_2*(a) \tag{2}$$

$$CO(g) + * \underset{K_{-2}}{\overset{K_2}{\rightleftharpoons}} CO*(a) \tag{3}$$

$$CO*(a) + O_2*(a) \xrightarrow{K_4'} CO_2(g) + O*(a) \tag{4}$$

$$CO*(a) + O*(a) \xrightarrow{K_5'} CO_2(g) + * \tag{5}$$

where * indicates an adsorption site. The last step is assumed to be very rapid.

Since θ_{O_2} cannot be determined in these experiments, a Langmuir adsorption isotherm has been assumed for O_2. Then, if oxygen adsorption, equation (2), is the rate limiting step, this mechanism leads to the following rate expression in terms of θ_{CO}:

$$R = 2K_1 P_{O_2}(1 - \theta_{CO}) \tag{6}$$

If surface reaction, equation (4), is the rate limiting step, the following rate expression results:

$$R = 2K_4' \frac{K_1}{K_{-1}} P_{O_2}(1 - \theta_{CO})\theta_{CO} \tag{7}$$

The alternative class of reaction mechanism frequently considered in the literature (particularly at high temperatures) is a Rideal-Eley reaction between chemisorbed oxygen and gas phase CO. As proposed by Sklyarov, et al. (7) the mechanism is summarized as follows:

$$O_2(g) + * \underset{K_{-1}}{\overset{K_1}{\rightleftharpoons}} O_2*(a) \tag{8}$$

$$CO(g) + * \underset{K_{-2}}{\overset{K_2}{\rightleftharpoons}} CO*(a) \tag{9}$$

$$CO(g) + O_2*(a) \xrightarrow{K_4} CO_2(g) + O*(a) \tag{10}$$

$$CO(g) + O*(a) \xrightarrow{K_5} CO_2(g) + * \tag{11}$$

Once again it is assumed that reaction with adsorbed atomic oxygen, equation (11), is very rapid. It is also assumed that desorption of O_2 is very slow. Thus, $K_5 >> K_4$ and $K_4P_{CO}\theta_{O_2} >> K_{-1}\theta_{O_2}$. Then in general under steady state conditions

$$R = \frac{2K_4P_{CO}(1 - \theta_{CO})}{(1 + K_4P_{CO}/K_1P_{O_2})} \tag{12}$$

For oxygen adsorption rate limiting, equation (6) again applies. For the Rideal-Eley step, equation (10), rate limiting, equation (12) reduces to

$$R = 2K_4P_{CO}(1 - \theta_{CO}) \tag{13}$$

The typical data of Figures 6 and 7 are utilized here to critically evaluate the various postulated mechanism types and rate-limiting steps which led to the kinetic expressions equations (6), (7), (12) and (13). Lack of agreement of a mechanistic model with the data is a strong refutation of the model for these particular experiments. Agreement between model and data, of course, only indicates the model is not impossible.

Examination of Figure 7 shows a definite downward curvature of A versus $R(P_{CO})$. (The quadratic term in the regression of the data of Figure 7 is significant to the 95% confidence level.) Thus, the results of Figure 4-6 are inconsistent with equation (6), oxygen adsorption rate limiting for either mechanism-type. Moreover, the derived sticking probability in this case would be $s \simeq 1.5 \times 10^{-8}$, an implausibly low value for a partially covered surface.

The kinetic expression, equation (7), which applies in the case of surface reaction rate limiting for the Hinshel-wood-type mechanism can be tested by plotting the data of Figure 7 in the following linearized form of equation (7):

$$A = 2K_4' \frac{K_1}{K_{-1}} A_{max} - A_{max}\left(\frac{R}{P_{O_2}A}\right) \tag{14}$$

where it is assumed that θ_{CO} equals band area, A, divided by band area for full coverage, A_{max}. The appropriate plot is shown in Figure 8. The agreement is evident. Thus, it is concluded that the observed results are consistent with equation (7), surface reaction between adsorbed CO and adsorbed O_2 rate limiting.

There is a clear inconsistency of the data with the general kinetic expression for the Rideal-Eley type mechanism, equation (12). Again, the data of Figure 7 are plotted according to the following linearized form of equation (12):

$$P_{CO}(1 - A/A_{max})/R = 0.5K_4 + 0.5K_1(P_{CO}/P_{O_2}) \tag{15}$$

The appropriate plot is Figure 9. The non-linearity of the results is apparent. It is concluded, therefore, that the results of this study are not consistent with the Rideal-Eley mechanism.

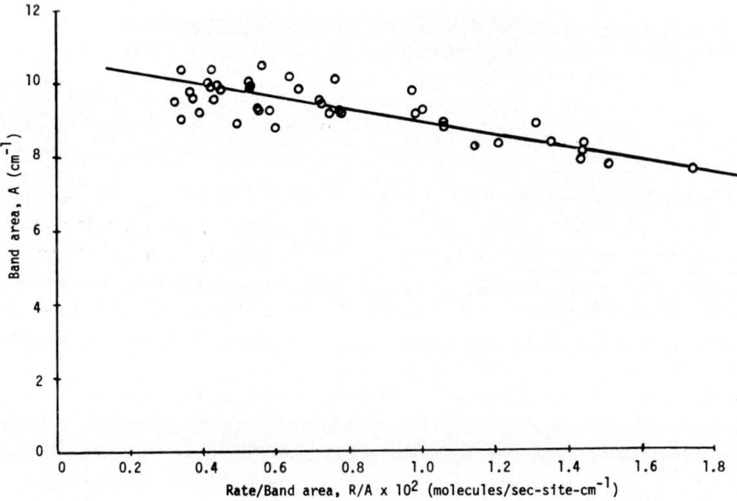

Figure 8. Test of the surface reaction model

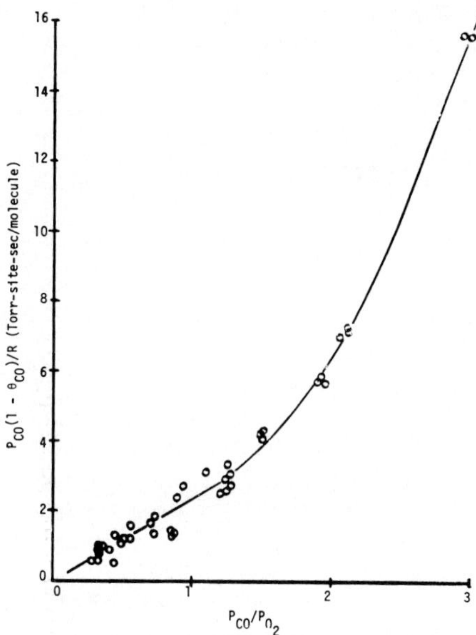

Figure 9. Test of Rideal-Eley model

In summary, then, the results of this investigation are not consistent with an oxygen adsorption rate limiting step; the results are not consistent with the Rideal-Eley type mechanism; the results are consistent with surface reaction between chemisorbed CO and chemisorbed O_2.

C. Low Coverage Regime and Transition in Kinetics

In the low coverage kinetics regime the observed results indicate the rate was limited by pore diffusion of CO to the catalyst. The expected CO mass transfer rate has been calculated to be about 5 x 10^{-7} gmol/sec-Torr CO; the observed rate, 7.5 x 10^{-7} gmol/sec-Torr CO.

The lowest surface coverage observed before transition to the low coverage regime was about $\theta_{CO} \simeq 0.6$. It is postulated that the transition might be the result of a phase change in the chemisorbed CO from a densely-packed, condensed phase at higher P_{CO} to a loosely-packed disperse phase at lower P_{CO}. In equation (3) K_4', the surface reaction rate constant, might be expected to be significantly higher for the disperse CO phase. Sickafus and Bonzel (11) have shown, for example, that just such a phase change occurs for CO chemisorbed on Pd at about $\theta_{CO} = 0.5$. Also, the infrared extinction coefficient for CO chemisorbed on silica-supported platinum was observed by Eischens, et al. (12) to change abruptly at $\theta_{CO} \simeq 0.6$, possibly reflecting such a phase change for the CO/Pt system. Insufficient evidence exists to test this hypothesis, but as a possible explanation for the transition in kinetics it appears not unreasonable.

IV. REFERENCES

1. Sills, R.A., "Simultaneous Infrared and Kinetic Studies of the Platinum Catalyzed Carbon Monoxide Oxidation," Ph.D. thesis, M.I.T., Cambridge, Mass., 1970.
2. Cochran, H.D., Donnelly, R.G., Modell, M., and Baddour, R. F., "The Carbon Monoxide Oxidation on a Platinum Catalyst," (to be published).
3. Hugo, P., and Jakubith, M., "Dynamic Behavior and Kinetics of Carbon Monoxide Oxidation on Platinum Catalysts," Chem.-Ing.-Tech., 44(6), 383 (1972).
4. Langmuir, I., "The Mechanism of the Catalytic Action of Platinum in the Reactions $2CO+O_2 = 2CO_2$ and $2H_2+O_2 = 2H_2O$," Trans. Far. Soc., 17, 621 (1922).
5. Heyne, H., and Tompkins, F.C., "Application of Infrared Spectroscopy and Surface Potential Measurements in a Study of the Oxidation of Carbon Monoxide on Platinum," Proc. Roy. Soc., Ser. A, 292, 460 (1966).
6. Wood, B.J., et al., "Competitive Sorption Kinetics of Oxygen and Carbon Monoxide on Platinum," J. Catal., 18, 70 (1970).
7. Sklyarov, A.V., et al., "Oxidation of Carbon Monoxide by Oxygen on Platinum Purified in Ultrahigh Vacuum," Dok. Akad. Nauk SSSR, 189(6), 1302 (1969) translated by Consultant Bureau, Inc., Doklady Physical Chemistry.
8. Oki, S., and Kaneko, Y., "On the Mechanism of Carbon Monoxide Oxidation," J. Res. Inst. Catal. Hokkaido, 18(2), 93 (1970).
9. Bonzel, H.P., and Ku, R., "Carbon Monoxide Oxidation on a Pt(110) Single Crystal Surface," J. Vac. Sci. and Tech., 9(2), 663 (1972).
10. Sickafus, E.N., and Bonzel, H.P., "Surface Analysis by Low-Energy Electron Diffraction and Auger Electron Spectroscopy," Prog. Surf. Membrane Sci., 4, 115 (1971).
11. Eischens, R.P., et al., "The Effect of Surface Coverage on the Spectra of Chemisorbed CO," J. Phys. Chem., 60, 194 (1956).

SURFACE ANALYSIS AND ADHESIVE BONDING. II. POLYIMIDES

David W. Dwight, Mary Ellen Counts and
James P. Wightman
Virginia Polytechnic Institute and State University

I. ABSTRACT

Surface analysis by SEM/EDAX shows effects upon fracture surfaces of adhesive formulation, process, and test variables. Empirical correlations between shear strength and structure of the fracture surface are promoted by qualitative fracture mechanics theory. Interfacial and bulk failure, viscoelastic, plastic and brittle mechanisms, and the distribution of initial flaws are the controlling factors. A series of experimental polyimide adhesives was tested on titanium 6-4 adherends, and voids in the glue line appear to be a prime factor limiting strength and reproducibility. Strength varies inversely with temperature of testing. SEM shows a corresponding increase in interfacial failure and decrease in plastic and brittle mechanisms. Improved high-temperature strength was obtained by filling the adhesive with aluminum powder; interfacial separation was eliminated and much finer features characterize the microtopography. High-temperature failure appears to proceed farther from the interface, judging by estimates of the residual adhesive thickness using EDAX. Changes in solvent, amine and anhydride in the adhesive formulation also are characterized by unique fracture surfaces.

II. INTRODUCTION

Adhesive bonding of aerospace systems and components is increasing at a rapid pace (1). Substantial cost savings are possible with the reduction in weight and in manufacturing costs. Other advantages over riveted or bolted structures are facile joining of thin and contoured sheets, and reduced stress concentration and galvanic corrosion; adhesives are the only practicable way to join fiber-reinforced composites and honeycomb structures. However, new service requirements (10,000 hours at 600°F [316°C]) exceed the property limits of most synthetic organic polymers. The total system also must withstand exposure to high humidity and severe temperature cycles. Furthermore, restrictions are placed upon polymer synthesis by the need for good processability with very low volatiles under stringent autoclave conditions.

Our interdisciplinary program has synthesis and strength-testing of novel high-temperature polymers as its primary focus (2), with detailed surface analyses of substrates and fracture surfaces to aid in interpreting the results (3). This report integrates some of the results of the preliminary studies with new data derived from changes in solvent, amine and anhydride during adhesive formulation. Also we have studied the effects of aging and strength-testing at high temperatures, and the improvement in these properties with aluminum powder adhesive filler. Interpretation and correlation of the strength-testing and SEM results is facilitated by reference to qualitative fracture mechanics theory (4). This approach identifies: (1) inherent flaws, (2) viscoelastic and plastic deformation, (3) crazing and crack propagation, and (4) interfacial failure as the main factors contributing to overall strength. Changes in parameters of adhesive formulation, processing, and testing alter the proportions of each of these failure mechanisms. Results from SEM combined with energy-dispersive x-ray fluorescence analysis (EDAX) suggest that, in some cases, useful estimates of the thickness of residual adhesive layers may be made. For improvements in strength, it is first important to know the details of crack-initiation, and the extent and morphology of the failure <u>zone</u> (5).

III. EXPERIMENTAL

Photomicrographs were obtained with the Advanced Metals Research (AMR) Corporation Model 900 Scanning Electron Microscope operating at 20 kV. The specimens were cut to approximately 1 x 1 cm with a high pressure cutting bar and fastened to SEM mounting pegs with adhesive-coated, conductive copper tape. To enhance conductivity, a thin (∿200Å) film of Au/Pd alloy was vacuum evaporated onto the samples. Photomicrographs were taken with the samples inclined 20° from incident electron beam.

Rapid elemental analysis was obtained by energy-dispersive x-ray fluorescence with the EDAX International Model 707A unit attached to the AMR-900 SEM.

Adhesive synthesis and testing has been described elsewhere (2).

<u>Abbreviations of Monomers and Solvents</u>

BTDA-Benzophenone Tetracarboxylic Acid Dianhydride
DABP-Diaminobenzophenone
PMDA-Pyromellitic Dianhydride

ODPA-Oxidiphthalic Anhydride
EHA-13-m,m''-Diamino Terbenzylone
DG-Dyglyme, DMAC-Dimethylacetamide, DMF-dimethylformamide

IV. RESULTS AND DISCUSSION

A. Aging and Testing at Elevated Temperatures

A series was prepared with a standard BTDA + m,m'-DABP/DG adhesive and exposed to 295°C for 30 days in air, and then shear tested at 25°, 225°, 250° and 270°C, respectively. Little thermal- or oxidative-degradation occurred during the severe exposure (joints were not stressed during aging), but strength dropped rapidly at higher test temperatures. SEM examination (Figures 1-5) indicates that two processes occur as the test temperature increases: (1) the percentage of interfacial failure increases, and (2) the amount of plastic deformation and brittle fracture decreases.

The same general features characterize the fracture surface of the sample tested at room temperature (Figure 1) as

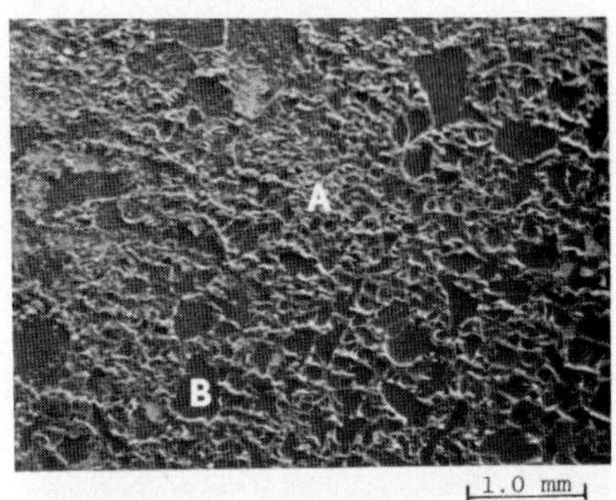

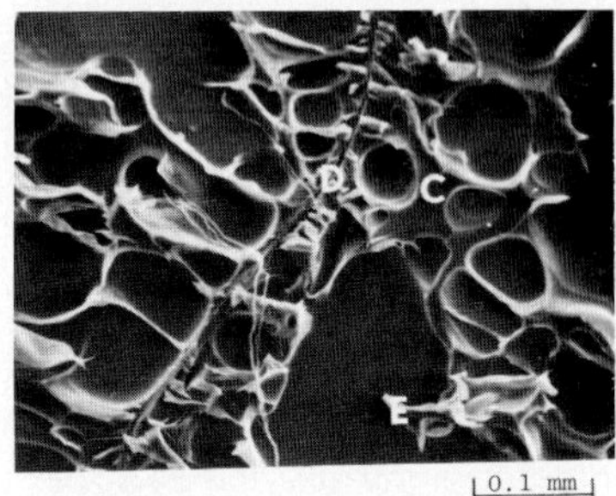

Fig. 1. Photomicrographs of the fracture surface of BTDA + m,m'-DABP/DG adhesive after aging 30 days at 295°C; 3570 psi shear strength at room temperature. Features similar to previous, (unaged) room temperature results (3).

have been seen in previous (unaged) high strength samples (3). At lower magnification there is a fairly uniform distribution of raised material that forms a filigree pattern "A" superimposed upon lower, smooth, oblong areas "B" that have dimensions ranging from 0.02 to 0.5 mm, approximately. The low areas are the bottoms of voids that were probably created during the formation of the joint and expanded during fracture. The filigree is composed of void-cell walls that have undergone plastic deformation and finally fractured, primarily by a brittle cleavage-crack propagation mechanism. It is only the deformed areas that provide strength. Some of the cracks are smooth and quite parallel to the substrate, such as in area "C" on the higher magnification photomicrograph, but also fine louvers, tilted at an angle to the adherend, appear along the curved line that comes down the center "D". Localized "hot spots" that could promote ductility indicated at "E" may arise from adiabatic fracture processes. Note that only a very small amount of interfacial failure appears, and many, thin walls result from plastic deformation.

Generally similar features are seen in the 225°C sample (Figure 2) but the proportion of interfacial failure "A", and

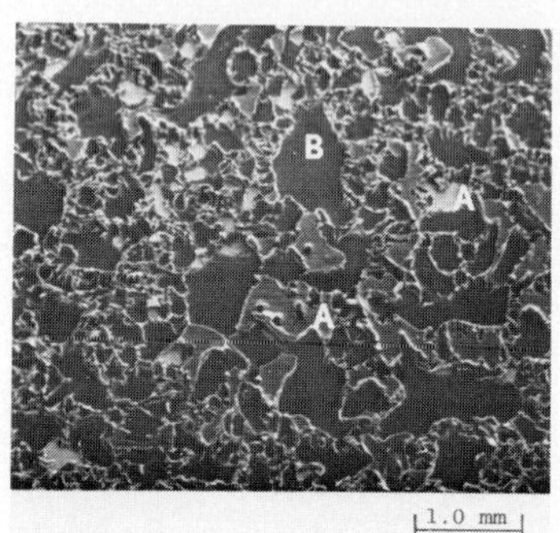

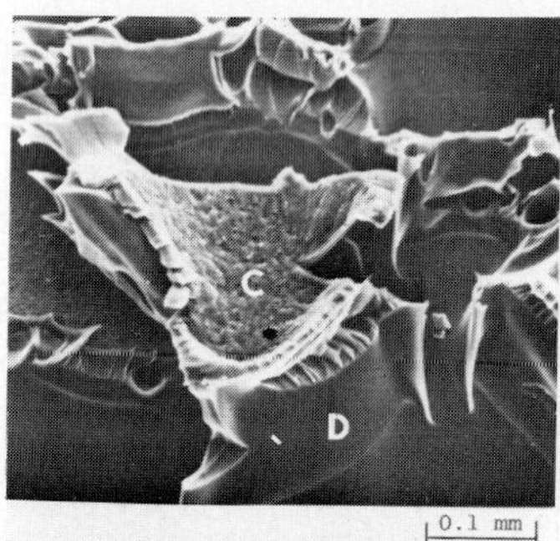

Fig. 2. Same as Fig. 1 except 2065 psi strength, tested at 225°C.

void area "B" has increased. Areas of adhesive that have detached from the opposite adherend show that original interfacial contact was good, because the polymer faithfully replicates the detailed surface features of the titanium (area "C"). Comparing Figures 1 and 2, it appears that the voids have coalesced in the latter, sometimes reaching dimensions over 1 mm. Correspondingly, void-cell walls are thicker; area "D" at higher magnification provides a striking example of a plastically-drawn void-cell wall, circumscribed by brittle cleavage cracks.

Figure 3 shows that interfacial failure and void area increase with temperature (250°C). Large void spaces are almost completely interconnected across the sample. At higher magnification, some strength across the interface is indicated by continuity where the brittle, cleavage-crack louvers meet the exposed substrate. Interfacial failure predominates in the 270°C test. The opposite sides of mating fracture surfaces are shown in Figure 4; it is quite simple and instructive to locate the matching features. There is very little plastic deformation, hence little strength. Brittle cleavage cracks without much louvering account for the failure that is

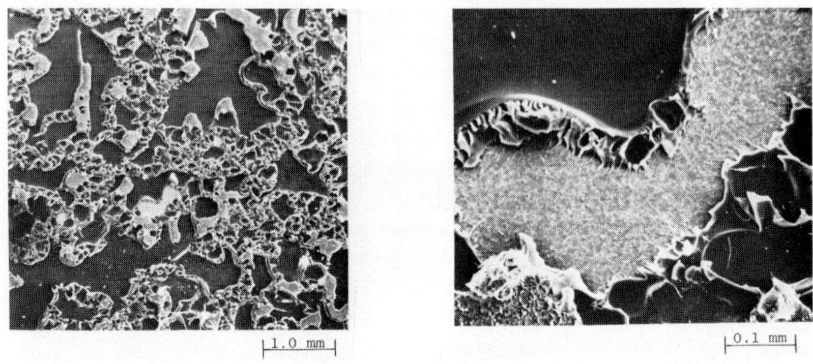

Fig. 3. Same as Fig. 1 except 1465 psi strength, tested at 250°C.

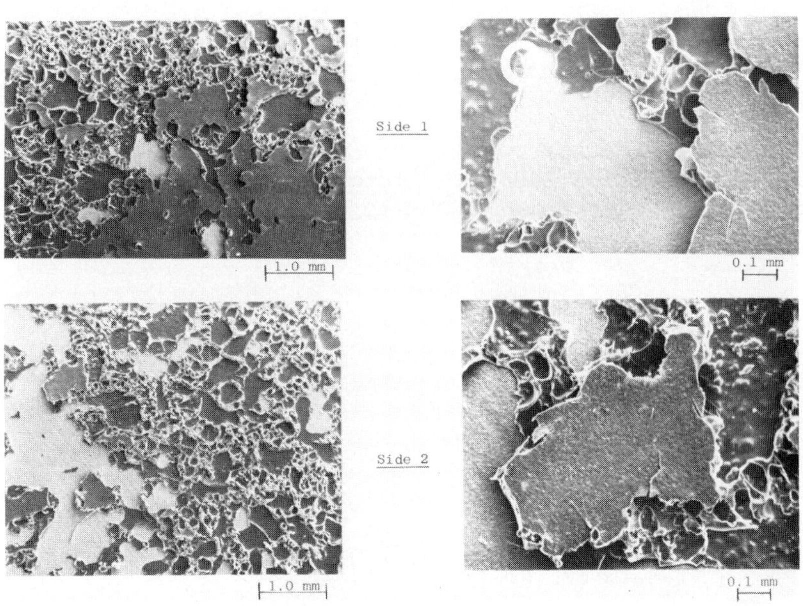

Fig. 4. Same as Fig. 1 except 760 psi strength, tested at 270°C; opposite sides of mating fracture surfaces.

not interfacial. The circled area of Side 1 is shown at higher magnification in Figure 5, illustrating a discontinuity (low forces) at the polymer/metal interface.

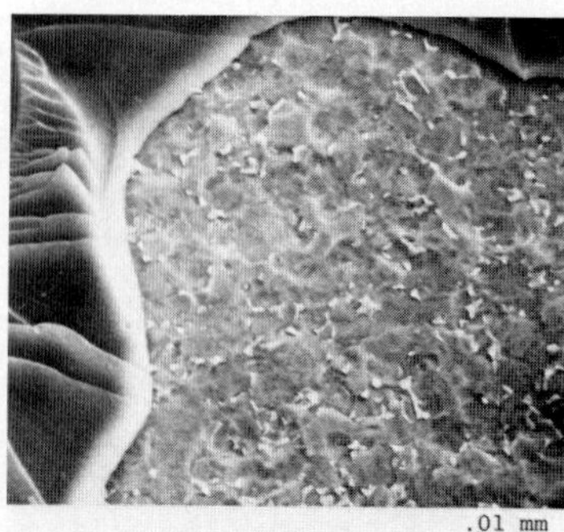

Fig. 5. High magnification view of area circled in Fig. 4. Note the brittle cleavage-crack louvers on the left and the discontinuity at the polymer/metal interface.

Apparently high temperature reduces interfacial forces, probably through differential thermal expansion, and the polymer experiences less stress. In other words, without sufficient bonding at the polymer/metal interface, mechanisms that provide adhesive strength (elastic deformation, crazing, etc.) will not come into play. From this point of view, it seems reasonable that overall strength would be the product of an interfacial bonding term and a term that sums the contributions to the strength of bulk materials.

Progar and St.Clair found a two-fold improvement in high-temperature strength (65/35 BTDA/PMDA + m,m'-DABP) was obtained by using aluminum powder as an adhesive filler (2). Figure 6 shows typical surface features of filled-adhesive samples fractured at room temperature. The aluminum filler is apparent as lumps about 1-10 μm in diameter, covered with at least a thin layer of polyimide. Fracture-surface features are much smaller than with unfilled adhesive. At high magnification it appears that the fracture initiates in a myriad of minute voids (or nucleation sites) existing in the walls of the larger void areas, and the fracture surface is composed of very thin, plastically-deformed microvoid-cell walls. This feature is unique to the filled adhesive.

No interfacial failure occurred, in contrast to the results at 250°C with unfilled adhesive. Perhaps the aluminum filler prevents interfacial failure at high temperature by adjusting relative thermal expansion. Another effect of filler is to increase the amount of fracture surface area by expanding the filigree pattern into a more continuous area.

EDAX provided information on the thickness of adhesive film remaining on the substrate after fracture, since the 20 kV electron beam should penetrate about 1 μm of organic polymer. EDAX results from two magnifications of the room temperature

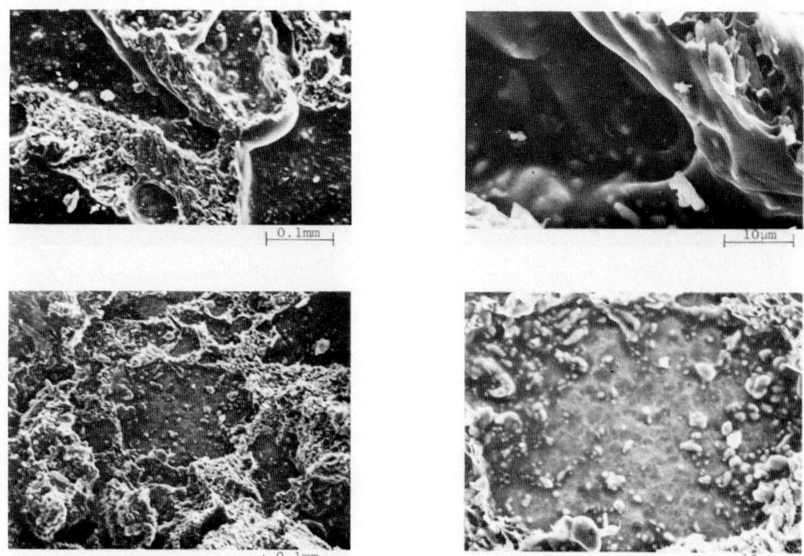

Fig. 6. Photomicrographs of the fracture surface of 65/35 BTDA/PMDA + m,m'-DABP/DG adhesive filled with 50% (top) and 70% (bottom) aluminum powder.

samples are shown in Figure 7. The gold and palladium peaks are due to the ∿200 Å conductive coating layer; they are useful for approximate internal standards. The aluminum filler actually gave the largest peak in the original spectra, omitted here for clarity. In line A (50% Al), all the adherend (Ti) signal was found to be coming from small holes, seen in Figure 6 (top) at high magnification; no Ti signal could be obtained at the bottom of voids. The opposite was true at 70% Al. More Ti appears on the survey scan, and it dominates the spectrum from void bottoms. Figure 6 (bottom) shows the area from which the last EDAX spectrum was taken. Clearly there is a layer of polymer there, but it seems to be so thin that titanium surface features show through vaguely. EDAX examination of both high temperature samples failed to uncover a Ti signal at any magnification. Thus it can be concluded that the room temperature fracture occasionally penetrates nearer to the adherend than the high temperature fracture, which can hardly come closer than 1μm.

The combination of SEM and EDAX provides most of the essential information on the locus and micromechanics of fracture. It is true that the SEM electron beam can penetrate several hundred Angstrom units of adhesive and give an EDAX signal from underlying adherend. However the combination between the EDAX spectra and the SEM photomicrographs usually leaves little doubt about the details of fracture, as illus-

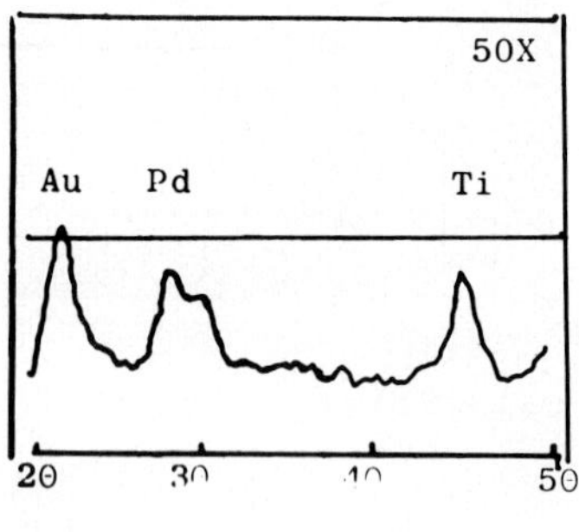

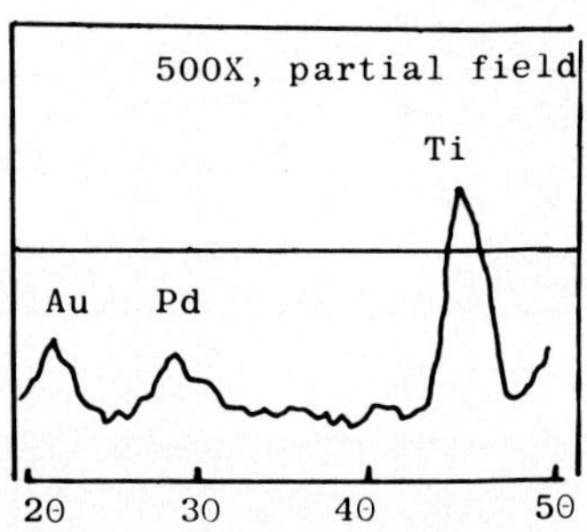

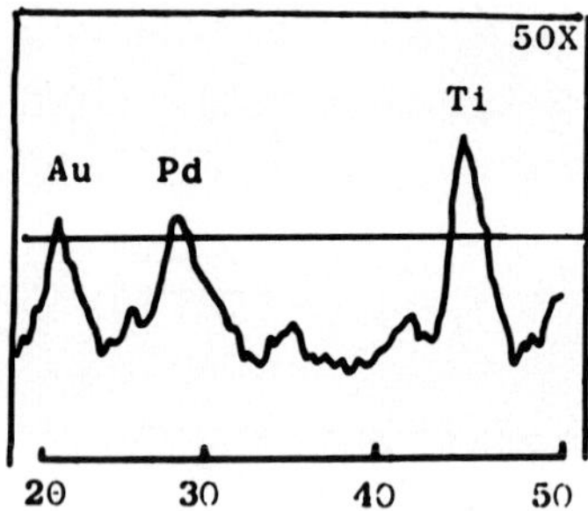

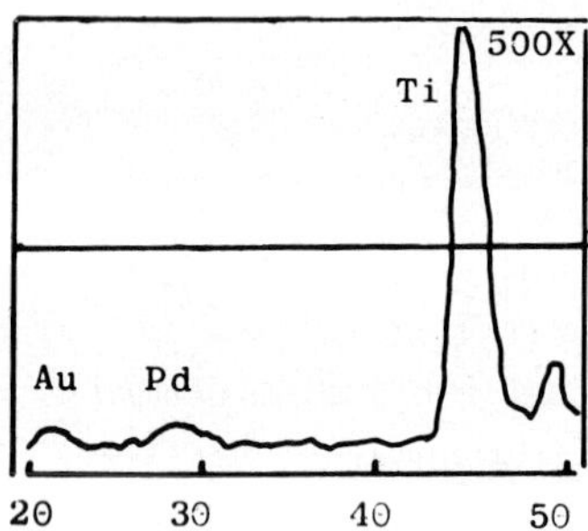

Fig. 7. Elemental analysis, using EDAX at two magnifications, of the same samples shown in Fig. 6. Top - Some adherend shows through on the 50X "survey" on the left, but all the Ti signal originated from holes like the one shown in Fig. 6. Bottom - More adherend signal at 50X, and it dominates the spectrum when the beam was focussed into the void bottom.

trated in Figures 6 and 7. Moreover, by variation of the beam voltage and use of calibration samples of known thickness, it may be possible to make more quantitative measurements of residual adhesive film thickness. An important advantage of SEM/EDAX is the ability to focus the electron beam and analyze only very small, selected areas.

B. Anhydride, Amine and Solvent

Several adhesive formulation parameters were varied, and good shear strengths obtained with changes in comonomer and solvent (except for the use of p,p'-DABP, which gave 1400 psi). Figures 8-10 display the micro-mechanics of fracture for these samples. The shear strength results can be rationalized in terms of the proportion of voids, interfacial and brittle failure, and plastic deformation.

The use of DMF as solvent for standard polyimide gives little interfacial failure, but a large proportion of connected-void area compared to fracture area, shown in Figure 8.

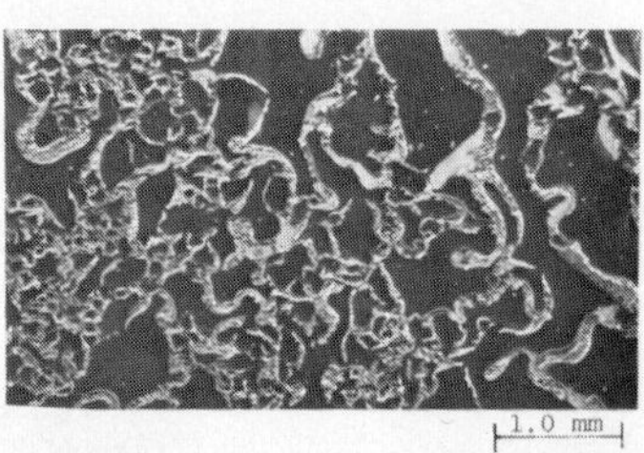

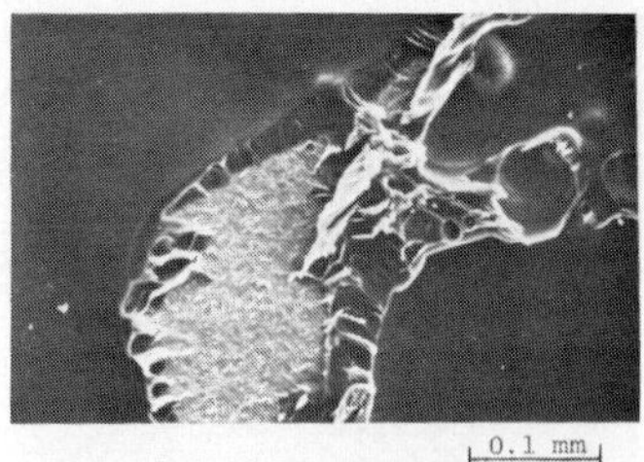

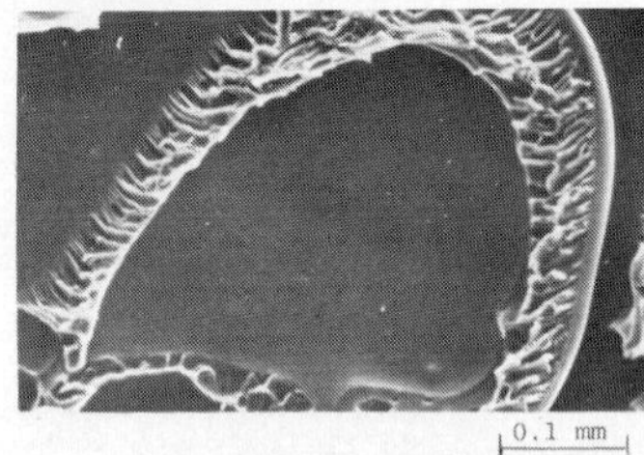

Fig. 8. Photomicrographs of the fracture surface of BTDA + m,m'-DABP adhesive in DMF solvent; 3700 psi shear strength.

Apparently the 3700 psi strength is developed by the initiation and annihilation of a large number of brittle cleavage cracks illustrated by numerous louvers in the higher magnification photomicrographs. Figure 9 shows a large proportion

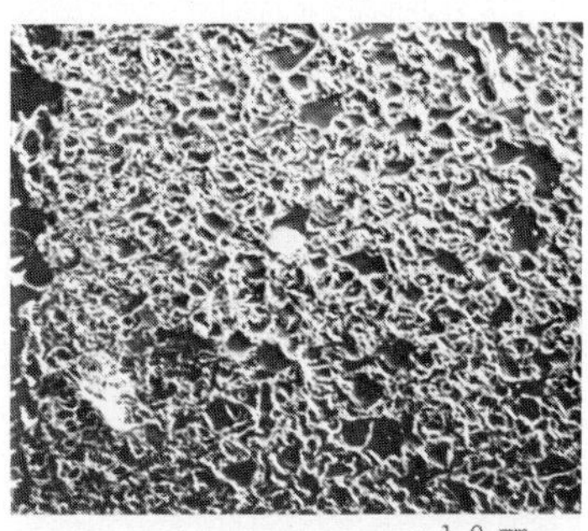

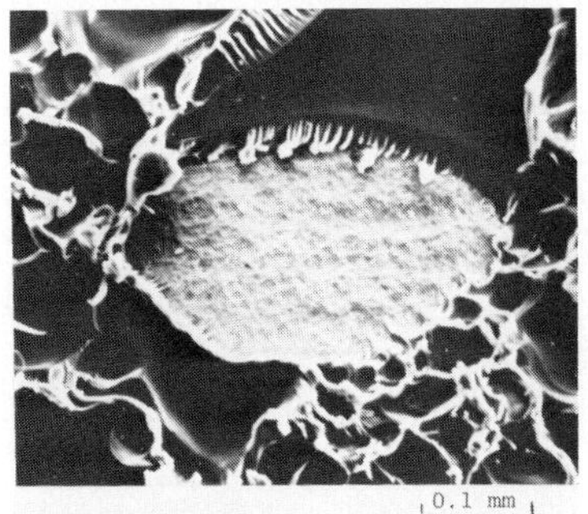

Fig. 9. Photomicrographs of the fracture surface of OPDA + m,m'-DABP/DG adhesive; 4500 psi shear strength.

of drawn and fractured polymer in a high-strength sample prepared with anhydride comonomer, ODPA. At higher magnification, a balance between moderate plastic deformation and brittle fracture is apparent, as well as exposed substrate and an indication of strong polymer/metal bond (continuity at the interface).

EAH-13 comonomer was used to provide a totally imidized, film adhesive with enough flow at high temperature so that good interfacial contact occurred during pressing at 200 psi and 300°C for an hour. The photomicrographs in Figure 10 show

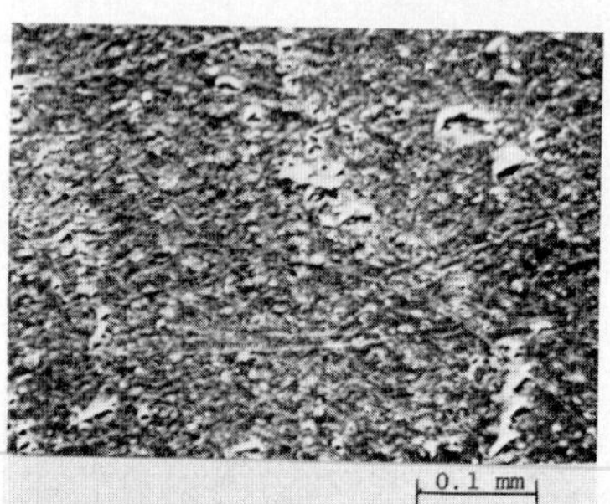

Fig. 10. Photomicrographs of the fracture surface of BTDA + EAH-13/DMAC, "imidized flow-bonding" adhesive; 3500 psi shear strength. Note the replication of the titanium adherend surface, and the absence of voids or plastic deformation: failure was completely interfacial.

that the polymer has formed a detailed replica of the titanium surface. Failure is totally interfacial on one side of the joint. It is unusual for good strength to be obtained without contributions from plastic and brittle mechanisms. Perhaps elastic deformation occurred. Otherwise, the true area over which polymer/metal polar and dispersion forces interact must be many times the geometrical joint area, due to the adherend roughness. Most important is the absence of the typical void structures so characteristic of joints prepared by imidization during bonding.

V. SUMMARY AND CONCLUSIONS

The basic findings to date of our interdisciplinary adhesive program are summarized in Table 1. Lap shear strength decreased when the test temperature increased; interfacial failure and void area increased and plastic and brittle mechanisms decreased. Interfacial bonding must be weakened by

TABLE 1
Variations in Adhesive Formulation, Test Parameters, Strength and Fracture Surfaces

Variable		Lap Shear Strength (psi) †	Fracture Surface Analysis: Figure No.	Fracture Surface Analysis: Comments
Temperature (BTDA + m,m'-DABP)				
Aging, °C	Testing, °C			
295	25	4320	1	∿50/50 filigree and and void, almost no interfacial failure; thin plastic deformation, brittle fracture with louvers.
295	225	2060	2	More interfacial and void area; thicker-wall deformation, brittle fracture with louvers.
295	250	1460	3	Similar except more void and interfacial area.
295	270	850	4,5	∿60/30/10 interfacial/void/filigree; small plastic deformation and brittle fracture surface area.
Aluminum Powder (65/35 BDTA/PMDA + m,m'-DABP DG)				
% Al	Testing, °C			
50	25	3780	6,7	∿60/40 void and filigree; much finer features at hi mag.-microvoids in walls, thin plastic deformation area large; lumps of Al seem covered with polymer. EDAX:Ti

TABLE 1 (continued)

Variable		Lap Shear Strength (psi)†	Fracture Surface Analysis	
			Figure No.	Comments
%Al	Testing, ^{o}C			
50	250	1210	--	Similar at 20X, but melting rather than drawing. EDAX:No Ti
70	25	3720	6,7	Fracture area continuous instead of filigree, low void and no interfacial area; detail similar to above EDAX:Ti
70	250	2340	--	Similar, except EDAX:no Ti
Solvent (BTDA+mm'-DABP)				
	DG	5280	6*,9*	∿50/50 filigree and void, almost no interfacial failure; thin plastic deformation, brittle fracture with louvers.
	DG	3860	10*, 11*	Similar except more interfacial and void area.
	DMAC	2510	7*,8*	Large void area and interfacial failure, no filigree; little deformation or brittle fracture surface
	DMF	3860	8	∿20/70/10 filigree/void/interfacial; little deformation but high area of brittle fracture with louvers.

TABLE 1 (continued)

Variable	Lap Shear Strength (psi)†	Fracture Surface Analysis Figure No.	Comments
Anhydride (m,m'-DABP/DG)			
BTDA	5280	6*,9*	Similar to line 1
PMDA	0	15*,16*	60/40 interfacial and void; small brittle fracture area.
ODPA	3250	9	High % filigree; short, thin deformation and brittle louvers.
Amine (BTDA/DG)			
m,m'-DABP	5280	6*,9*	Similar to line 1
m,p'-DABP	2070	12*,13*, 14*	50/50 void and interfacial; some hackled brittle fracture
EAH-13/DMAC	3500	10	100% interfacial; no voids, deformation, or brittle failure.

*Reference 3. †Average of three samples.

differential thermal expansion. Filling the adhesive with aluminum powder doubled the high-temperature shear strength; interfacial failure was eliminated and features of fracture surfaces were an order of magnitude smaller.

DG and DMF promoted good contact with the adherend, but DG resulted in plastic deformation while DMF gave high-area, brittle fracture. DMAC appeared to give poor adherend contact and little deformation or brittle fracture surface. Changes in solvent may effect polymer physical properties through variations in chain entanglement.

Adhesive formulations with either the anhydride PMDA or para-structure in the amine appeared more brittle than when anhydrides BTDA or ODPA or m-amines were used. Plastic deformation decreased and interfacial failure and low-area brittle cracks (without extensive louvering) increased in the former cases. The totally pre-imidized "flow-bonding" adhesive film, employing the amine EAH-13, produced no voids in the glue line during the bonding step, and failed interfacially (with good strength in spite of the absence of brittle and plastic mechanisms.)

In all other cases, voids were generated during cure either by residual solvent or water of imidization. The distribution of void sizes and locations within the glue line seems to be random and probably was the main cause of poor reproducability and lowered shear strengths. Not only is the area of drawing and fracture limited by void areas, but also they serve as nucleation sites for crack initiation. When strength results varied anomalously on replicate samples, there was usually a corresponding variation in void area.

Currently work is in progress to eliminate voids, synthesize more effective polymers, employ high-modulus fiber composites as adherends, expand durability tests, and refine the semi-empirical approach to understand adhesive bonding by quantitative analysis of fracture-surface data and rheological (WLF) measurements.

VI. ACKNOWLEDGMENTS

NASA support (Grant No. NSG-1124), program coordination by Norm Johnston, and expert SEM/EDAX operation of Frank Mitsianis are gratefully acknowledged.

VII. REFERENCES

1. Dauksys, R. J., SAMPE Quarterly, 1 October 1973.
2. a. St.Clair, T. L. and Progar, D. J., in "Adhesion Science and Technology" (L.-H. Lee, Ed.), p. 187, Plenum Press, New York, 1975.
   b. St.Clair, A.K. and St.Clair, T.L., SAMPE Mat. Rev., 7, 53 (1975).
   c. Progar, D. J. and St.Clair, T. L., Organic Coatings and Plastics Chemistry Preprints 35, 185 (1975).
3. Bush, T. A., Counts, M. E. and Wightman, J. P., in "Adhesion Science and Technology" (L.-H. Lee, Ed.), p. 365, Plenum Press, New York, 1975.
4. Dwight, D. W., in "Kendall Award Symposium, 1976", to be published in J. Colloid Interface Sci.
5. Good, R. J., J. Adhesion, 4, 133 (1972).

DETERMINATION OF MADELUNG CONSTANTS FOR INFINITE AND SEMI-INFINITE IONIC LATTICES BY DIRECT SUMMATION*

J.V. Calara and J.D. Miller
University of Utah

ABSTRACT

A direct summation technique was developed for the calculation of Madelung Constants for infinite and semi-infinite ionic lattices. For a binary ionic lattice X_mY_n, *the Madelung Constant , A, (referenced to the cation) is related to the direct sums of potentials about the cation,* V^+, *and the anion,* V^-, *by the following expression*

$$A = \frac{1}{2}(V^+ + \frac{n}{m}V^-)$$

Rapid convergence and excellent agreement with published Madelung Constants for NaCl, CsCl, and CaF_2 *was obtained. Surface Madelung Constants (summation for a semi-infinite lattice) is readily determined and good agreement was achieved with the value for the NaCl <100> plane reported in the literature. With this direct summation technique the Surface Madelung Constant for monoionic planes can be calculated, as demonstrated for the <111> plane of* CaF_2.

I. INTRODUCTION

Lattice sums for the electrostatic interactions in ionic crystals, first evaluated by Madelung (1), are now known to high degrees of accuracy for many crystal types (see Sakamoto (2)). The problem of conditional convergence has been solved either by manipulations on a mathematical model of the lattice (Madelung (1), Hund (3), Emersleben (4), Ewald (5)) or by directly summing over neutral groups of the lattice (Evjen (6), Frank (7)). The direct summation methds of Evjen (6) and Frank (7) are not generally applicable to crystals of moderate complexity and are even less tractable when dealing with semi-infinite lattices.

The direct summation scheme presented in this communication allows calculation of the Madelung Constants for infinite and semi-infinite lattices of ionic crystals which can be decomposed into neutral linear elements. This computational

---

*The authors wish to acknowledge the financial support provided by the National Science Development Board of the Phillipines.

technique was developed as part of a flotation research program designed to explain the surface potentials developed by ionic solids, specifically CaF_2(8).

II. DISCUSSION

A. Infinite Lattice

Sum of potentials over individual charges of an ionic lattice lead to a conditionally convergent series. To transform the series into an absolutely convergent one, the ionic lattice must be decomposed into identical groups or repeating units of high multipole order, as was done by Evjen (6) and Frank (7). In our scheme, linear, symmetric groups were selected for the calculation of both Bulk and Surface Madelung Constants.

Since the potential contribution of a 2^n-pole decays with $r^{-(n+1)}$, where r is the distance of a test charge, summation over a three-dimentional lattice requires that a symmetric linear repeat unit be at least a 2^4-pole, a hexadecapole (symmetry removes all n-odd pole components). Figure 1 depicts a linear array of fractional charges selected for the NaCl lattice to assure an absolutely convergent summation.

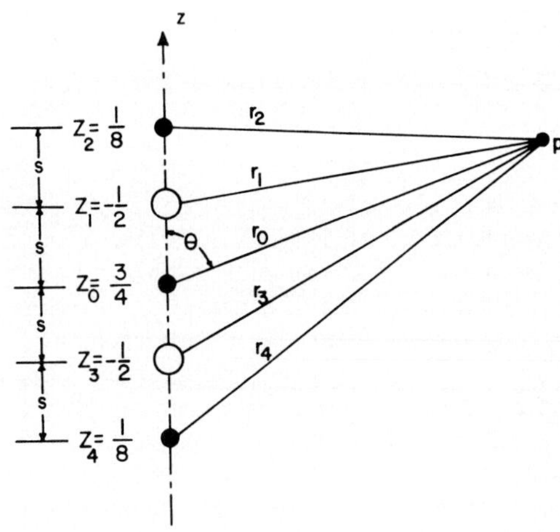

Fig. 1. High multipole order repeat unit suitable for the sodium chloride lattice summation. Overlapping of this repeat unit reproduces the bulk sodium chloride lattice.

The contribution to the potential, ε, at point p of each charge is $Z_i e/r_i$ where e is the electronic charge. If the charge array's spacing is such that $r_o >> s$, all the distances to p may be expanded as a power series in (s/r_o). Neglecting higher order terms, the potential ε can be shown to be (8)

$$\varepsilon = \frac{3es^4}{r_o^5} \left(\frac{35\cos^4\theta - 30\cos^3\theta + 3}{8}\right). \tag{1}$$

An assembly of overlapping units of the configuration presented in Fig. 1 will reproduce the interior of the NaCl lattice, and the sum of potentials V^+ around a central cation, or V^- around a central anion will be absolutely convergent because of the rapid decay of potential contribution from each repeat with increasing distance $\varepsilon = f(1/r_o^5)$.

Figure 2 illustrates the pseudo-cubic unit cells of three different crystal types: NaCl, CsCl, and CaF_2. The dashed lines indicate the charges belonging to the selected repeat unit for each crystal type and also gives their crystallographic orientation. The potential sum using these repeat units is absolutely convergent as dictated by previous arguments and the limit of the sum is therefore independent of the geometry of growth imposed by the summation procedure. But, as noted by Harris (9), although the limit is unique for a given repeating unit of a crystal, it is not necessarily the correct Madelung sum.

Any finite sample assembled from repeating units has a surface possessing a dipole moment per unit area, although the repeat units themselves have no such moment. The magnitude and sign of the surface dipole moment is characteristic of the repeating unit (9). Now, the potential ε due to the dipolar surface approaches a limit as the sample grows indefinitely because the $1/r_o^2$ potential decay from a unit surface dipole area is just compensated by the r_o^2 dependence of the surface area. For a sufficiently large sample such that the summation has sufficiently converged, the difference between the direct sum V^+ (around the cation, say) and the surface contribution ξ must then be the potential due to an infinite lattice, infinite in the sense of having no surface. This difference, A, is the Madelung Constant (Eqn. 2).

$$A = V^+ - \xi \tag{2}$$

Using sodium chloride as an example, V^+ is the direct sum around the sodium site, A is the Madelung Constant, ξ is the surface contribution. For the sodium chloride lattice, the Madelung Constant is the same whether evaluated around a Na^+ or Cl^- site, whereas the surface contribution, ξ, must be

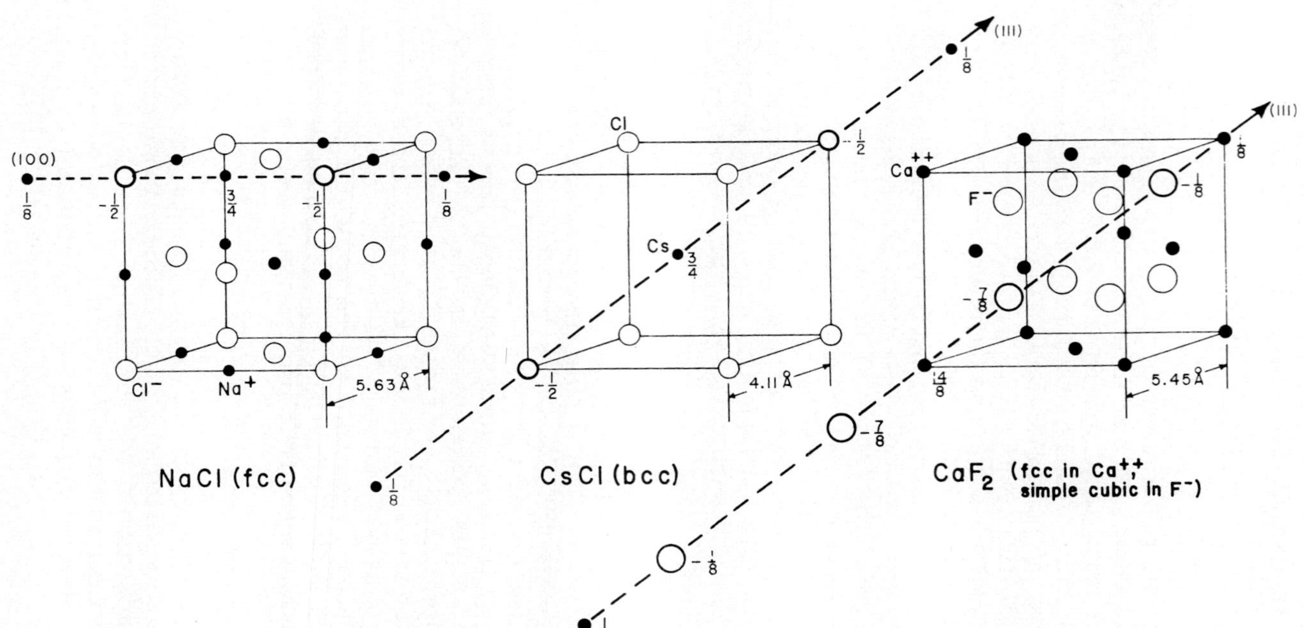

Fig. 2 Crystal structures for NaCl, CsCl and CaF_2 illustrating the selected linear repeating units, their orientations and the charge distribution in their repeating unit array.

equal but opposite in sign for the two reference sites. Hence, if V^- is the direct sum around the Cl^- site, then

$$A = V^- + \xi \tag{3}$$

Adding Eqns. (2) and (3),

$$A = \frac{1}{2}(V^+ + V^-) \tag{4}$$

For an non-symmetrical ionic solid such as CaF_2, parallel arguments lead to:

$$A = \frac{1}{2}(V^+ + 2V^-) \tag{5}$$

In general, for a binary ionic lattice X_mY_n where ions of the same charge have identical environments, the Madelung Constant for a cation site is,

$$A = \frac{1}{2}(V^+ + \frac{n}{m}V^-) \tag{6}$$

provided m and n have no common factor. Thus, the Madelung Constant is just half the total of potential sums around each constituent of an ionic "molecule".

The repeating units presented in Fig. 2 were arranged around central reference ions in concentric cubical shells for NaCl and CsCl and rhombohedral shells for CaF_2. The choice of growth shape is immaterial for the summation for the infinite lattice. A computer was used to obtain the direct sums, V^+ and V^-, and the Madelung Constants were evaluated according to Eqn. (6). Figure 3 illustrates the convergence of the potential sums V^+ and V^- for the NaCl lattice and the convergence of the Madelung Constant (dashed line). It should be noted that V^+ and V^- converge fairly rapidly and that their average converges even faster to the correct Madelung sum. After nine shells, the computed constant for NaCl was 1.747560, which compares well with the accepted value of 1.747564 (Sakamoto (2)). For CaF_2, the computed constant was 11.63659, accurate to six figures. Sometimes the convergence was quite rapid. Four shell summation for CsCl gave 1.7625, accurate to four figures.

B. Semi-Infinite Lattice

Consider the case of NaCl. After obtaining the direct sums V^+ and V^-, the surface contribution ξ is calculated from Eqns. (2) and (3):

$$\xi = \frac{1}{2}(V^+ - V^-) \tag{7}$$

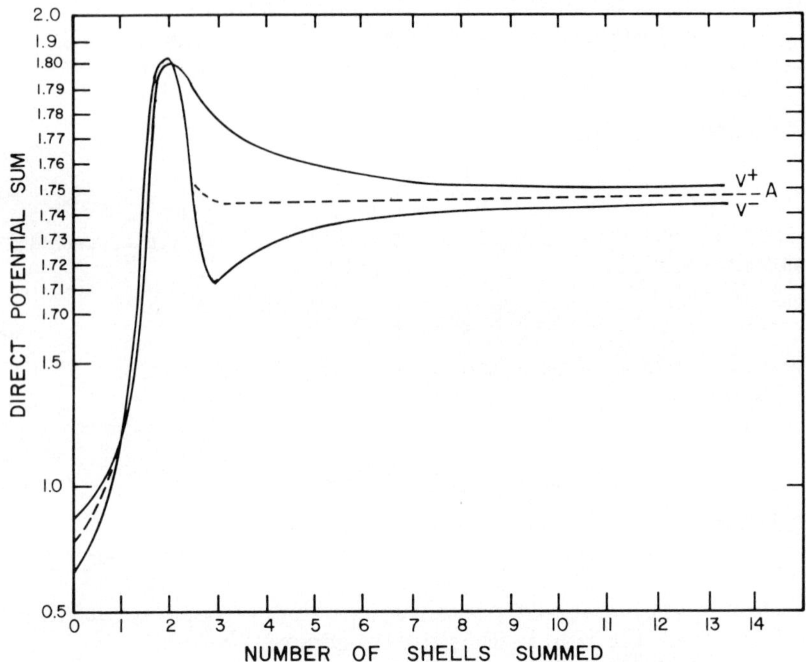

Fig. 3 Convergence of cationic and anionic potential sums and the Madelung Constant for an infinite sodium chloride lattice as a function of the number of shells used in the summation.

Now, let a <100> plane pass through the sodium reference ion. Proceeding as before, a sum of potentials, V_s^+, around this surface sodium ion which excludes the contributions from all points lying on one side of the plane can be determined. When sufficient convergence is attained, this procedure will exclude exactly half of the dipolar surface contributions, ξ. Subtracting $\frac{1}{2}\xi$ from V_s^+ yields the Madelung Constant A_s for a <100> surface sodium ion on a semi-infinite lattice:

$$A_s = V_s^+ - \frac{1}{2}\xi \tag{8}$$

The orientation of the repeat units with respect to the cleavage plane and the growth geometry selection must be such that truncation of repeat units produces a neutral surface. Table 1 summarizes Madelung Constants calculated for the NaCl, CsCl, and CaF_2 lattices and selected cleavage planes for these crystals.

TABLE 1
Madelung Constants for Infinite and Semi-Infinite Lattices. All Values Calculated for the Cationic Site and Referred to the Cubic or Pseudo-Cubic Cell's Edge.

Crystal	*Direct Sums*	*Madelung Constants* *Calculated*	*Literature*	*Ref.*
NaCl, bulk	$V^+ = 3.503644$	3.495122	3.495129	(2)
	$V^- = 3.486580$			
NaCl, <100>	$V_s^+ = 3.3278$	3.3288	3.34	(10)
NaCl, <100>	$V_s^+ = 3.05752$	3.08093	3.01	(10)
CsCl, bulk	$V^+ = 2.02586$	2.03515	2.03536	(2)
	$V^- = 2.04474$			
CaF_2, bulk	$V^+ = 15.13676$	11.63658	11.63675	(2)
	$V^- = 4.06820$			
CaF_2, <111>	$V_s^+ = 13.817803$	12.04467	none	(2)

As can be seen in Table 1, the Bulk Madelung Constants obtained by this summation technique reproduce the published values. The accuracy increases with the number of shells summed. Few results for Surface Madelung Constants are available in the literature. The calculated value for the <100> NaCl cleavage plane agrees with that reported by Levine and Mark (10). Their reported value for the <110> NaCl cleavage plane differs significantly from the value calculated by our direct summation method. However using their method, we calculate a Surface Madelung Constant identical to that obtained with our direct summation method. The method of Levine and Mark (10) is not applicable to a monoionic surface such as the <111> CaF_2 plane, which can be evaluated, using our direct summation technique. See Table 1. There are no other direct summation methods reported that accomplish the same end.

III. CONCLUSIONS

Charge arrays (repeating units) with vanishing quadrupole and lower moments results in absolute convergence of lattice potential sums and easy evaluation of Madelung constants by a direct summation technique. Calculated values for simple and complex crystals are in excellent agreement with those reported in the literature.

IV. REFERENCES

1. Madelung, E., Z. Physik, 19, 524 (1918).
2. Sakamoto, Y., J. Chem. Phys., 28, 164 (1958).
3. Hund, F., Z. Physik, 94, 11 (1935).
4. Emersleben, O., Z. Physik, 127, 588 (1950).
5. Ewald, P.P., Ann. Physik, 64, 253 (1921).
6. Evjen, H.M., "On the Stability of Certain Heteropolar Crystals", Phys. Rev., 39, 675 (1932).
7. Frank, F.C., Phil. Mag., 41, 128 (1950).
8. Miller, J.D. and Calara, J.V., "Analysis of the Surface Potential Developed by Non-Reactive Ionic Solids", *Flotation Vol. 1, SME/AIME* (1976).
9. Harris, F.E., *Theoretical Chemistry: Advances and Perspectives*. (H. Eyring and D. Henderson, eds.) Academic Press, New York, Vol. 1, 147 (1975).
10. Levine, J.D. and Mark, P., "Theory and Observations of Intrinsic Surface States on Ionic Crystals", Phys. Rev., 144, 751 (1966).

DEFORMATIONAL INSTABILITY OF A PLANE INTERFACE WITH TRANSFER OF MATTER

Marcel Hennenberg[$], Torben Smith Sørensen[*], Flemming Yssing Hansen[*] and Albert Sanfeld[$]

$ Faculté des sciences, Université Libre de Bruxelles
** Fysisk-Kemisk Institut, Technical University of Denmark*

I. INTRODUCTION

In previous publications three of the authors (M.H., T.S.S. and A.S.) have studied the deformational stability of plane and spherical interfaces between two immiscible liquids when autocatalytic or cooperative chemical reactions take place at the interface (1 - 6). In the present paper and in two preceding papers (7,8) we turn our attention towards the fascinating phenomenon often called "interfacial turbulence" - being a diffusion-induced type of Marangoni effect.

The first theoretical approach to this stability problem is the one given by Sternling and Scriven in 1959 (9). They studied two immiscible liquids with transfer of a third component by diffusion from one phase to the other. The concentration profile of the third component is assumed linear and it is assumed that the interface is in *local equilibrium* with the adjacent liquids. Furthermore, that the interface stays flat and that the hydrodynamic motion can be described as two-dimensional roll-cells.

Experiments done by Orell and Westwater (O & W) (10) showed - by means of Schlieren photography - that in the system ethylene glycol - ethyl acetate with acetic acid as the diffusing component the interface is not at all smooth and flat during interfacial turbulence, but rugged and full of protrusions and indentations. Furthermore, a linear profile is quite unrealistic for experiments wherein two "semi-infinite" liquid phases are brought into contact. Functions of the *erf* - type would be more suitable.

Since we have used in our previous articles the general tensor formulation of Scriven (11) and Aris (12) for the interfacial mass and momentum balance over a deformed element of surface, we are in a position to give an exact formulation of the deformational stability problem, and to obtain certain analytical results for as well a linear as an exponential concentration profile (7,8), the latter simulating very well the boundary conditions in experiments such as performed by

O & W. By using the best estimations we can make about the parameters in the ethylene glycol- ethyl acetate- acetic acid system, we shall be able to see - by analytic and computer calculations - if our model yields roughly the measured results in the O & W experiments, *e.g.* for the dimension of the convection cells.

By doing so we hope to be able to add an interesting chapter to the science of surface chemistry as well as to contribute to the theory of hydrodynamic stability analysis (extension of the Rayleigh-Taylor stability problem of two superposed fluids in a gravity field (13)) and to expand the concept of "dissipative structures" (14) pertinent to thermodynamic systems far from equilibrium into the realm of surface science.

Apart from the academic and interdisciplinary interest in the present problem our model might ultimately also gain economic importance since mass transfer rates are greatly enhanced by the set-in of interfacial turbulence in technological liquid-liquid extraction processes.

II. DIFFUSION-CONVECTION EQUATIONS AND INTERFACIAL CONDITIONS

We shall restrict ourselves to a *linear* stability analysis, and therefore we linearise all partial d.e. we shall meet in the problem at hand. The linearised eqn. of change for a three dimensional bulk concentration perturbation δc from a one dimensional reference profile (z-direction, perpendicular to the undeformed interface) is the diffusion-convection eqn.

$$\frac{\partial \delta c^{(\ell)}}{\partial t} = -\delta W_z^{(\ell)} \frac{d}{dz} c_o^{(\ell)} + D_\ell \nabla^2 \delta c^{(\ell)} \qquad (\ell = 1,2) \qquad (1)$$

The meaning of the symbols used here and in the following eqns is explained in the list of symbols in the end of this paper. Eqn. (1) is exact for a linear profile

$$c_o^{(1)} = c_{oo}^{(1)} - \beta_1 z \; ; \; c_o^{(2)} = c_{oo}^{(2)} - \beta_2 z \qquad (2)$$

but for an exponential profile ($\beta_1 > 0$, $\beta_2 > 0$)

$$c_o^{(1)} = c_\infty^{(1)} + c_{oo}^{(1)} e^{\beta_1 z}; \; c_o^{(2)} = c_\infty^{(2)} + c_{oo}^{(2)} e^{-\beta_2 z} \qquad (3)$$

we have neglected the slow time dependence of the reference concentration profile $c_o(z)$ due to diffusion. Thus, we assume that *vorticity* diffusion is much faster than diffusion of *matter*, *i.e.*

$$\nu_\ell >> D_\ell \tag{4}$$

which fact greatly simplifies the eqns. later on.

To solve eqn. (1) we must first have a knowledge of the z-component of the hydrodynamic motion which can be obtained by solving the linearised Navier-Stokes eqns. for the two bulk liquids subject to the incompressibility conditions

$$\nabla \cdot \delta \mathbf{W}^{(\ell)} = 0 \tag{5}$$

and no-slip interfacial conditions. The solution is for a single Fourier component (13)

$$\tilde{W}_z^{(1)} = \{-\frac{k+q_1}{2k} \cdot B_1 + \frac{k-q_2}{2k} \cdot B_2\} \cdot e^{kz} + B_1 e^{q_1 z} \tag{6a}$$

$$\tilde{W}_z^{(2)} = \{\frac{k-q_1}{2k} \cdot B_1 - \frac{k+q_2}{2k} \cdot B_2\} \cdot e^{-kz} + B_2 e^{-q_2 z} \tag{6b}$$

where B_1 and B_2 are unknown parameters which have to satisfy a compatibility eqn. later on. In eqns. (6a&b) k is the magnitude of the wavevector in the expansion

$$\delta W_z^{(\ell)} = \tilde{W}_z^{(\ell)}(z) \cdot e^{i(k_x \cdot x + k_y \cdot y)} \cdot e^{\omega t} \tag{7}$$

and the parameters q_ℓ are given by

$$q_\ell^2 = k^2 + \frac{\omega}{\nu_\ell} \quad ; \quad \mathrm{Re}(\, q_\ell \,) > 0 \tag{8}$$

In the interfacial momentum balance for the z-direction we shall also use the solution for the pressure perturbation δp. By taking the divergence on both sides of the linearised Navier -Stokes eqn. we see that δp for incompressible liquids must satisfy a Laplace eqn. and we obtain

$$\delta p^{(\ell)} = \tilde{\mathbf{p}}^{(\ell)}(z) \cdot e^{i(k_x \cdot x + k_y \cdot y)} \cdot e^{\omega t} \tag{9}$$

with

$$\tilde{p}^{(1)} = -\omega \rho_1 \{-\frac{k+q_1}{2k^2} \cdot B_1 + \frac{k-q_2}{2k^2} \cdot B_2\} e^{kz} \tag{10a}$$

$$\tilde{p}^{(2)} = \omega \rho_2 \{\frac{k-q_1}{2k^2} \cdot B_1 - \frac{k+q_2}{2k^2} \cdot B_2\} e^{-kz} \tag{10b}$$

The solution to the linearised diffusion-convection eqns. becomes in case of a linear profile

$$\tilde{c}_\ell(z) = C_1^{(\ell)} e^{\pm r_\ell z} + C_2^{(\ell)} e^{\pm kz} + C_3^{(\ell)} e^{\pm q_\ell z} \quad (11)$$

(+ for $\ell = 1$, - for $\ell = 2$) with the expansion

$$\delta c^{(\ell)} = \tilde{c}_\ell(z) e^{i(k_x \cdot x + k_y \cdot y)} e^{\omega t} \quad (12)$$

and

$$r_\ell^2 = k^2 + \frac{\omega}{D_\ell} \quad ; \quad \mathrm{Re}(r_\ell) > 0 \quad (13)$$

The coefficients for liquid 1 are given by

$$C_2^{(1)} = \frac{\beta_1}{2k\omega} \{ (k-q_2)B_2 - (k+q_1)B_1 \} \quad (14a)$$

$$C_3^{(1)} = \frac{\beta_1}{\omega} \cdot B_1 \quad (14b)$$

and we have symmetrical relations for liquid 2. The coefficients $C_1^{(1)}$ and $C_1^{(2)}$ can be expressed through B_1, B_2 and a parameter $\tilde{\Gamma}$ by using the *local equilibrium* condition and a linear adsorption isoterm for the adsorption of surfactant

$$\tilde{\Gamma} = K_\ell \cdot \tilde{c}^{(\ell)}(z=0) \quad (15)$$

with the expansion of the surface excess concentration

$$\delta\Gamma = \tilde{\Gamma} \cdot e^{i(k_x \cdot x + k_y \cdot y)} e^{\omega t} \quad (16)$$

For the *exponential profile* we get instead of eqn. (11)

$$\tilde{c}_\ell(z) = C_1^{(\ell)} e^{\pm r_\ell z} + C_2^{(\ell)} e^{\pm(k+\beta_\ell)z} + C_3^{(\ell)} e^{\pm(q_\ell+\beta_\ell)z} \quad (17)$$

and instead of (14a & b)

$$C_2^{(1)} = \frac{c_{oo}^{(1)} \beta_1}{D_1} \{ (k+\beta_1)^2 - r_1^2 \}^{-1} \{ \frac{k-q_2}{2k} \cdot B_2 - \frac{k+q_1}{2k} \cdot B_1 \} \quad (18a)$$

$$C_3^{(1)} = \frac{c_{oo}^{(1)} \beta_1}{D_1} \{ (q_1+\beta_1)^2 - r_1^2 \}^{-1} \cdot B_1 \quad (18b)$$

III. COMPATIBILITY CONDITION

The three unknown coefficients $\tilde{\Gamma}$, B_1 and B_2 must satisfy three linear, homogeneous eqns. giving rise to a compatibility eqn. (coefficient determinant equal to zero) between the parameters of the system amounting to a secular eqn. between the wavevector k and the complex time constant ω. The three eqns. are the linearised interfacial mass balance and the linearised interfacial momentum balance in the z-direction and in the x- and y-directions (combined to a relation for the z-component of the velocity through the eqn. of incompressibility) (5,6,7,8).

The linearised mass balance on an element of surface is given by (ΔX signifies the property X taken in liquid 2 adjacent to the interface minus the corresponding value in liquid 1 and s denotes the surface z = 0)

$$\frac{\partial \delta\Gamma}{\partial t} = \Gamma^o\left(\frac{\partial W_z}{\partial z}\right)_s + D_s\left\{\frac{\partial^2}{\partial x^2} + \frac{\partial^2}{\partial y^2}\right\}\delta\Gamma + \Delta\left\{D\,\frac{\partial \delta c}{\partial z}\right\} \tag{19}$$

where we have used that the reference state itself has no accumulation of the third component at the interface, *i.e.* for the linear case

$$\beta_1\, D_1 = \beta_2\, D_2 \tag{20}$$

and for the exponential

$$\beta_1\, c_{oo}^{(1)}\, D_1 = -\,\beta_2 c_{oo}^{(2)}\, D_2 \tag{21}$$

Introducing the Fourier-expansions and the solutions of the hydrodynamic eqns. and the diffusion-convection eqns. given in the preceding section we obtain

$$M_{11}\,\tilde{\Gamma} + M_{12}\, B_1 + M_{13}\, B_2 = 0 \tag{22}$$

with the coefficients (for brevity we remove the ℓ-indices and let Σ signify the summation over ℓ = 1,2)

$$M_{11} = \omega + k^2 D_s + \Sigma(r\cdot D/K) \tag{23a}$$

$$M_{12}^{lin} = (k-q_1)\left\{\frac{\Gamma^o}{2} + \frac{\beta_1 D_1}{\omega}\left\{\frac{k-r_2}{2k} + \frac{(r_1-1)(k+q_1)}{2k(k-q_1)} + \frac{(q_1-r_1)\nu_1}{(k-q_1)(\nu_1-D_1)}\right\}\right\} \tag{23b}$$

$$M_{12}^{exp} = (k-q_1)\frac{\Gamma^o}{2} + c_{oo}^{(1)}\beta_1\{\frac{1}{q_1+\beta_1+r_1} - \frac{k+q_1}{2k(k+\beta_1+r_1}\} + \frac{c_{oo}^{(2)}\beta_2(q_1-k)}{2k(k+\beta_2+r_2)} \tag{23c}$$

$$M_{13}^{lin} = (k-q_2)\{-\frac{\Gamma^o}{2} + \frac{\beta_1 D_1}{\omega}[\frac{k-r_1}{2k} + \frac{(r_2-k)(k+q_2)}{2k(k-q_2)} + \frac{(q_2-r_2)\nu_2}{(k-q_2)(\nu_2-D_2)}]\} \tag{23d}$$

$$M_{13}^{exp} = (q_2-k)\frac{\Gamma^o}{2} + \frac{c_{oo}^{(1)}\beta_1(k-q_2)}{2k(k+\beta_1+r_1)} + c_{oo}^{(2)}\beta_2\{\frac{k+q_2}{2k}\cdot\frac{1}{k+\beta_2+r_2} - \frac{1}{q_2+\beta_2+r_2}\} \tag{23e}$$

In the formulae (23b) and (23d) the approximation (4) has *not* been applied in the last term, because a detailled calculation shows that it gives false results to apply the approximation *before* the first order expansion in ω (to be carried out later) instead of *after* (7).

The two interfacial momentum balances give rise to two eqns. of the same form as (22) independent of the profile assumption (5,7,8). The coupling between mass and momentum balances enters solely due to the connection between the perturbation in surface excess concentration of the surfactive third component and the interfacial tension

$$\delta\sigma = -\alpha\delta\Gamma \tag{24}$$

The coefficients become

$$M_{21} = g \tag{25a}$$

$$M_{22} = (k-q_1)\{\frac{1}{2k}(\Gamma^o\omega - g\frac{\Delta\rho}{\omega} + \frac{k^2\sigma^o}{\omega}) + \mu_2 + \frac{q_1}{k}\mu_1 + \frac{\omega}{2k^2}\Sigma\rho\} \tag{25b}$$

$$M_{23} = (k-q_2)\{\frac{1}{2k}(\Gamma^o\omega - g\frac{\Delta\rho}{\omega} + \frac{k^2\sigma^o}{\omega}) + \mu_1 + \frac{q_2}{k}\mu_2 + \frac{\omega}{2k^2}\Sigma\rho\} \tag{25c}$$

$$M_{31} = \alpha k \tag{25d}$$

$$M_{32} = (k-q_1)\{-\frac{1}{2k}(\Gamma^o\omega + k^2(\kappa+\varepsilon)) - \frac{q_1}{k}\mu_1 - \mu_2 \tag{25e}$$

$$M_{33} = (k-q_2)\{\frac{1}{2k}(\Gamma^o\omega + k^2(\kappa+\varepsilon)) + \mu_1 + \frac{q_2}{k}\mu_2\} \tag{25f}$$

The compatibility eqn. is then given by

$$\det(M) = 0 \tag{26}$$

IV. FIRST ORDER EXPANSION OF THE COMPATIBILITY CONDITION

It is possible to obtain certain *analytical* results of great help for later numerical computations by means of a first order expansion of the compatibility eqn. (26). In this way we obtain an *exact* solution for the non-oscillatory critical boundary ("exchange of stabilities" in the terminology of Chandrasekhar) where $\omega_r = \omega_i = 0$ and for the stability in the immediate vicinity of this boundary, since we have here that $\omega_r \to 0$. We obtain (7,8)

$$B\omega^* + C = 0 \tag{27}$$

with the dimensionless time constant given by ($\Sigma = \sum_{\ell=1}^{2}$)

$$\omega^* = \frac{\Sigma\mu}{k\sigma_o}\,\omega \tag{28}$$

and the dimensionless parameters

$$B = -\,\phi_g\{\phi_{\kappa+\varepsilon}\phi_K + \phi_{D,1}\phi_{\rho,1}\} - 2\phi_{D,1}\phi_{\kappa+\varepsilon} - L_\alpha\{\phi_g L_{D,3} + 2L_{D,2}\} + L_{D,1}L_g\phi_{\kappa+\varepsilon} \tag{29a}$$

$$C = -\,\phi_g\{\phi_{D,1}\phi_{\kappa+\varepsilon} + L_{D,2}L_\alpha\} \tag{29b}$$

The ϕ's are dimensionless numbers common to the linear and the exponential concentration profiles and are given by

$$\phi_g = 1 - \frac{g\Delta\rho}{k^2\sigma_o} \tag{30a}$$

$$\phi_{\kappa+\varepsilon} = 2 + \frac{k(\kappa+\varepsilon)}{\Sigma\mu} \tag{30b}$$

$$\phi_{\rho,1} = \frac{\sigma_o}{(\Sigma\mu)^2}\{\Gamma^o + \frac{\Sigma\rho}{2k}\} \tag{30c}$$

$$\phi_K = 1 + \frac{1}{2k}\,\Sigma\,\frac{1}{K} \tag{30d}$$

$$\phi_{D,1} = \frac{\Sigma\mu}{\sigma_o}\{kD_s + \Sigma\frac{D}{K}\} \tag{30e}$$

The L's are also dimensionless numbers, but differ from the linear to the exponential profiles. In the linear case we have

$$L_\alpha^{lin} = \frac{\alpha\beta_1 D_1\Sigma\mu}{k\sigma_o^2} \tag{31a}$$

$$L_g^{lin} = \frac{g\beta_1 D_1 \Sigma\mu}{k^2\sigma_o^2} \tag{31b}$$

$$L_{D,1}^{lin} = -\frac{3\sigma_o}{4k\Sigma\mu}\Sigma\frac{1}{D} \tag{31c}$$

$$L_{D,2}^{lin} = \frac{k\sigma_o}{\Sigma\mu}\{\frac{\Gamma^o}{\beta_1 D_1} + \frac{1}{4k^2}\Delta(\frac{1}{D})\} \tag{31d}$$

$$L_{D,3}^{lin} = -\frac{\sigma_o^2}{8(\Sigma\mu)^2 k^2}\Delta(\frac{1}{D^2}) \tag{31e}$$

The analogous eqns. for the exponential case are

$$L_\alpha^{exp} = \frac{\alpha\beta_1 c_{oo}^{(1)} D_1 \Sigma\mu}{k\sigma_o^2} \tag{32a}$$

$$L_g^{exp} = \frac{g\beta_1 c_{oo}^{(1)} D_1 \Sigma\mu}{k^2\sigma_o^2} \tag{32b}$$

$$L_{D,1}^{exp} = \frac{\sigma_o}{\Sigma\mu}\Sigma\frac{3k+\beta}{D(2k+\beta)^2} \tag{32c}$$

$$L_{D,2}^{exp} = \frac{k\sigma_o}{\Sigma\mu}\{\frac{\Gamma^o}{\beta_1 c_{oo}^{(1)} D_1} - \Delta(\frac{1}{D(2k+\beta)^2})\} \tag{32d}$$

$$L_{D,3}^{exp} = (\frac{k\sigma_o}{\Sigma\mu})^2 \Delta\{\frac{1}{kD^2(2k+\beta)^3}\} \tag{32e}$$

From the above eqns. we can get expressions for the solution in the vicinity of k = 0 by taking the limit $k \rightarrow 0$. In the linear case we obtain

$$\omega^* \rightarrow \frac{2k\ \Sigma\mu}{\sigma_o\ \Sigma\frac{1}{D}}, \qquad k \rightarrow 0 \tag{33a}$$

and in the exponential

$$\omega^* \rightarrow \frac{k}{\sigma_o} \cdot \frac{\alpha\beta_1 c_{oo}^{(1)} D_1 \Delta(\frac{1}{D\beta^2}) - 2(\Sigma\mu)(\Sigma\frac{D}{K}) - \Gamma_\alpha^o}{\Sigma\frac{1}{K} + \frac{1}{2}\frac{\{\Sigma(D/K)\}\Sigma\rho}{\Sigma\mu}} \tag{33b}$$

The critical value(s) of k for which $\omega_r = \omega_i = 0$ one obtains from the eqn.

$$C(k) = 0 \tag{34}$$

We shall assume that $\phi_g \neq 0$, since this will always be the case for negative $\Delta\rho$. The case $\phi_g = 0$ corresponds to the instability boundary in the well-known Rayleigh-Taylor stability problem of a heavy fluid superposed on a lighter one (13). For the slope $d\omega^*/dk$ taken at a critical value of k we obtain

$$\left(\frac{d\omega^*}{dk}\right)_{k_{cr}} = \frac{\phi_g(k_{cr})}{B(k_{cr})\sigma_o}\{2D_s\Sigma\mu + (\kappa+\varepsilon)\Sigma\frac{D}{K} + 2k_{cr}(\kappa+\varepsilon)D_s+X\} \tag{35a}$$

where X in the linear case is given by

$$X = -\frac{\alpha\beta_1 D_1}{2k_{cr}^3}\Delta\left(\frac{1}{D}\right) \tag{35b}$$

and in the exponential case

$$X = 4\alpha\beta_1 c_{oo}^{(1)} D_1 \Delta\left(\frac{1}{D(2k+\beta)^3}\right) \tag{35c}$$

V. ETHYLENE GLYCOL - ETHYL ACETATE - ACETIC ACID SYSTEM

Orell and Westwater[10] studied a two-phase system where phase 1 was mainly ethylene glycol and phase 2 mainly ethyl acetate. With concentrations from 0.1 to 10 w% acetic acid in the lower phase (phase 1), "dissipative structures" in the form of more or less regular polygonal convection cells (stationary and travelling) and later on also surface ripples with a larger wavelength appeared near the interface. A considerable interfacial deformation was also observed.

We want to test our model of spontaneous interfacial convection qualitatively and quantitatively by using the best parameters we can estimate for the ethylene glycol - ethyl acetate - acetic acid system. We shall use the following values:

$$\mu_1 = 0.103 \text{ Poise} \qquad D_1 = 0.21\cdot 10^{-5}\ \text{cm}^2/\text{s}$$

$$\mu_2 = 6.36\cdot 10^{-3} \text{ Poise} \qquad D_2 = 3.4\cdot 10^{-5}\ \text{cm}^2/\text{s}$$

$\rho_1 = 1.064\ g/cm^3$ $\quad D_s = 1.8\cdot 10^{-5}\ cm^2/s$

$\rho_2 = 0.9142\ g/cm^3$ $\quad \sigma_o = 1.44$ dynes/cm

$\Gamma^o = 2.67\cdot 10^{-10}\ g/cm^2$ $\quad K_1 = 1.27\cdot 10^{-7}$ cm

$\alpha = 4.13\cdot 10^8\ cm^2/s^2$ $\quad K_d = K_2/K_1 = 1.52$

We have made extensively use of fig. 12 in the Orell and Westwater paper which is an equilibrium three phase diagram for the studied system at 25 C. Up to around 5 w% acetic acid there is an immiscibility gap between a phase poor in ethyl acetate and a phase rich in that component. Tie-lines are shown across the immiscibility gap with corresponding values of the interfacial tension written on each tie-line. The values of ρ_1 and ρ_2 are calculated from the weight percentages in the two phases when only a trace of acetic acid is present - assuming that the volume of mixing is null. Viscosities are calculated from the approximate Eyring-formula for liquid mixtures (15)

$$\log \mu = X_{EG} \log \mu_{EG} + X_{EA} \log \mu_{EA} \tag{36}$$

(the presence of a trace of acetic acid is neglected).

From fig. 12 in O & W's paper the interfacial tension in the system may be given as function of the concentration of acetic acid in the inferior or the superior phase - again neglecting the volume of mixing. The result is shown in fig.1.

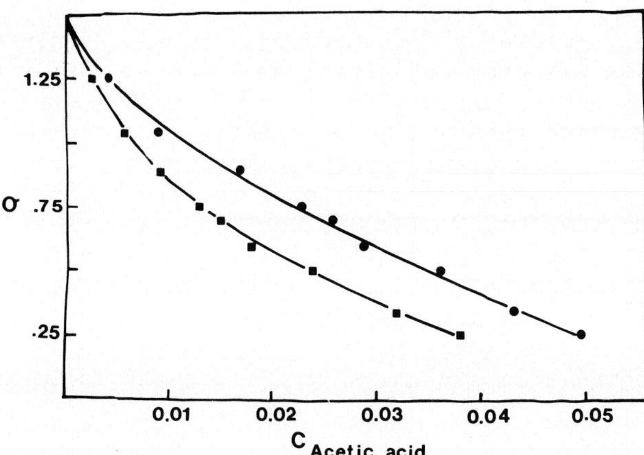

Fig. 1. The influence of acetic acid on the interfacial tension in the ethyle ıe glycol - ethyl acetate system. c: g/cm³ ● *inferior phase,* ■ *superior phase. σ in dynes/cm.*

The functional relationship between σ and the sub- or superphase concentration of acetic acid is very well fitted by the two exponentials

$$\sigma = 1.55 \exp(-33.639\cdot c^{(1)}) \tag{37a}$$

$$\sigma = 1.55 \exp(-51.375\cdot c^{(2)}) \tag{37b}$$

When only a trace of acetic acid is present, the adsorbed acetic acid on the interface behaves as an ideal two-dimensional gas with $\alpha = RT/M = 4.13\cdot 10^8\ cm^2/s^2$. At the same time the adsorption isotherm becomes linear, and from the initial slopes $d\sigma/dc$ calculated from (37 a&b) and the value of α the values of K_1 and K_2 can be estimated. We have taken a concentration of acetic acid of 0.5 w% $\sim$ 0.0042 g/cm^3 in phase 1 as representative of a very dilute system, and we have assumed that the concentration $c_{(1)}$ immediately adjacent to the interface is half of that value ($c_\infty^{(2)}$ is zero in all the experiments done by O & W). From that figure σ_o is calculated by eqn. (37a) and Γ^o from the value of K_1 according to the assumption of local equilibrium.

Finally, the value of the diffusion coefficient at 25C of acetic acid in pure ethylene glycol is found to be $0.13\cdot 10^{-5}\ cm^2/s$ in Landolt-Börnstein's tables (5. Teil, Bandteil a, p.674 , 1969 ed.). From that figure D_1 and D_2 are calculated by means of inverse viscosity ratios (Stokes law hypothesis). The surface diffusion coefficient is taken as the mean value between D_1 and D_2 in the lack of a better figure. Anyway, the terms containing D_s show up to be negligible in comparison with the terms in D_1 and D_2 in the expression (23a) for M_{11}.

From eqn. (34) with $\Phi_g \neq 0$ we obtain in the case of a linear concentration profile the following possibilities

$$\beta_1 - \beta_2 = \beta_1\left(1 - \frac{D_1}{D_2}\right)\begin{cases} > 0 & \text{1 root } k=k_{cr} \\ < 0 & \text{no root} \end{cases} \tag{38}$$

and we see that a critical k will only exist if the direction of diffusion is from the phase with the smallest value of the diffusion coefficient to the phase with the greatest. This is actually the case in the O & W experiments. It is also in accordance with the observations of Thiessen (16) on spontaneous interfacial convection during mass transfer through spherical films between two liquid phases.

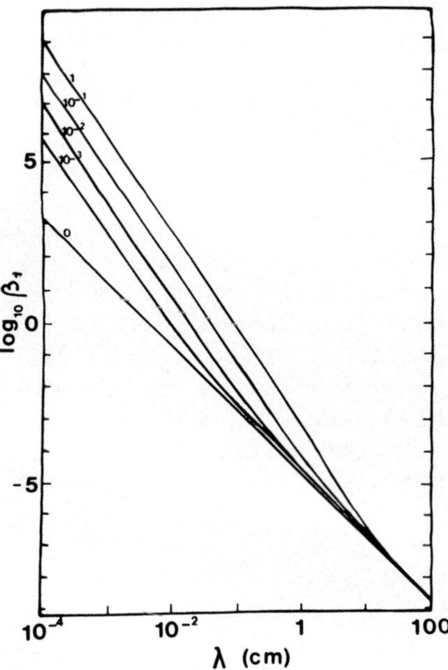

Fig. 2. The critical wavelength for non-oscillatory instability for a linear concentration profile for various concentration gradients (β_1) *and surface viscosity* $\kappa+\varepsilon$ *0, 0.001, 0.01, 0.1 and 1 surface Poise. Unit of* β_1*: g/cm*^{4}.

In fig. 2 the critical wavelength for exchange of stabilities is calculated from eqn. (34) for a linear concentration profile as a function of the steepness of the concentration profile β_1. The critical wavelength increases with decreasing steepness. The effect of surface viscosity is also shown: the critical wavelength is displaced towards higher values for increasing surface viscosity. A physical interpretation of the behaviour is given in the next section.

We have also performed some preliminary computer calculations directly on the determinantal eqn. (26). We use the fact that

$$\det(M)^{*}\cdot\det(M) \geqq 0 \tag{38}$$

and use a minimalisation programme to find the absolute minima of that function - varying only ω_i and ω_r.

The results are given on fig. 3 for positive values of ω_r. The curves are for $\omega_i = 0$ corresponding to *stationary* convection cells. The points $k_{cr} \neq 0$ where $\omega_r = 0$ are in accordance with the analytically calculated values and so are the slopes just around k_{cr} with the analytical expression given by eqns. (35a & b).

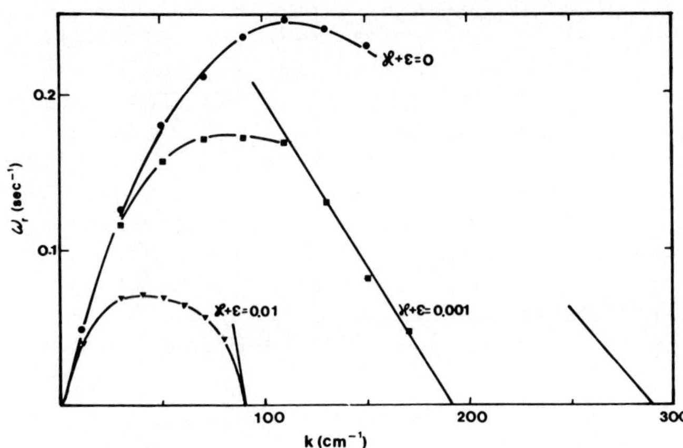

Fig. 3. Time constant for growth of stationary convection cells with a linear concentration profile for surface viscosities 0.01, 0.001 and 0 surface Poise evaluated by computer. Straight lines indicate analytically calculated slopes at the critical values of k. $\beta_1 = 0.042$ *g/cm*^{4}.

The convection cells most probable to be the dominating ones when the perturbation grow into the non-linear region are the ones of maximal values of ω_r on fig.3. For the three surface viscosities 0.01, 0.001 and 0 surface Poise we have k_{max} = 34, 82, and 112 cm^{-1} corresponding to wavelengths λ_{max} = 0.185, 0.0766, and 0.0561 cm. The slope β_1 used in fig.3 corresponds to a drop from the subphase concentration to zero over a distance of 1 mm. On fig.7 in the Orell and Westwater paper the experimental wavelength is shown to be 0.02 to 0.03 cm at the moment of contact between the two phases. Thereafter, it is steadily growing with increasing contact time. For a 0.5 w% solution of acetic acid in the subphase the cell dimension increases to 0.05 cm after 3 h of contact. The figures are well in accordance with the λ_{max} for the two lowest surface viscosities on fig.3, and the fact that cell dimensions increase with decreasing steepnes of the concentration profile is also in accordance with figs.2 and 3, although it must be remembered that our theory strictly describes only the *onset* of instability.

A difficulty is, however, that Orell and Westwater only found *travelling* cells at the lowest concentrations of acetic acid, *i.e.* ω_i would be expected to be $\neq 0$. We have scanned the region ω_i (0 - 0.2 sec^{-1}) for $\kappa + \varepsilon = 0.001$ and 0 surface Poise and $\beta_1 = 0.042$ g/cm^4, but we have not found any sign of a zero in the positive, definite function (38). However, we had some numerical difficulties with the minimalisation programme for $\kappa + \varepsilon = 0$ around k_{cr} (part of curve not shown in fig.3) because the function (38) had a lot of minima both in the ω_r-direction and in the ω_i-direction, and it was very difficult to distinguish the absolute minima from other minima. Therefore, oscillatory solutions may still hide in this region.

According to eqn.(33a) the slope $(d\omega_r^*/dk)_{k=0}$ should always have a finite positive value, and with the definition (28) of ω^* it is seen that $(d\omega_r/dk)_{k=0}$ must be zero and that the first term in k in the Taylor-expansion of ω_r must be parabolic. This is not evident from fig.3 because of the coarseness of the scale, but on fig.4 we see how the parabola is asymptotically approached when k becomes very small.

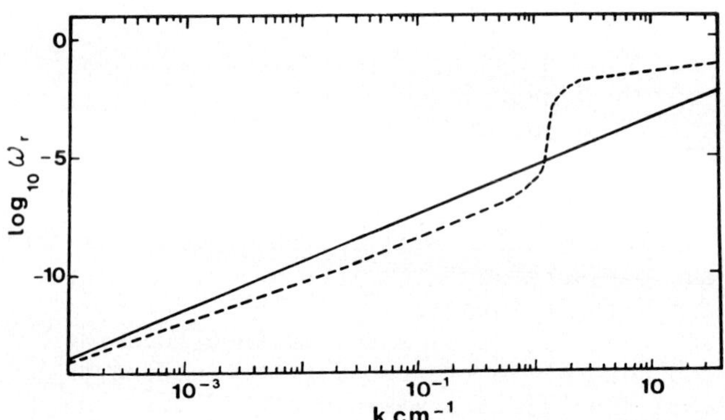

Fig. 4. The asymptotic behaviour of ω_r for small wave-vectors. $\kappa+\varepsilon=0.001$ surface Poise, $\beta_1=0.042$ g/cm^4. Dashed line: computed values. Full line: asymptotic parabola.

An exponential profile is of course much more realistic than a linear one for the type of experiments done by Orell and Westwater. The analytical eqn. for the exchange of stabilities boundary (34) may here be rewritten

$$1+\frac{1}{\alpha\Gamma^o}\{2\Sigma\mu + k(\kappa+\varepsilon)\}\cdot\{kD_S + \Sigma\frac{D}{K}\} = \frac{\beta_1 c_{oo}^{(1)} D_1}{\Gamma^o}\Delta\{\frac{1}{D(2k+\beta)^2}\} \qquad (39)$$

We choose the same surface concentration of acetic acid as in the linear case ($\Gamma^{o} = 2.67 \cdot 10^{-10}$ g/cm^2), and then by the local equilibrium assumption and the eqn.(21) for zero accumulation at the interface in the reference state we get immediately

$$c_{oo}^{(1)} = -0.0021 \text{ g/cm}^3 \quad c_{oo}^{(2)} = +0.00138 \text{ g/cm}^3 \quad \beta_1 = 9.40 \cdot 10^{-2} \cdot \beta_1$$

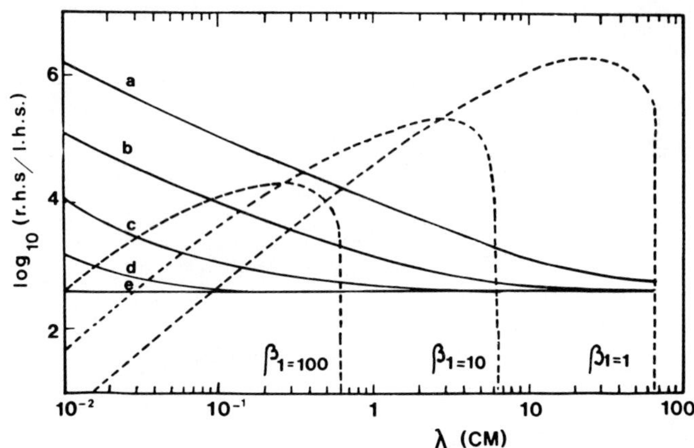

Fig. 5. Graphs for determination of critical values of k from eqn.(39). Full line: l.h.s. a: 1 b:0.1 c:0.01, d:0.001 e:0 surface Poise. Dashed line: r.h.s. Unit of β_1 *: g/cm*^{4}.

On fig.5 we have plotted r.h.s. and l.h.s. of eqn.(39) against λ to find the critical values of λ from the crossing points. We have now two critical wavelengths for stationary cells, and by calculation of $(d\omega^*/dk)_{k=0}$, $(d\omega^*/dk)_{k=k_{cr}^{(1)}}$ and $(d\omega^*/dk)_{k=k_{cr}^{(2)}}$ from the analytical expressions (33b),(35a) and (35c) we see that the first slope (k=0) is negative, the next positive and the last negative. This corresponds to instability between the two critical wavevectors and stability outside that interval.

From fig.5 we observe, that there will be no roots if the concentration profile is too steep or the surface viscosity too high. When the steepness decreases due to diffusion, we arrive for a certain surface viscosity to a situation with one critical wavevector corresponding to convection cells with a definite dimension (touching between full and dashed curves in fig. 5). When the steepness further decreases, we get two roots and an instability interval of still increasing values of the critical wavelengths. This is in accordance with fig.6 in the O. & W. paper where the size distribution of the polygonal cells are shown as normal distributions with mean values and standard deviations increasing with time.

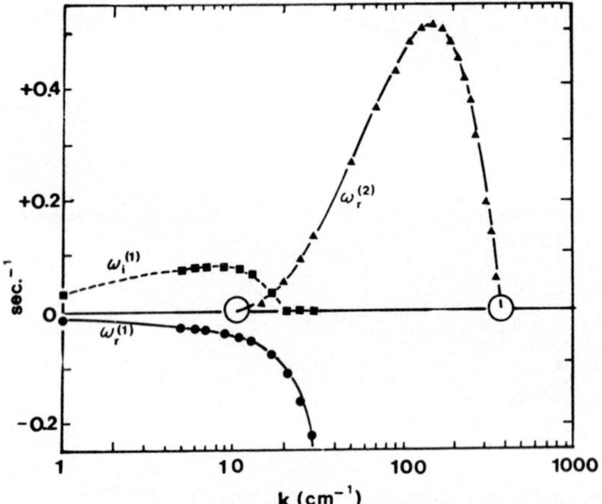

Fig. 6. Real and imaginary components of time constant for an exponential concentration profile. Solution (2) is purely real. Circles indicate analytically calculated values of critical wavevectors. $\kappa+\varepsilon$ *= 0.001 surface Poise.* β_1 *= 100 cm*$^{-1}$

On fig. 6 we have plotted the real and imaginary components of ω computed by the minimalisation programme. Soln. (1) corresponds to damped capillary waves. At about k = 20cm^{-1} the damping becomes so great - due to increased rate of shear for small wavelengths - that the solution becomes aperiodic. Soln. (2) is purely real and unstable between the encircled critical values of k (10.47 and 381 cm^{-1}) corresponding to the analytically calculated values. The k_{max} is about 150 cm^{-1} corresponding to a λ equal to 0.042cm in very good agreement with the initial cell dimensions cited in the O.& W. paper. It is seen that the travelling waves observed by O.& W. cannot be explained by a linear instability analysis, since the oscillatory mode is damped. The region of overlapping of the two solutions seems to indicate, however, that non-linear interaction between the two modes could appear, so that hybrid unstable and oscillatory solns. could appear in the non-linear region. It should be stressed in that connection that a decrease in surface viscosity would cause the damping coefficient to rise for greater k-values and the region of overlapping would be enlarged. Therefore, the surface viscosity is probably less than 0.001 surface Poise in the O. & W. case.

VI. EXERGY RELEASE - EXCESS DISSIPATION PRINCIPLE

Some of the important features shown in the preceding analysis of stability may tentatively be explained by a principle for the formation of "dissipative structures" which we shall baptise the *exergy release - excess dissipation principle*.

The principle has been stated by Chandrasekhar[13](Chap. II, § 14) for the instability problem of a fluid heated from below in the field of gravity. He states that instability occurs at the minimum temperature gradient at which a balance can steadily be maintained between the internal energy released by the buoyancy force and the kinetic energy dissipated by the viscous processes. To be precise, it is not internal energy which is "released" (internal energy is conserved), but *exergy*. Exergy is the maximum available work which can be extracted from a thermodynamical system, and the precise relation between exergy and internal energy has recently been pointed out in a paper by Sørensen (17). We see that in Chandrasekhar's example thermal convection sets in when a certain convection pattern releases just enough buoyancy exergy to overcome the *excess* dissipation involved in the hydrodynamic motion (*i.e.* the dissipation due to thermal conduction is kept out of the book-keeping).

In the case at hand the exergy could be extracted by equalising the concentration of acetic acid to their Nernstian values in the two liquid phases by means of reversible processes involving ideal semi-permeable membranes. Thus, the exergy here is a *chemical potential exergy*. The criterion for the appearance of a dissipative structure of convection is now, that the exergy released by the convection processes is just able to overcome the viscous dissipation in the convection process. With a linear profile we get more and more exergy released the longer we take the wavelength of the hydrodynamical perturbation, since the depth of the perturbation simultaneously increases. We should not be astonished to find,therefore, that the system is unstable above a certain critical wavelength. For the more realistic exponential profile,however, an increase in wavelength will first bring about an increase in exergy release, but only until a certain wavelength where the perturbation penetrates too far into the region of homogeneous concentrations. Then the exergy release does not at all increase at the rate exhibited by a linear profile, and it is very likely that viscous dissipation again takes over. Therefore, the system becomes stable again at a second critical wavelength. We hope to be able at a later stage to give a complete *quantitative* illustration of the exergy release - excess dissipation principle.

VII. LIST OF SYMBOLS

Latin letters

o	index indicating unperturbed reference state
t	time
x,y	Cartesian coordinates
z	Cartesian coordinate peperdicular to unperturbed interface z = o
δz	interfacial deformation
ℓ	phase index = 1,2
$c^{(\ell)}$	concentration of third component in phase ℓ
$c_o^{(\ell)}$	unperturbed concentration distribution of third component
$c_{oo}^{(\ell)}$	defined in eqns.(2) for a linear and in eqns.(3) for an exponential profile
$c_\infty^{(\ell)}$	defined in eqns.(3) for an exponential profile
$W_z^{(\ell)}$	z-component of solvent velocity in phase ℓ
q_ℓ	parameter defined in eqn.(8)
r_ℓ	parameter defined in eqn.(13)
p	hydrostatic pressure
K_ℓ	adsorption coefficients for the third component to the interface
K_d	Nernstian distribution coefficient
B_1,B_2	hydrodynamic coefficients, eqns.(6a & b)
$C_i^{(\ell)}$	(i = 1,2,3) diffusion-convection coefficients eqn.(11) for linear, eqn.(17) for exponential profile
B,C	dimensionless constants in eqn.(27)
k_x,k_y	components of wavevector k
g	gravitational acceleration
M_{ij}	(i,j = 1,2,3) matrix elements in determinantal compatibility eqn.
$L_{subscript}$	dimensionless numbers different for linear and exponential profiles.

Greek letters and other symbols

δ perturbation from reference state

ω_r, ω_i real and imaginary components of time constant ω

ω^* dimensionless time constant, eqn.(28)

λ wavelength = $2\pi/k$

∇ gradient operator

ΔX $X^{(2)} - X^{(1)}$ taken at z = o

Σ summation over ℓ-index

$\tilde{X}(k)$ Fourier component of parameter X(x,y)

β_ℓ parameter characterising unperturbed concentration profile. Linear profile: eqns.(2), exponential Profile: eqns.(3)

α defined in eqn.(24)

σ interfacial tension

κ, ε interfacial dilatational and shear viscosity

Γ surface excess concentration of third component

μ_ℓ, ν_ℓ viscosity and kinematic viscosity for the bulk liquids

ρ_ℓ density of bulk liquid

$\phi_{subscript}$ dimensionless numbers common for linear and exponential profile.

VII. REFERENCES

1. Steinchen, A., and Sanfeld, A., Chemical Physics 1, 156 (1973)
2. Sanfeld, A., and Steinchen, A., Biophysical Chemistry 3, 99 (1975)
3. Sørensen, T.S., Hennenberg, M., Steinchen, A., and Sanfeld, A., "Coupling between Macromolecular Reactions and Hydrodynamic Motion at an Interface". International Symposium on Macromolecules. Jerusalem. July 1975.
4. Sørensen, T.S., Hennenberg, M., Steinchen, A., and Sanfeld, A., "Surface Chemical and Hydrodynamical Stability". International Conference on Colloid and Surface Science. Budapest. September 1975.
5. Hennenberg, M., Sørensen, T.S., Steinchen, A., and Sanfeld, A., J. Chim. Phys. 72, 1202 (1975)
6. Sørensen, T.S., Hennenberg, M., Steinchen, A., and Sanfeld, A., "Chemical and Hydrodynamical Analysis of

Stability of a Spherical Interface", J. Colloid Interface Sci.(in press, 1976).
7. Hennenberg, M., Sørensen, T.S., and Sanfeld, A., "Deformational Instability of a Plane Interface with Transfer of Matter". Paper submitted for publication in J. Chem. Soc. Faraday Trans. II (1976)
8. Sørensen, T.S., Hennenberg, M., and Sanfeld, A., "Deformational Instability of a Plane Interface with Perpendicular Linear and Exponential Concentration Gradients". Paper submitted for publication in J. Colloid Interface Sci. (1976).
9. Sternling, C.V., and Scriven, L.E., A.I.Ch.E. Journal 5, 514 (1959)
10. Orell, A., and Westwater, J.W., A.I.Ch.E. Journal 8, 350 (1962)
11. Scriven, L.E., Chem.Eng.Sci. 12, 98 (1960)
12. Aris, R., "Vectors, Tensors, and the Basic Equations of Fluid Mechanics". Prentice-Hall, Englewood Cliffs,N.J., 1962.
13. Chandrasekhar, S., "Hydrodynamic and Hydromagnetic Stability", Chap. X. Oxford University Press 1961.
14. Glansdorff, P., and Prigogine, I., "Thermodynamic Theory of Structure, Stability and Fluctuations". Wiley-Interscience, London- New York - Sydney - Toronto,1971.
15. Hirschfelder,J.O., Curtiss, C.F., and Bird, R.B., "Molecular Theory of Gases and Liquids", John Wiley & Sons, New York- London 1954, § 9.2, p.630.
16. Thiessen, D., Z. Physik. Chem. 232, 27 (1966)
17. Sørensen, T.S., "Brønstedian Energetics, Classical Thermodynamics and the Exergy", Acta Chem. Scand. (in press 1976)

ACKNOWLEDGEMENT

M.Hennenberg gratefully acknowledges the support from a Belgo-Danish exchange fellowship from the Danish Ministry of Education and T.S.Sørensen the support from travel grants from International Conference on Colloids and Surfaces, Technical University of Denmark and Otto Mønsted's Fond.

The authors also wish to thank prof. M.Dupeyrat, Lab. de Chim. Phys., Université de Paris VI, for a dramatic demonstration of spontaneous interfacial deformation and prof. M.G. Velarde, Departamento de Fisica-C-3, Universidad Autonoma de Madrid, for useful discussions and bibliographical advice.

INTERACTIONS BETWEEN MOLECULES SORBED IN A CYLINDRICAL PORE OF ZEOLITE

T.Takaishi,
Institute for Atomic Energy, Rikkyo University.

The three-body effect was studied for a system composed of a zeolite and two molecules sorbed in its cylindrical pore. Numerical computations were carried out for a circular cylinder with a small radius comparable to that of sorbed molecules. The numerical results show that the three-body effect is essentially a short range force, and approximated by a simple function,

$$a\{r(12)\}^{-6} - b\{r(12)\}^{-7}$$

where a and b are positive factors, and r(12) the mutual distance between sorbed molecules. In this expression the first term, which is the leading one, is repulsive.

If this potential is added to an unperturbed pair potential, for instance, the 6-12 potential, the resultant potential has a very shallow valley. In cases of rough approximations, an L-shape or the hard sphere potential might be used as a pair potential for sorbed molecules in the pore.

The three-body effect in the present system is much larger than that in adsorbed molecules on a plane surface, as far as the mutual distance between molecules is less than 3.6 times of the molecular radius, while smaller at larger separations.

The applicability of these results is discussed for molecules sorbed in mordenite whose pore has a form of elliptic straight cylinder. Through a geometrical consideration, we arrived at the conclusion that the above potential form is a good approximation for such larger molecules as krypton and xenon in mordenite.

I. INTRODUCTION

It has been well recognized that the interaction potential between a pair of physisorbed molecules on a solid surface differs considerably from that between an isolated pair in vacuum (1). The difference is ascribed to a third-order perturbation energy in the system of three bodies, i.e., two molecules and the solid adsorbent (2), which is called the three-body effect or the non-additive term. This third-order perturbation energy amounts to 20—30% of the total interaction energy between physisorbed molecules on solid surfaces, as shown by Sinanoğlu and Pitzer (2), McLachlan (3), and MacRury and Linder (4). Experimental results are, more or less, well explained in terms of these theories (1).

It is concievable that the three-body effect may become large for molecules sorbed in a pore of zeolite, since they are environed by the pore-wall and heavily perturbed. With this in view, we calculate in this paper the three-body effect term for molecules in a cylinder pore. The pore in mordenite has a form of an elliptical straight cylinder with maximum and minimum diameters of 7.0 and 5.8Å, respectively. The present model of circular cylinder differs from the real geometry of mordenite, but may represent an essential feature of the problem.

II. FORMULATION

In this paper, we concern only non-polar spherical molecules, for theoretical simplicity. Consider two molecules of the same kind, specified by symbols *1* and *2*,and atoms(or ions) constituting zeolite specified by symbols *i* and *j*. The dispersion interaction energy of this system, U_{disp}, is given by, to the third-order,

$$U_{\text{disp}} = \{\sum_i E(1i) + \sum_i \sum_{j\neq i} E(1ij)\} + \{\sum_i E(2i) + \sum_i \sum_{j\neq i} E(2ij)\}$$

$$+ \{E(12) + \sum_i E(12i)\} , \qquad [1]$$

where $E(kl)$ denotes the second-order perturbation energy between the k- and l-th particles, and $E(klm)$ the three-body effect energy between the k-, l-, and m-th particles. The term in the first wavy parethesis gives the dispersive adsorption energy, and $E(1i)$ is much larger than $E(1ij)$. At the present stage, we are reluctantly obliged to introduce a drastic approximation in estimating the value of $E(1i)$, and hence the inclusion of the term $E(1ij)$ is of less significance. Furthermore, our main interest concerns the magnitude of the ratio $E(12i)/E(12)$, i.e., the relative importance of the three body term in the interaction energy between a pair of sorbed molecules, so that $E(1ij)$ and $E(2ij)$ are neglected in the following.

The second-order energy, $E(1i)$, is divided into two parts

$$\sum_i E(1i) = E(\mathrm{I}) + E(1s;\mathrm{v.d.W.}) \qquad [2]$$

where $E(\mathrm{I})$ denotes the electrostatic interaction energy between the surface field and the induced dipole by it, and $E(1s;\mathrm{v.d.W.})$ the van der Waals potential in the adsorption. The latter term is given by,

$$E(1s;\mathrm{v.d.W.}) = \frac{3}{2}\alpha_1\Delta_1 \sum_i \frac{\alpha_i\Delta_i}{(\Delta_1 + \Delta_i)} \{r(1i)\}^{-6}$$

where α's denote the polarizabilities of particles, Δ's the mean exitation energies of particles, $r(1i)$ the distance between the 1- and i-th particles, and summation is carried out over all atoms (or ions) constituting the zeolite. The following simplifications are introduced as usual :

(i) Δ_i's and α_i's can be replaced by thier averaged values of the solid, Δ_S and α_S, respectively,

(ii) the summation can be replaced by a proper integration,

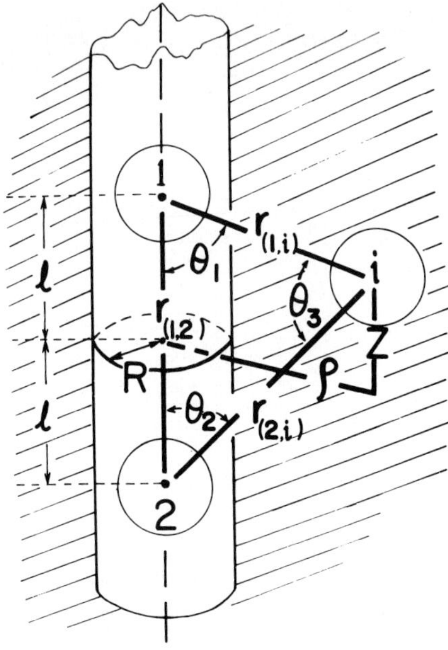

Fig. 1. The geometry of sorbed molecules in a circular cylinder.

and (iii) the admolecules lie on the center axis of the pore. By adopting a geometry shown in Fig.1, the above equation is reduced to

$$E(12s;\mathrm{fl}) = -\frac{3}{2}\frac{\Delta_1\Delta_s\alpha_1\alpha_s}{\Delta_1+\Delta_s}\int\frac{d\tau}{r(12)^6} = -\frac{3}{8}n_s\pi^2\frac{\Delta_1\Delta_s\alpha_1\alpha_s}{\Delta_1+\Delta_s}R^{-3}, \qquad [3]$$

where n_s denotes the number of atoms (or ions) contained in unit volume of zeolite.

Similarly, we have

$$\sum_i E(12i) = E(12s;\mathrm{es}) + E(12s;\mathrm{fl}), \qquad [4]$$

where $E(12s;\mathrm{es})$ is the electrostatic part of the three-body term, and $E(12s;\mathrm{fl})$ the interaction energy between fluctuating dipoles in the particles. The latter term is given as, (5)

$$E(12s;\mathrm{fl})$$

$$= \frac{3}{4}\Delta_1\alpha_1^2\sum_i\frac{\{1+3\cos\theta_1\cos\theta_2\cos\theta_3\}(2\Delta_1+\Delta_i)}{\{r(12)\cdot r(1i)\cdot r(2i)\}^3(\Delta_1+\Delta_i)^2}\alpha_i\Delta_i$$

$$= \frac{3}{4}\frac{(2\Delta_1+\Delta_s)\Delta_1\Delta_s}{(\Delta_1+\Delta_s)^2}\alpha_s\alpha_1^2\sum_i\frac{1+3\cos\theta_1\cos\theta_2\cos\theta_3}{\{r(12)\cdot r(12)\cdot r(12)\}^3}$$

where the approximation (i) is used, and $r(1i)$, $r(2i)$, θ_1, θ_2, and θ_3 are defined in Fig.1. By assuming that two admolecules lie on the center axis of the pore, and replacing the summation by integration, we have

$$E(12s;\mathrm{fl}) = \frac{3}{4}n_s\frac{(2\Delta_1+\Delta_s)\Delta_1\Delta_s}{(\Delta_1+\Delta_s)^2}\alpha_s\alpha_1^2G(l,R), \qquad [5]$$

with

$$G(l,R) = \frac{4}{r(12)^3}\int_0^\infty dz\int_R^\infty \rho d\rho\frac{1+3\cos\theta_1\cos\theta_2\cos\theta_3}{\{r(1i)\cdot r(2i)\}^3}, \qquad [6]$$

$$r(12) = 2l, \qquad r(1i) = [(l-z)^2+\rho^2]^{1/2},$$

$$r(2i) = [(l + z)^2 + \rho^2]^{1/2}, \quad \cos\theta_1 = (l - z)/r(1i),$$
$$\cos\theta_2 = (l - z)/r(1i), \text{and} \quad \cos\theta_3 = \cos(\pi - \theta_1 - \theta_2). \qquad [7]$$

Introducing Eq.[7] into Eq.[6], and integrating in terms of ρ, we have,

$$G(l,R) = \pi/2\cdot\{r(12)\cdot R\}^{-3}F(\xi), \qquad [8]$$

with,

$$F(\xi) = 1/2\cdot(1 + \xi)^{3/2}\int_0^\infty \zeta^{-4}I(\xi,\zeta)d\zeta \qquad [9]$$

where
$$I(\xi,\zeta) = -1 + \frac{1 + 3\xi\zeta^2 + (4 + \xi)\zeta^4 + \zeta^6}{(1 + 2\xi\zeta^2 + \zeta^4)^{3/2}}, \qquad [10]$$

$$\xi = (R^2 - l^2)/(R^2 + l^2),$$

and
$$\zeta = z\,(R^2 + l^2)^{-1/2}. \qquad [11]$$

In a region of $\zeta << 1$, Eq.[10] is approximated as

$$I(\xi,\zeta) = (1 + \xi)[5 - 3\xi + (5\xi^2 - 8\xi - 1)\zeta^2 - \zeta^4(\cdot\cdot)\cdots].$$

Numerical values of $F(\xi)$ are plotted in Fig.2. The curve is fairly well approximated by a simple function,Eq.[12]. The relative deviation,

$$\delta F/F = [F(\xi) - 1.5(R/l)^3 + 0.86(R/l)^4]/F(\xi),$$

is plotted also in the figure. The deviation is large in the region of $l/R >> 1$ and of $l/R < 1$, but small in the intermediate region. In the former region, however, the absolute value of F is very small and the error introduced by the approximation has no serious effect. In the present problem, we concern exclusively to small pore whose radii are comparable to the radius of sorbate molecules. Then, in the region of $l/R < 1$, a large repulsive force operates between cores of sorbate molecules, and the relative weight of F becomes very small. Thus, we can safely adopt the approximation,

$$F = 1.5(R/l)^3 - 0.86(R/l)^4. \qquad [12]$$

Introducing Eq.[6] and [12] into Eq.[5], we have

$$E(12s;\text{fl}) = 3\pi n_s \frac{\Delta_1\Delta_s(2\Delta_1 + \Delta_s)\alpha_s\alpha_1^2}{(\Delta_1 + \Delta_s)^2\{r(12)\}^6}\left[1.5 - \frac{1.72R}{r(12)}\right], \qquad [13]$$

which is essentially a short range force. The above expression is practically useful for its simplicity.

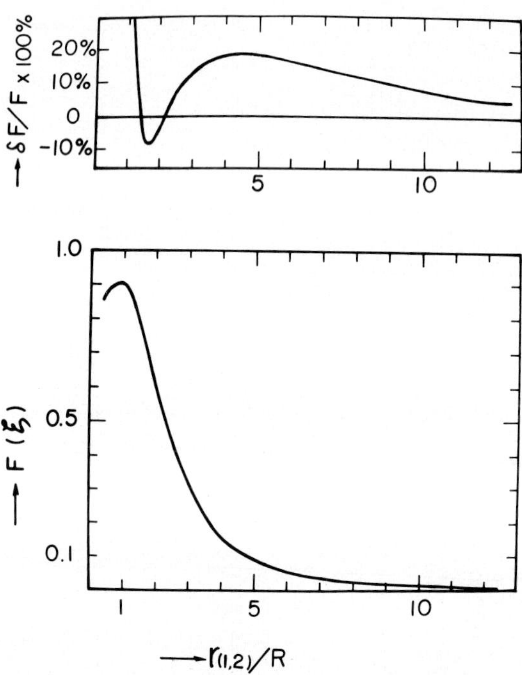

Fig. 2. Curve of F(ξ), nad the relative deviation of an approximate function to F(ξ) against the mutual distance between sorbates. R, radius of the pore; r(12), mutual distace between a pair of molecules.

III. DISCUSSION

Equation [13] is compared with the Sinanoğlu-Pitzer potential which concerns adsorbed molecules on a plane surface. This is approximately expressed as (2),

$$E(\text{plane};\text{fl}) = \frac{3}{4}\, n_S \frac{\Delta_1\Delta_s(2\Delta_1 + \Delta_s)\alpha_s\alpha_1^2}{(\Delta_1 + \Delta_s)^2\{r(12)\cdot z\}^3}\, 0.085\pi \qquad [14]$$

where z denotes the distance between the sorbates and the surface plane. Dividing Eq. [13] by Eq. [14], we have

$$\frac{E(12s;\mathrm{fl})}{E(\mathrm{plane};\mathrm{fl})} = 5.9\left[\frac{2\ R}{r(12)}\right]^3 \left[1.5 - \frac{0.86R}{r(12)}\right], \qquad [15]$$

if we put $z = R$. Equation [15] is plotted against $R/r(12)$ in Fig. 3, which shows that $E(12s;\mathrm{fl})$ is larger than $E(\mathrm{plane};\mathrm{fl})$ in a region of $r(12) < 3.6R$, and that $E(12s;\mathrm{fl})$ is a short range force in contrast to $E(\mathrm{plane};\mathrm{fl})$.

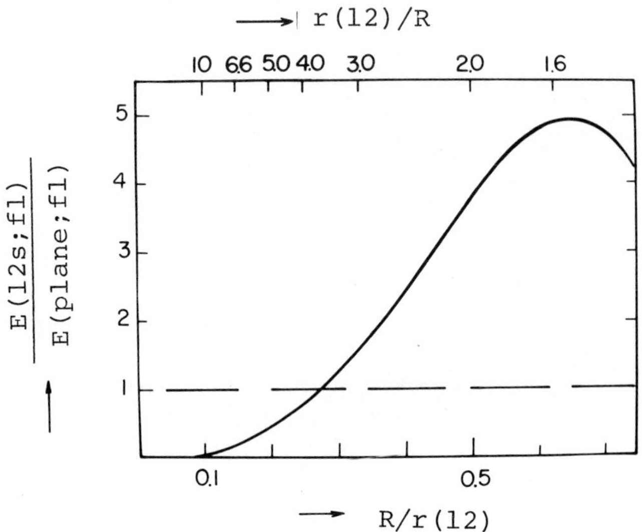

Fig. 3. Comparison of the magnitude of the three-body effect term for molecules in a cylindrical pore, $E(12s;\mathrm{fl})$*, and that on a plane surface,* $E(\mathrm{plane};\mathrm{fl})$

Now let us investigate the magnitude of $E(12s;\mathrm{fl})/E(12)$. Combining the relation,

$$E(12) = -\ (3/4)\ \alpha_1^2\ \Delta_1\{r(12)\}^{-6}$$

with Eq. [13], we have

$$E(12s;\mathrm{fl})/E(12) = -\ 3n_s\pi\alpha_s[1.5 - 0.86\{2R/r(12)\}], \qquad [16]$$

if we put $\Delta_1 = \Delta_s$ for simplification. The value for $n_s\alpha_s$ is estimated for a representative case, say, for mordenite. Its unit cell has a dimension of 18.13Å X 20.48Å X 7.52Å, containing $Na_8(AlO_2)_8(SiO_2)_{40}$, according to Meier (6). By definition we have

$$n_s\alpha_s = \sum_i \alpha_i = n(cell) \sum_{cell} \alpha_i$$

where the summation concerning i designates the summation over all constituting ions contained in unit volume of zeolite, the second summation sign means the summation over those in unit cell, and $n(cell)$ is the number of unit cells contained in unit volume of zeolite. By neglecting small contributions of cations, we have

$$\sum_{cell} \alpha_i = 96\alpha(O^{--}) = 96 \times 3.88 \text{ Å}^3 ,$$

where Pauling's value for $\alpha(O^{--})$ is used (7). Thus, we have

$$n_s\alpha_s = 0.133$$

A pair potential between sorbed molecules, $u'(12)$, is calculated, if the use is made of the 6-12 potential,

$$u(12) = 4\varepsilon_o[\{\sigma/r(12)\}^{12} - \{\sigma/r(12)\}^6] , \quad [17]$$

as the unperturbed pair potential. Adding Eq.[13] to the

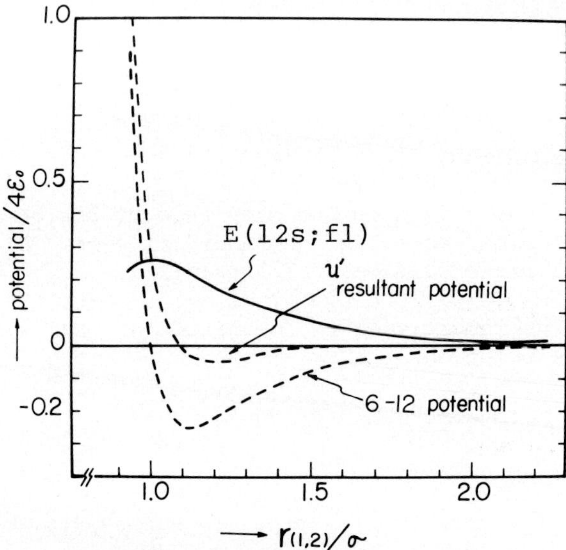

Fig. 4. A pair potential between sorbed molecules. E(12s;fl), three-body effect term; u', resultant pair potential; the energy is expressed in the unit of $4\varepsilon_o$ defined by Eq.[17].

above, we have

$$\frac{u'(12)}{4\varepsilon_o} = \left[\frac{\sigma}{r(12)}\right]^{12} + 0.88\left[\frac{\sigma}{r(12)}\right]^{6} - 1.62\left[\frac{\sigma}{r(12)}\right]^{7}$$

where we put provisionally $\sigma = 2R/1.5$. Numerical values of $u'(12)$ are shown in Fig.4. The valley in the curve is very shallow, and may be neglected at temperature not so low. In other wards, the potential may roughly be approximated by an L-shape or the hard sphere potential.

In the last place, an elliptical cylindrical pore is discussed, which is found in mordenite. The factor deciding the sign of the right hand side of Eq.[6], $(1 + 3\cos\theta_1\cos\theta_2\cos\theta_3)$, becomes negative, if any one of angles, θ_1, θ_2, and θ_3, is larger than 117°. Points giving negative values for this factor are located in two cones or rotation of the arc shown in Fig. 5. Figure 6 shows situations that two molecules are compactly packed in a circular cylinder (abbreviated as C-cylinder) or an elliptical cylinder (abbreviated as E-cylinder). Apart from remote regions from the molecules, the larger the hatched region contained in the pore, the smaller the attractive part of $E(12s;\mathrm{fl})$. The same relation exists between the unhatched region and the repulsive part. The figure shows qualitatively that the attractive part increases with the increase of the angle ϕ, while the repulsive part decreases. If $r(12)$ is increased without changing the relative position of the sorbates against the pore-wall, ϕ-value decreases. This means that the difference between $E(12s;\mathrm{fl})$-values in the C-cylinder and E-cylinder is large only for small $r(12)$-values. Hence, $E(12s;\mathrm{fl})$ in the E-cylinder may be a short range force as well as that in the C-cylinder, and may be approximated as

$$E(12s;\mathrm{fl}) = 3\pi n_s \frac{(2\Delta_1 + \Delta_s)\Delta_1\Delta_s\alpha_s\alpha_1^{\,2}}{(\Delta_1 + \Delta_s)^2\{r(12)\}^6}\left[a - b\frac{2\bar{R}}{r(12)}\right], \qquad [18]$$

where a and b are adjustable parameters subjecting to the conditions, respectively,

$$1.2 \geq a > 0, \qquad b \geq 0.86 ,$$

and $2\bar{R}$ the arithmetic mean of the maximum and minimum diameters of the ellips.

The value for ϕ can be estimated by the relation given in the preceding paper (8),

$$\sigma(1D;\mathrm{obs}) = \sigma\cos\phi$$

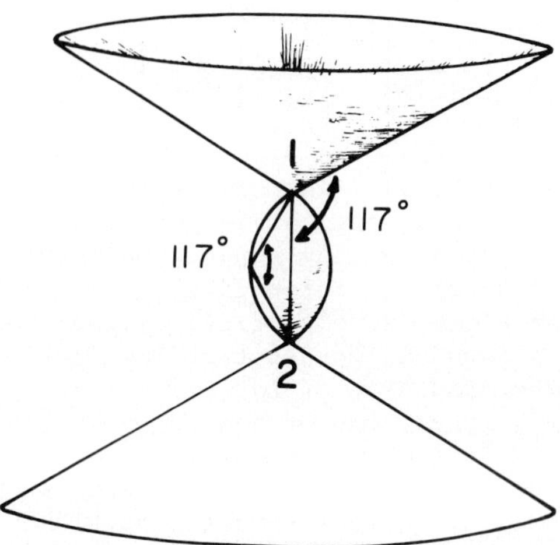

Fig. 5. Cones and a rotation of arc, any point in which satisfies the condition: $1 + 3\cos\theta_1\cos\theta_2\cos\theta_3 < 0$.

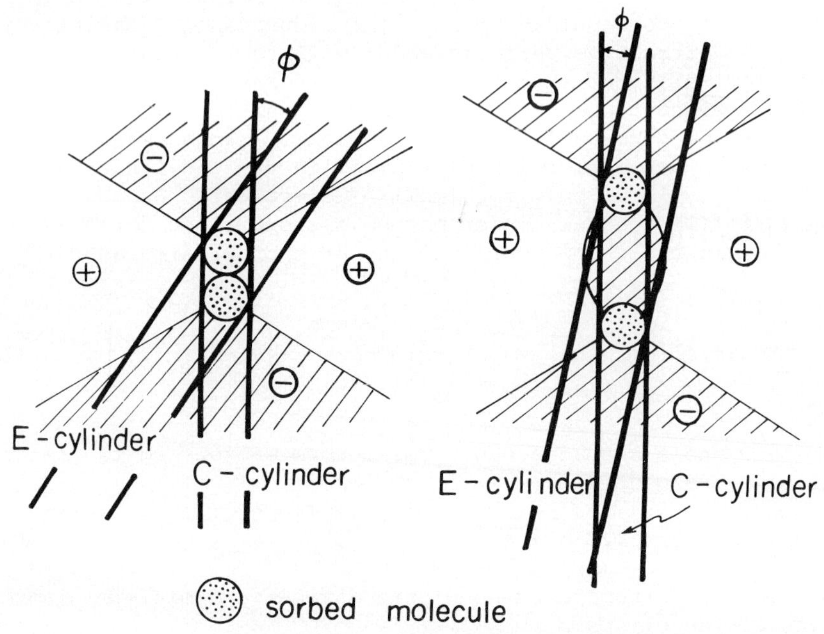

Fig. 6. Geometries that two molecules are compactly packed in a circular cylinder and an elliptic cylinder.
⊖, the region in which $1 + 3\cos\theta_1\cos\theta_2\cos\theta_3 < 0$;
⊕, the region in which $1 + 3\cos\theta_1\cos\theta_2\cos\theta_3 > 0$.

where σ is the molecular diameter defined by Eq.[17], and σ(1D;obs) the obserbed effective diameter for one-dimensional gas in the pore.[1] For argon in the pore of mordenite, we have had

$$\sigma(\text{1D;obs}) = 1.84\text{Å}, \text{ and } \quad \sigma = 3.405\text{Å}$$

and hence $\phi = 57°$.

The larger the molecular diameter, the smaller ϕ, and consequently the better the approximation [18]. Thus, Eq.[18] may be applied with more success to xenon and krypton than to argon, on account of thier large molecular diameters or resulting smaller ϕ-values.

IV. ACKNOWLEDGEMENTS

The author thanks Dr.S. Harasawa, of the Rikkyo University, for his assistance in mumerical computation. This work was partially supported by a Grant-in-Aid for Individual Research from the Ministry of Education of the Japanese Goverment, No.855017 (1973).

V. REFERENCES

1. Takaishi,T., Prog. Surf. Sci. 6, 43 (1975).
2. Sinanoğlu, O., and Pitzer, K.S., J .Chem. Phys. 32, 1279 (1960).
3. McLachlan, A.D., Mol.Phys. 7, 381 (1964).
4. MacRury, T.B., and Linder, B., J. Chem. Phys. 54, 2056 (1971); *ibid*. 56, 4356 (1972).
5. Kihara, T., in " Advances in Chemical Physics " (I. Prigogine, Eds.), Vol. 1,p.267. Interscience Publisher, New York, 1956.
6. Meier, W.T., Z. Kristallogr. 115, 439 (1961).
7. Pauling, L., Proc. Roy. Soc.(London) A 114, 181 (1927).
8. Takaishi, T., Yusa, A., and Amakasu, F., Trans. Faraday Soc. 67, 3565 (1971).

---

1) In Fig.9 of the reference (8), $\sigma\cos\theta$ was miss-written as $\sigma\sin\theta$, and the relation must be corrected as above.

THERMODYNAMIC STABILITY CRITERIA FOR SURFACES

A. STEINCHEN and A. SANFELD,
Free University of Brussels.

When dealing with stability problems, three types of situations have to be considered : the equilibrium states, the linear region near equilibrium, and the nonequilibrium situations beyond the linear region.

Spontaneous microscopic fluctuations always occur in all systems. When these fluctuations are damped in time, and do not grow to change the macroscopic state, the system is stable.

At equilibrium (far from a phase transition), and near equilibrium, fluctuations play a minor role. They give "corrections" to macroscopic results.

In classical thermodynamics, the response of the system to external perturbations is related to the well known Le Châtelier-Braun theorem (1).

Nevertheless, equilibrium instabilities occur in the region of phase transitions (1), (2), for example the nucleation of a new phase, and the demixtion. These instabilities give rise to the formation of equilibrium structures that are maintained in reversible processes, or for processes that remove slightly the system from equilibrium.

The first region beyond equilibrium to be studied was the socalled linear region, where the Onsager relations (3) and the theorem of the minimum entropy production were established (4). The behavior of matter in this region corresponds to the natural "continuation" of equilibrium.

More recently, Glansdorff and Prigogine (5) derived new stability criteria and an evolution criterion for the linear region and the non linear region as well.

They showed that the linear phenomena have always a stabilizing contribution. The most interesting situations are to be found beyond the linear region, in far from equilibrium systems.

For example, the autocatalytic reactions, far from equilibrium that often occur in biological systems and that can give rise to chemical instabilities (5) such as chemical oscillations, chemical waves, dissipative space structures once coupled with diffusion.

Another type of interesting instability is the onset of convection cells, in the Bénard instability of a fluid submitted to a thermal constraint (6), (7). The transition between the laminar flow to the turbulent flow is another example of instability in non linear phenomena (5), (8), (9).

In open systems, far from equilibrium, beyond the validity of the Onsager laws, completely new situations can be obtained, after an instability threshold, i.e. after a critical point beyond which the system may present phenomena of self organization. They correspond to a new ordering of matter, that was called by Prigogine (5) "dissipative structures" because these structures are maintained by external fluxes and correspond to an increase of the dissipation. The fluctuations drive the macroscopic regime and confer to the evolution an essentially statistical character. For that reason Glansdorff and Prigogine introduced the concept of "order through the fluctuations".

The space time structure beyond an instability depends on the growth pattern of fluctuations which size plays a central role. Dissipative structures are generated through a nucleation process involving fluctuations beyond a critical size (10), (11).

As concerns the thermodynamic stability criteria, the basic assumption is that of the "local equilibrium" given by the Gibbs formula for a continuous volume phase

$$\delta s = T^{-1}\delta u + p T^{-1}\delta v - \sum \mu_\gamma T^{-1}\delta x_\gamma \qquad [1]$$

where s, u and v are respectively the entropy, the internal energy, and the volume by unit mass, μ_γ is the chemical potential and x_γ the mass fraction of component γ.

By an extension of the Duhem (2) stability criterion of equilibrium, for small perturbations, Glansdorff and Prigogine (5) showed that the stability criterion of equilibrium was related to the negative sign of the second derivative of entropy.

$$\delta^2 s = -\frac{\rho}{T}\left[\frac{C_v}{T}(\delta T)^2 + \frac{\rho}{\chi}(\delta v)^2_{x_\gamma} + \sum\left(\frac{\partial \mu_\gamma}{\partial x_{\gamma'}}\right)\delta x_\gamma \delta x_{\gamma'}\right] \leqslant 0 \qquad [2]$$

with C_v the heat capacity and χ the compressibility.

That quadratic form can be split into three separate stability conditions for equilibrium that are

$C_v > 0$ thermal stability [3]

$\chi > 0$ mechanical stability [4]

$\sum\left(\frac{\partial \mu_\gamma}{\partial x_{\gamma'}}\right)\delta x_\gamma \delta x_{\gamma'} > 0$ stability in respect to diffusion [5]

For polar systems (12) in an electric field at equilibrium, Sanfeld and Steinchen (12) (13) have shown that the stability criterion contains an additional term related to the fluctuations of polarization

$$-\frac{\rho}{k}(\delta \bar{p})^2 < 0 \qquad [6]$$

where $\bar{p}$ is the polarization vector by the unit mass and $k = \frac{\varepsilon - 1}{4\pi}$ the electric susceptibility.

The dipolar stability condition for equilibrium then reads $\varepsilon \geqslant 1$ [7].

Glansdorff and Prigogine have extended the thermodynamic theory of stability to non-equilibrium conditions. They assumed that in the complete range where a macroscopic description is possible, and where the basic hypothesis of local equilibrium subsists, the conditions [3] - [5] and therefore [2] remain valid. The quantity $\delta^2 s$ is therefore a negative quadratic form. However this does not guaranty the stability of the system. If a fluctuation takes place in a system out of equilibrium, that gives rise to $\delta^2 s \leq 0$ the time variation $\partial_t \delta^2 s$ has to be ≥ 0 for the system to be stable. The basic concepts of the Lyapounov theory of stability (15) lead to use of $\delta^2 s$ as a Lyapounov function. According to this theory, if $\partial_t \delta^2 s \geq 0$ together with $\delta^2 s \leq 0$, for all values of t, the reference state is stable.

The choice of $\delta^2 s$ as Lyapounov function instead of any other quadratic function finds its justification in its physical significance in terms of probability of fluctuations in Einstein's theory (5).

To take into account the fluctuations of the local centre-of-mass motion (barycentric velocity V_i), the stability problem requires a larger set of independent variables, to treat the convective effect. These variables are u, v, x_γ and V_i where V_i denotes the i velocity component.

The new Lyapounov function is then the negative quadratic expression (5)

$$\delta^2 \rho z = \delta^2\left(\rho s - T_0^{-1} \rho \frac{V^2}{2}\right) = \delta^2 \rho s - T_0^{-1} \rho (\delta \vec{v})^2 \leq 0 \qquad [7]$$

and the stability criterion reads

$$\begin{aligned} \frac{1}{2} \partial_t \delta^2 \rho z = & \sum_k \delta J^k \delta X^k \\ & - \left[\delta(\rho u) \delta v_j + \delta P^{ij} \delta v_i + \frac{1}{2} \rho v_j (\delta v)^2\right] T^{-1}_{,j} \\ & - \sum_\gamma \delta \rho_\gamma \delta v_j \left[F_{\gamma j} T^{-1} - (\mu_\gamma T^{-1})_{,j}\right] \\ & - \left[\delta P^{ij} \delta T^{-1} - T^{-1} \delta(\rho v_j) \delta v_i\right] v_{i,j} \\ & + \frac{1}{2} T^{-1} (\delta v)^2 (\rho v_j)_{,j} \end{aligned}$$

$$+\left[\delta(\rho u)\delta T^{-1}_{,j} - \sum \delta\rho_\gamma \delta(\mu_\gamma T^{-1})_{,j}\right] v^{o}_j$$

$$-\Big\{\delta J_j \delta T^{-1} - \sum \delta(\rho_\gamma \Delta_{\gamma j})\delta(\mu_\gamma T^{-1}) - T^{-1}\big[\delta P^{ij}\delta v_i$$

$$+\tfrac{1}{2}\rho v_j (\delta v)^2\big] - v_j \delta^2(\rho s)\Big\}_{,j} \geqslant 0 \qquad [8]$$

where P^{ij} is the tensor of pressure, $F_{\gamma j}$ the external force acting on component γ, J_j the heat flux, $\Delta_{\gamma j}$ the diffusion velocity of component γ, the symbol, j is the classical covariant derivative and the superscript ^{o} refers to the reference stationary state.

The first term of the r.h.s. of [8] is the excess entropy production (product of fluctuations of fluxes δJ_k and fluctuations of generalized forces δX_k) related to dissipative effects (heat conduction, diffusion, viscous dissipation and chemical reactions).

The next terms concern the dissipative phenomena together with mechanical effects.

The last bracket term is the divergence of a flux of excess entropy.

Once the relation between fluxes J_k and forces X_k are nonlinear, for example in chemical autocatalytic processes (5), the term $\delta J_k \; \delta X_k$ can become $\leqslant 0$ and may lead to a violation of the unequality [8], giving rise to a "chemical instability". The system may leave the thermodynamic branch to follow another branch, which may correspond to a quite different structure.

The second type of terms for example can be responsible for hydrodynamic instability such as the onset of convection in a fluid in a gradient of temperature, in the Bénard problem (6).

The convection cells observed after a critical threshold of the temperature gradient are stationary patterns corresponding to a self organization of the system maintained by a thermal flux. For charged systems in an electric field,

Sanfeld and Steinchen (13), (14) have demonstrated the stability criteria of nonequilibrium systems

$$\delta^2(\rho\zeta) = \delta^2\left(\rho z - T_0^{-1}\frac{E^2}{4\pi}\right) \leq 0 \qquad [9]$$

and

$$\frac{1}{2}\partial_t\,\delta^2(\rho\zeta) = \frac{1}{2}\partial_t\,\delta^2(\rho z) + \delta(I_j T^{-1})\,\delta E_j \geq 0 \qquad [10]$$

We see that [10] involves an additional term due to the fluctuations of the total electric current I_j and to the fluctuations of the electric field E_j. The electric term is able to destabilize the system if the electric field $\underline{E}$ is related to the total current $\underline{I}$ by a "negative resistance".

That unstable behavior of systems with a negative resistance is well known in electrochemistry (16), where oscillations in current tension curves were observed in the region of negative resistance.

The thermodynamic criteria of Glansdorff and Prigogine were dealing only with continuous one-phase systems. We want now to extend them to multiphase systems with interfaces.

The equilibrium stability of the whole system with an interface is governed by $\delta^2 S \leq 0$ where S can be split into three parts S' the entropy of volume phase' and S'' the entropy of volume phase " and S^s the entropy of the surface in a Gibbs' surface model (17). The equilibrium stability condition for each volume phase reads $\delta^2 S' \leq 0$ and $\delta^2 S'' \leq 0$. As the volume phases can be reduced in size in such a way that the second derivative of entropy of the whole system reduces to the second derivative of entropy of the surface, this last quantity has to be ≤ 0 separately for the system to be stable. The equilibrium stability criterion for a plane surface reads

$$\delta^2 s^s = -\frac{T}{T_s}\left[\frac{C_\Omega}{T_s}(\delta T_s)^2 + \frac{T}{E}(\delta\omega)^2_{x_\gamma} + \sum \frac{\partial \mu^s_\gamma}{\partial x^s_{\gamma'}}\,\delta x^s_\gamma\,\delta x^s_{\gamma'}\right] \leq 0 \qquad [11]$$

with s^s the surface entropy by unit mass,
Γ the surface density or total adsorption
c_Ω the surface heat capacity
μ_γ^s the chemical potential of γ in the surface
x_γ^s the surface mass fraction of γ
s the subscript s refers to the surface
$E = \Omega \frac{\partial \sigma}{\partial \Omega}$ the Gibbs' elasticity $\frac{\omega}{\sigma}$ variation of surface tension σ with the area Ω
ω the area by unit mass

The equilibrium stability criterion for the surface can be split into three conditions :

$$c_\Omega > 0 \quad \text{thermal surface stability} \qquad [12]$$

$$E > 0 \quad \text{mechanical stability of the surface} \qquad [13]$$

$$\sum \frac{\partial \mu_\gamma^s}{\partial x_\gamma^s} \delta x_{\gamma'}^s \delta x_\gamma^s > 0 \quad \text{stability in respect to diffusion in the surface} \qquad [14]$$

For nonequilibrium systems. The problem is somewhat more complicated for surfaces than for bulk phases because the basic assumption of local equilibrium is not generally demonstrated. (17), (18).

To take into account the so-called non-autonomy of the surface, Defay (17) introduced supplementary variables (the concentrations of adjacent phases) for nonequilibrium surfaces

$$\delta s^s = T_s^{-1} \delta u^s - \sum_\gamma \mu_\gamma^s T_s^{-1} \delta \Gamma_\gamma - \sum_\gamma \varepsilon_\gamma' T_s^{-1} \delta \rho_\gamma' - \sum_\gamma \varepsilon_\gamma'' T_s^{-1} \delta \rho_\gamma'' \qquad [15]$$

with ε'_γ and ε''_γ the cross-chemical potentials
Γ_γ^s the adsorption of γ

However, for systems where the concentrations in the bulks are maintained constant, or for surface without exchange of matter with the bulk such as insoluble layers, the two last terms of [15] vanish and the hypothesis of local equilibrium

$$\delta s^s = T_s^{-1}\delta u^s - \sum_\gamma \mu_\gamma^s T_s^{-1} \delta \Gamma_\gamma \qquad [16]$$

can be applied. We will restrict ourselves here to that case.

The stability criterion for non equilibrium systems then reads

$$\frac{1}{2}\partial_t \delta^2(T s^s) = \sum_k \delta J_s^k \delta X_s^k + \Delta_s[\delta T_s^{-1} \delta J_z]$$

$$-\left\{\delta T_s^{-1} \delta j_\alpha - \sum_\gamma \delta(\mu_\gamma^s T_s^{-1}) \delta(\Gamma_\gamma \Delta_\gamma^\alpha)\right\}_{,\alpha} \geq 0 \qquad [17]$$

where the surface excess entropy production

$$\sum_k \delta J_s^k \delta X_s^k = \delta T_{s,\alpha}^{-1} \delta j_\alpha + \sum_\gamma \delta(\mu_\gamma^s T_s^{-1})_{,\alpha} \delta(\Gamma_\gamma \Delta_\gamma^\alpha)$$

$$+ \sum \delta(F_\gamma^\alpha T_s^{-1}) \delta(\Gamma_\gamma \Delta_\gamma^\alpha) - \sum_\rho \sum_\gamma \delta(\mu_\gamma^s T_s^{-1}) \nu_{\gamma\rho} M_\gamma \delta V_\rho^s \qquad [18]$$

with J_z the heat flux normal to the interface, j_α the heat flux by unit length in the surface, z the normal coordinate, α the coordinates in the surface, Δ_γ^α the diffusion velocity in the surface, V_ρ the reaction rate, M_γ the molecular weight. The symbol Δ_s is the difference through the surface of the bracket quantity, $\nu_{\gamma\rho}$

the stoechimetric coefficient, $;\alpha$ is the classical covariant derivative in the surface.

The linear phenomena of heat conduction and diffusion in the surface in [18] are always $\geqslant 0$, and give thus a stabilizing contribution to [17]. These linear transport phenomena in the surface have been treated on the basis of the minimum entropy production theorem by Bedaux and al. (19). For convective systems, these authors (19) discussed the concept of surface temperature and the Fourier's surface law (see also Waldmann (20)).

The only term of [17], able to be responsible for a surface instability is the chemical reaction term $\sum_{\rho} \sum_{\gamma} \delta(\mu^{s}_{\gamma} T_{s}^{-1}) \nu_{\gamma\rho} M_{\gamma} \delta v_{\rho}$ that can become < 0 for autocatalytic processes, and that can give rise to the same type of structures as observed in bulk phases. For example, in the reaction of oxydation of malonic acid by bromate in the presence of ferroin and of Ce^{IV} ions, surface concentration patterns were observed (21).

The term coming from the net heat fluxes from the bulks $\Delta_s \delta T_s^{-1} \delta J_z$ is always > 0 for fixed temperature in the bulks.

When integrating equation [17] over the whole surface, the boundary conditions can be chosen in such a way that the divergence term $\{ \quad \}_{,\alpha}$ vanishes (fixed boundary conditions on the line limiting the surface).

For convective surfaces, the Lyapounov function is

$$\delta^2(T_z s) = \delta^2(T_s s - T_s^{-1} T v^2) \leqslant 0 \qquad [20]$$

with v, the barycentric velocity in the surface, and the stability criterion reads :

$$\frac{1}{2}\partial_t \delta^2(\Gamma z^s) = \frac{1}{2}\partial_t \delta^2(\Gamma s^s) + \Delta_s[\delta T_s^{-1} u\, \delta v_z]$$

$$+ \delta v^\alpha_{,\beta}\ \delta(\Pi^{\alpha\beta} T_s^{-1})$$

$$- \delta v_\alpha \delta u^s T^{-1}_{s,\alpha} - \delta v_\alpha \sum_\gamma \delta \Gamma_\gamma [F_\alpha T_s^{-1} - (\mu_\gamma^s T_s^{-1})_{,\alpha}]$$

$$-\Big\{ T_s^{-1} \delta v_z \Delta_s \delta p^{zz} + T_s^{-1} \delta v_z\, \delta(\Gamma F_z)$$

$$+ T_s^{-1} \delta v_z\, \delta\Big[\sigma\Big(\frac{1}{R_1} + \frac{1}{R_2}\Big)\Big]$$

$$+ T_s^{-1} F_z \Delta_s \rho\, \delta v_z\, \delta z^s \Big\}$$

$$-\Big\{ T_s^{-1} \delta v_\alpha \Delta_s \delta P^{\alpha z} + T_s^{-1} (\delta v_\alpha\, \delta \sigma^{\alpha\beta})_{,\beta} \Big\}$$

$$+ \Big\{ T_s^{-1} [\delta P^{\alpha\beta} \delta v_\beta] \Big\}_{,\alpha} \geq 0 \qquad [20]$$

with u the internal energy by unit mass in the bulk phases, z the normal coordinate to the surface, α and β the coordinates in the deformed surface, $\Pi^{\alpha\beta}$ the viscous stress tensor in the surface, F the external force, p^{zz} the normal viscous pressure, R_1 and R_2 the principal radii of curvature, δz^s the height of the perturbation of the interface, σ the hydrostatic surface tension, ρ the bulk density, $P^{\alpha z}$ the shear viscous pressure in the bulk phases, $\sigma^{\alpha\beta}$ the tensor of surface tension. The term $\Delta_s [\delta T_s^{-1} u\, \delta v_z]$ is related to the fluxes of internal energy from the bulks. The term $\delta v^\alpha_{,\beta}\, \delta(\Pi^{\alpha\beta} T_s^{-1})$ is the viscous dissipation function in the surface, always ≥ 0 . The term $-\delta v^{\alpha} \delta u^{s} T^{-1}_{s,\alpha}$ accounts for the coupling between the onset of convection in the surface and the fluctuation of internal energy under a thermal constraint in the surface.

The term $-\delta v^{\alpha} \sum_{\gamma} \delta T_{\gamma} [F_{\alpha} T_{s}^{-1} - (\mu_{\gamma}^{s} T_{s}^{-1})_{,\alpha}]$

accounts for the coupling between the fluctuations of the surface concentrations and the convection, under a generalized diffusion constraint in the surface. These terms are the surface analogs of the Bénard terms in multicomponent systems, see [9].

The next four terms under bracket explain the well known Rayleigh-Taylor effect (6), while the next term

$$\{T_{s}^{-1} \delta v^{\alpha} \Delta_{s} \delta P^{\alpha z} + T_{s}^{-1} (\delta v^{\alpha} \delta \sigma^{\alpha\beta})_{,\beta}\}$$

is responsible for the Marangoni instability.

Rayleigh-Taylor instabilities have been already observed experimentally and were analyzed theoretically (22), (23). The kinetic analysis of this type of instability has shown that it corresponds to an exchange of stability (6), with a critical wavelength (space structure).

The Marangoni effect and the stability of longitudinal surface waves was studied in details in the last ten years (24), (25), (26), (27). For that effect also, surface stationary patterns were observed (24) and analyzed by a kinetic approach (7).

The analysis of the coupling between reactions in the surface and the onset of convection in the surface is of the first importance for biological systems, where chemical mechanisms in the biosurfaces were shown to induce surface deformation, motion, surface phase transitions between multiple steady states, and surface space-structures (28) - (32).

In conclusion, the surface stability criterion shows the existence of destabilizing terms, for a spread monolayer at the interface between two immiscible fluids, when an auto or crosscatalytic chemical reaction occurs at the surface.

Terms arising from the Rayleigh-Taylor and from the Marangoni effects may also exhibit a destabilizing contribution.

Experimental evidences and kinetic analysis corroborate the thermodynamic previsions.

REFERENCES

1. Prigogine, I., and Defay, R., "Chemical Thermodynamics", Longmans, Green, New York, 1962.
2. Duhem, P., "Traité d'Energétique", Tomes 1 et 2, Gauthier-Villars, Paris, 1911.
3. Onsager, L., Phys. Rev. 37, 405 (1931).
4. Prigogine, I., "Introduction to Thermodynamics of Irreversible Processes", Wiley, New York, 1967.
5. Glansdorff, P., and Prigogine, I., "Thermodynamics of Structure, Stability and Fluctuations", Wiley, New York, 1971.
6. Chandrasekhar, S., "Hydrodynamic and Hydromagnetic Stability", Clarendon Press, Oxford, 1961.
7. Velarde, M.G., in "Hydrodynamics" (R. Balian Ed.), Gordon and Breach, New York, 1974.
8. Lin, C.C., "The Theory of Hydrodynamic Stability", Cambridge University Press, 1966.
9. Davies, J.T., "Turbulence Phenomena", Academic Press, New York and London, 1972.
10. Nicolis, G., and Prigogine, I., "Self Organization Phenomena in Non-Equilibrium Systems", Wiley, New York, in press.
11. Nicolis, G., Malek-Mansour, M., Van Nypelseer, A., and Kitahara, K., J. Stat. Phys. 14, 417 (1976).
12. Sanfeld, A., "Introduction to Thermodynamics of charged and polarized Layers", Wiley, London, 1968.
13. Sanfeld, A., and Steinchen-Sanfeld, Bull. Ac. Sc. Belg., LVII, 684 (1971).
14. Steinchen, A., and Sanfeld, A., Experi. Suppl. 18, 599 (1971).
15. La Salle, J., and Lefschetz, S., "Stability by Liapunov's direct method", Acad. Press, New York, 1961.
16. Vetter, K.J., "Electrochemical Kinetics", Academic Press, New York, 1967.
17. Defay, R., Prigogine, I., Bellemans, A., and Everett, D.H., "Surface tension and adsorption", Longmans Green, London, 1966.
18. Kerszberg, M., Dissertation, Free University of Brussels, 1975.
19. Bedeaux, D., Albano, A.M., and Mazur, P., Physica, 82A, 438 (1976).
20. Waldmann, L., Z. Naturforschung, 22 a, 1269 (1967).
21. Winfree, A.T., Scientific American, 82, June (1974).
22. Sterling, C.V., and Scriven, L.E., A.I.Ch.E.J. 5, 514 (1959).

23. Schwartz, E., and Linde, L., Phys. of Fluids 4, 535 (1963)
24. Thiessen, D., C.I.D. Proc. IVth Int. Congr., Gordon and Breach, Brussels, 1964.
25. Lucassen-Reynders, E.H., Lucassen, J., in "Advances in Interface Science", p. 347, Elsevier, Amsterdam, 1969.
26. Saraga, L., and Ivanov, I.P., submitted to J. Coll. Int. Sc.
27. Joos, P., and Van Bockstaele, M., An. Quim. 71, 889 (1975).
28. Sanfeld, A., and Steinchen, A., Biophys. Chem. 247, 156 (1974).
29. Hennenberg, M., Sørensen, T.S., Steinchen, A., and Sanfeld, A., J. Chim. Phys. 72, 1202 (1975).
30. Sørensen, T.S., Hennenberg, M., Steinchen, A., and Sanfeld, A., J. Coll. Int. Sci., in press.

31. Gouda, J., and Joos, P., Chem. Eng. Sci., 30, 521 (1975).
32. Steinchen-Sanfeld, A. and Sanfeld, A., Chem. Phys. 1, 156 (1973).

ADSORPTION OF HEAVY METAL TRACES ON PARTICULATE MATTER IN SEA WATER

H. Bilinski, S. Kozar and M. Branica
The "Rudjer Bošković" Institute
Zagreb, Croatia, Yugoslavia

Adsorption of ions of trace metals Zn, Pb, Cu and Cd in sea water samples is studied on colloidal particles of SiO_2 *(Ludox SM 30) and* $\gamma\ Al_2O_3$ *(Alon), as modal substances for common particulate matter of oxides in marine environment. Experimental data are presented in forms of Langmuir adsorption isotherms. Surface coordination model and hydrogen bonding model are discussed in connection with the experiments.*

Order of adsorption on SiO_2 *is* $Zn \gg Pb > Cu \gg Cd$ *and on* $\gamma\ Al_2O_3$ *is* $Zn > Cu > Pb \gg Cd$*. Approximate values of adsorption constants of these metal ions on* SiO_2 *and* $\gamma\ Al_2O_3$ *in sea water are presented in detail.*

I. INTRODUCTION

The knowledge of the removal processes of trace metal ions from the sea into the sediment is very limited. For most heavy metals sea water is undersaturated with respect to any likely solid phase, and Krauskopf (1) has suggested that the concentration is reduced due to adsorption processes.

Suspended matter in sea water ranges in size from colloidal to very large particles, with corresponding different specific surface area and residence time. It is composed of both inorganic and organic matter. The inorganic fraction is slightly predominating, as quoted by Riley and Chester (2) and Horne (3).

Neihof and Loeb (4) have observed that different solids which exibit both positive and negative charges in artificial sea water, all became negatively charged in natural sea water, due to the adsorption of organic material.

Therefore it seems to be very difficult, as emphasized by Parks (5) to predict the behaviour of trace metals on the base of existing adsorption models.

The aim of this experimental work is to study (in natural sea water samples) the adsorption of some trace metal ions on colloidal particles of SiO_2 (Ludox SM 30) and γ Al_2O_3 (Alon), as model substances for common oxides in marine environment.

This work is continuation of earlier work in electrolyte solution containing 0.1 M KNO_3 (6,7). Experimental data are presented in tables and in forms of Langmuir adsorption isotherms. From several existing adsorption models which are quoted by Parks (5), only surface coordination model (8,9) and hydrogen bonding model (10,11,12) will be discussed in connection with our experimental data.

II. EXPERIMENTAL

A. Materials and Methods

For adsorption studies the commercial product "Alon" (Cabot Corp., Boston) was used, which consists predominantly of γ Al_2O_3. Its specific surface area is 117 m^2g^{-1}, as determined by Huang and Stumm (13). The stock "Alon" suspension was prepared as described earlier (13). In other series of experiments the commercial product Ludox SM 30 (SiO_2) (Du Pont de Nemours, GmbH) was used. Its specific area is 210-230 m^2g^{-1}. An aliquot of one or another adsorbent was pipetted into a polyethylene polarographic vessel, which contained 50 ml of natural sea water (Banjole, Adria, Yugoslavia, S = 37,3 ‰) and added trace metal ion of total concentration $C_T < 2x10^{-6}$M. Stock solutions of Zn, Cu, Pb and Cd nitrates were prepared from Merck p.a. chemicals and tetra distilled water, and diluted solutions were freshly prepared.

Analysis was performed in the following way: Trace metal ion, with $C_T < 2x10^{-6}$, was added to 50 ml sample of natural sea water and the concentration was measured 1-2 hours until it was constant. Upon addition of an adsorbent, the concentration was measured several hours, until it remained constant.

Fig. 1 shows three typical adsorption experiments. Peak height of I vs. E(V) curve was measured until a constant value was obtained. In most cases 5-6 hours were sufficient for adsorption equilibrium and if not measurement was continued for additional several hours at room temperature. Extra pure nitrogen gas was purified through glass tube containing BTS catalyzator at 200^oC (BASF Ludwigshafen, Germany) and through washing bottles (one containing $MgCO_3$ (S), $Na_2B_4O_7$ x x 10 H_2O 10.4 g/l and 0.1 M KH_2PO_4 and another with natural sea water), to polyethylene (Kartell-Milano) reaction vessel.

B. Instruments

Pulse polarograph, Southern Harwell Merck II was used, with constant voltage sweep 0.00213 $Vsec^{-1}$ and constant HMDE surface. The paper was shifted with speed 2.54 cm/min.

Preelectrolysis were 5 min with stirring and 30 seconds of hydrodynamic stabilizing period. Preelectrolysis potentials were for Zn (- 1.3 V), for Pb (- 0.7 V), for Cd (- 1.0 V) and for Cu (- 0.6 V). Ag/AgCl electrode was used as a reference electrode. Anodic peak potentials were: Zn - 0.98 V, Cd -0.62 V, Pb -0.45 V and Cu -0.04 vs. Ag/AgCl electrode.

For copper which has anodic peak potential near to the one of mercury, the most scattered data were obtained. pH measurements were performed by pH - meter Radiometer and were aproximatelly constant during the experiment, pH = 8.1 ± 0.2.

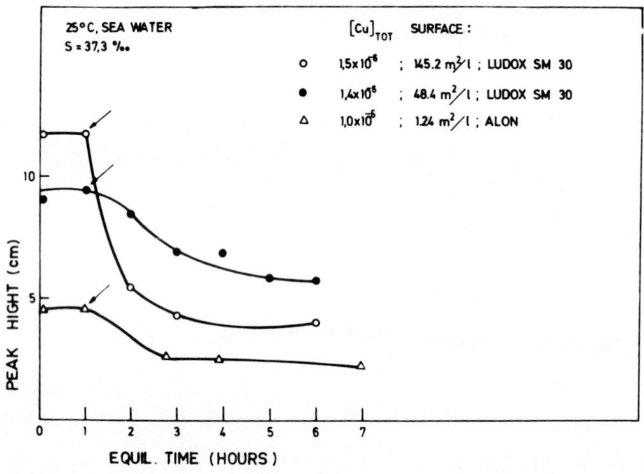

Fig. 1. Typical adsorption experiments by derivative pulse polarography. Peak height is plotted vs. equilibration time for three different $[Cu]_{TOT}$ *and surface areas of adsorbent.*

III. RESULTS

The explanation of adsorption data up to now is not unique inspite of various theories. We shall plot our original experimental data in Tables I and II, so that they can be recalculated from the point of view of the present and in in the future adsorption models[4] were developed.

The data of adsorption of each metal ion are plotted in the following order: $[Me]_{TOT}$ before ads., $[Me]_{EQ}$ after ads. Surface of adsorbent in $[cm^2 l^{-1}]$ and X, concentr, of adsorbent in $kg l^{-1}$.

TABLE I

Experimental data of adsorption of bivalent Zn, Cu, Pb *and* Cd *on* SiO_2 *(ludox SM 30), added to natural sea water after equilibration (more than 5 hours) at 20°C.*

Zn(II), SiO_2 *(ludox SM 30)*

Tot. Me(II), 10^{-7} *M*	*Equil.* Me(II), 10^{-7} *M*	*P* SiO_2 10^4 cm^2 l^{-1}	*Conc.* SiO_2 10^{-5} *kg* l^{-1}
5	2	4.84	1.21
3.5	0.5	9.68	2.42
16	5.4	9.68	2.42
17	1.3	19.36	4.84
23.5	0.86	48.4	12.1
17	0.1	484	121
8.3	2.2	9.68	2.42
13	2.6	9.68	2.42
15	5.2	9.68	2.42
11.5	5.4	6.78	1.69
7.1	2.6	4.84	1.21

Cu(II), SiO_2 *(ludox SM 30)*

13.7	9.2	48.4	12.1
10	4.7	96.8	24.2
10	4.7	96.8	24.2
16.1	2	53.24	133
14	7.05	96.8	24.2

TABLE I *(continued)*

Tot. Me(II), $10^{-7}M$	*Equil.* Me(II) $10^{-7}M$	*P* SiO_2 $10^4 cm^2 l^{-1}$	*Conc.* SiO_2 $10^{-5} kg\ l^{-1}$
19.3	8.6	96.8	24.2
12.5	7.9	48.4	12.1
15.3	5.6	145.2	36.3
20	8.2	96.8	24.2
16	12.5	38.72	9.68
18	11	58.8	14.5
15	11.8	38.72	9.68
15.8	13	38.72	9.68
5.6	3.5	48.4	12.1
Pb(II), SiO_2 *(ludox SM 30)*			
4	3.1	9.68	2.42
5.1	2.9	19.36	4.84
3.5	1.3	48.4	12.1
6.8	4.9	9.68	2.42
4.5	3.6	9.68	2.42
1.45	0.96	9.68	2.42
7.5	2.8	48.4	12.1
3.3	2.5	9.68	2.42
4.1	2.4	19.36	4.84
4.2	1.8	29.04	7.26
6.3	5.2	9.68	2.42
5.4	3.2	19.36	4.84
3.0	1.0	48.4	12.1
5.0	1.2	96.8	24.2
4.0	2.3	19.36	4.84
4.0	1.4	48.4	12.1
2.0	0.14	38.72	9.68
2.0	0.56	48.4	12.1

TABLE I *(continued)*

Tot. Me(II) $10^{-7}M$	*Equil.* Me(II) $10^{-7}M$	*P* SiO_2 $10^4 cm^2 l^{-1}$	*Conc.* SiO_2 $10^{-5} kg\ l^{-1}$
2.0	0.4	193.6	48.4
2.0	1.27	9.68	2.42
0.8	0.14	193.6	48.4
1.0	0.86	9.68	2.42
0.87	0.6	9.68	2.42
1.0	0.77	19.36	4.84
0.72	0.31	96.8	24.2
0.77	0.32	48.4	12.1
1.8	0.59	96.8	24.2
0.9	0.25	193.6	48.4
Cd(II), SiO_2 *(ludox SM 30)*			
20	17	96.8	24.2
18.5	17.2	484.2	121
7.0	6.3	96.8	24.2
9.0	7.4	96.8	24.2
10.5	9.6	484	121

TABLE II

Experimental data of adsorption of bivalent Zn, Cu, Pb *and* Cd *on* γ Al_2O_3 *(Alon), added to natural sea water after equilibrium (more than 5 hours) at 20°C.*

Zn(II), γ Al_2O_3 *(Alon)*

Tot. Me(II) $10^{-7}M$	*Equil.* Me(II) $10^{-7}M$	*P* γ Al_2O_3 $10^4 cm^2 l^{-1}$	*Conc.* γ Al_2O_3 $10^{-5} kg\ l^{-1}$
9.0	0.35	6.19	4.85
8.5	2.6	1.86	1.45
10.0	3.6	2.48	1.94
10.0	0.68	4.95	3.87

TABLE II *(continued)*

Tot. Me(II) $10^{-7}M$	*Equil.* Me(II) $10^{-7}M$	*P* γ Al_2O_3 $10^4 cm^2 l^{-1}$	*Conc.* γ Al_2O_3 $10^{-5} kg\ l^{-1}$
10.0	1.07	4.33	3.39
10.0	1.9	3.10	2.42
10.0	6.4	1.24	0.968
Cu(II), γ Al_2O_3 *(Alon)*			
9.6	3.4	3.53	2.76
7.2	2.2	6.19	4.84
7.2	4.0	1.24	0.968
8.0	2.2	12.38	9.68
7.2	1.9	9.29	7.26
5.5	1.7	4.95	3.87
9.6	8.8	0.43	0.339
9.6	8.4	0.62	0.484
10.0	8.5	0.62	0.484
6.9	5.0	1.24	0.968
10.7	8.0	0.93	0.726
Pb(II), γ Al_2O_3 *(Alon)*			
10.5	6.6	4.33	3.39
10.0	3.0	6.19	4.84
10.5	5.4	4.95	3.87
10.0	5.6	4.33	3.39
7.8	4.6	3.10	2.42
10.0	2.0	1.86	1.45
10.0	2.15	15.48	12.1
9.0	5.4	3.10	2.42

TABLE II *(continued)*

Tot. Me(II) $10^{-7}M$	*Equil.* Me(II) $10^{-7}M$	*P* γ Al_2O_3 $10^4 cm^2 l^{-1}$	*Conc.* γ Al_2O_3 $10^{-5} kg\ l^{-1}$
10.0	8.0	6.19	4.84
8.7	8.1	9.29	7.26
10.0	9.8	15.48	12.1
8.6	8.1	30.95	24.2
7.6	6.0	61.90	48.4
2.8	1.75	30.45	24.2
10.0	7.7	154.8	121.0

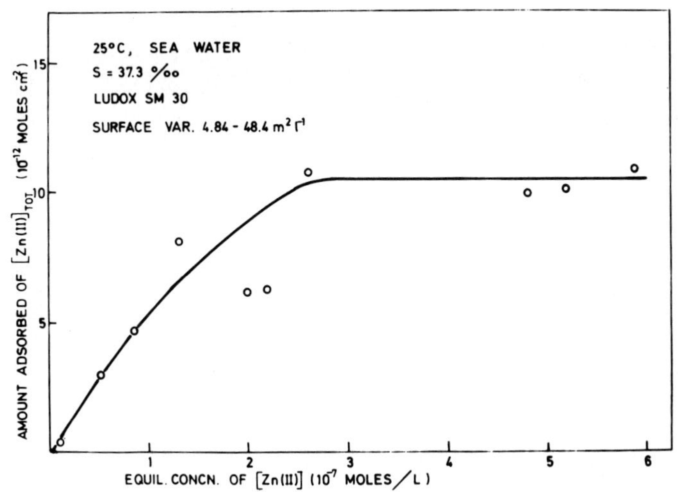

Fig. 2. Adsorption isotherm of Zn(II) *on* SiO_2 *(ludox SM 30) in natural sea water. Data are plotted according to eq. (1).*

On the basis of adsorption data presented in Tables I and II and Fig. 2-5, one can conclude the Langmuir equation can be satisfactorily applied. The corresponding curve is described by equation:

$$\Gamma = \Gamma_{\infty} \cdot C_{eq.} \; / \; (K + C_{eq.}) \qquad (1)$$

where: Γ is adsorption density of adsorbate on solid (moles cm^{-2});

Γ_{∞} is adsorption capacity (moles cm^{-2});

K is equilibrium adsorption constant.

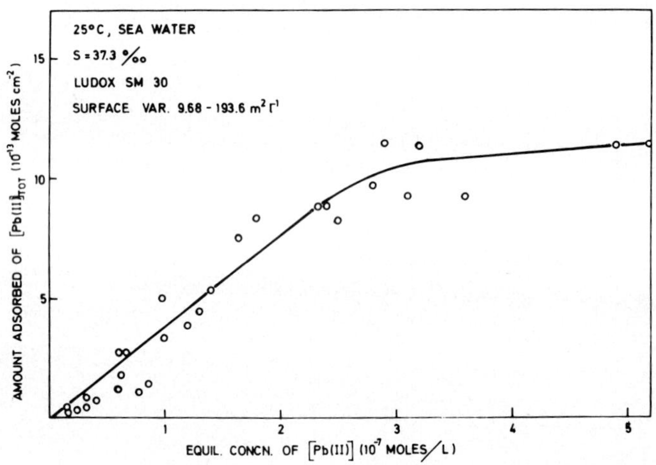

Fig. 3. Adsorption isotherm of Pb(II) *on* SiO_2 *(ludox SM 30) in natural sea water. Data are plotted according to eq. (1).*

The amount of metal ion adsorbed was calculated from experimental data as:

$$[Me]_{ADS.} = [Me]_{TOT} - [Me]_{EQ.} \qquad (2)$$

The value of Γ (mole cm^{-2}) is calculated from:

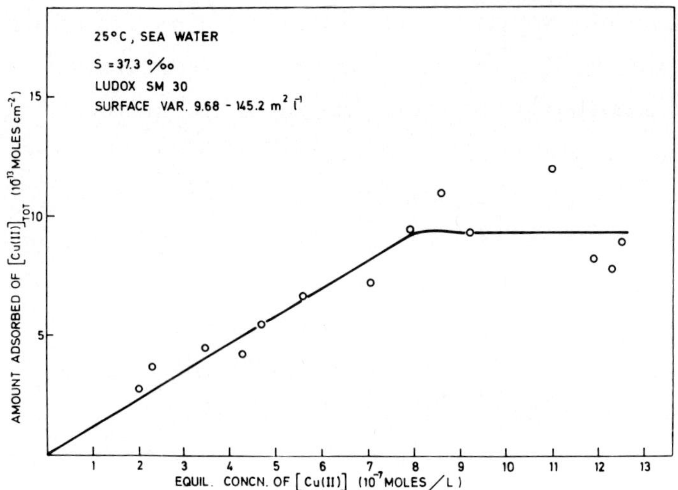

Figs. 4. and 5. Adsorption isotherms of Cu(II) *and* Cd(II) *on* SiO_2 *(ludox SM 30) in natural sea water. Data are plotted according to eq. (1).*

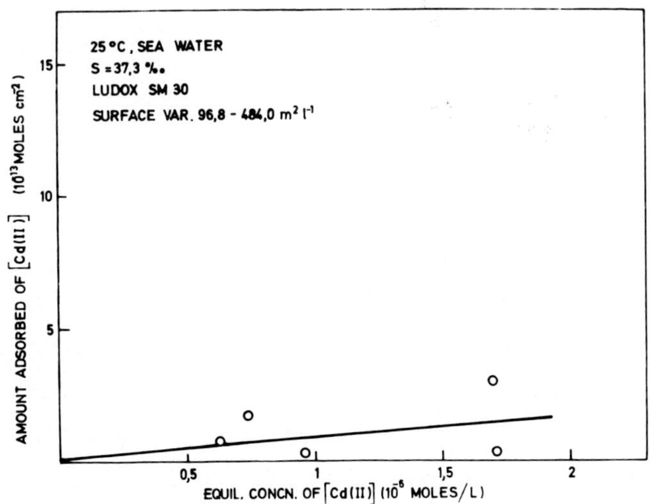

$$\Gamma = [Me]_{ADS.} / Surface \quad (3)$$

One should emphasize that a satisfactory fit of the experimental points to eq. (1) does not necessarily imply that the conditions that form the basis and mechanism of the theoretical Langmuir model are fulfilled (14).

From adsorption curves in Figs. 2-5 the value of Γ_∞ and $K \cdot \Gamma_\infty^{-1}$ were calculated as well as the specific surface occupied by one molecule of adsorbate. The results, which are plotted in Table III, show quantitatively the order in adsorption of metal ions in sea water on SiO_2 (Ludox SM 30), being Zn >> Pb > Cu >> Cd.

TABLE III

Adsorption constants and specific surface of SiO_2 *(Ludox SM 30) occupied by one molecule of metal ions of* Zn, Pb, Cu, Cd, *in natural sea water sample, after equilibration (more than 5 hours) at* 20^oC.

Metal	*Adsorbens*	Γ_∞ 10^{-12}	$K \cdot \Gamma_\infty^{-1}$ 10^4	*Specific Surface*
Zn	SiO_2 *(Ludox SM 30)*	13	12.8	$(35.7\ Å)^2$
Pb	"	1.2	4.7	$(117.6\ Å)^2$
Cu	"	0.9	0.22	$(135.9\ Å)^2$
Cd	"	*practically not adsorbed.*		

Fig. 6. shows experimental adsorption data for Zn(II), Cu(II), Pb(II) and Cd(II) on γ Al_2O_3 (Alon), plotted with form of Langmuir isotherm. A satisfactory fit of the experimental points is obtained for Zn(II) and Pb(II), while for Cu(II) a deviation exists due to greater analytical experimental error.

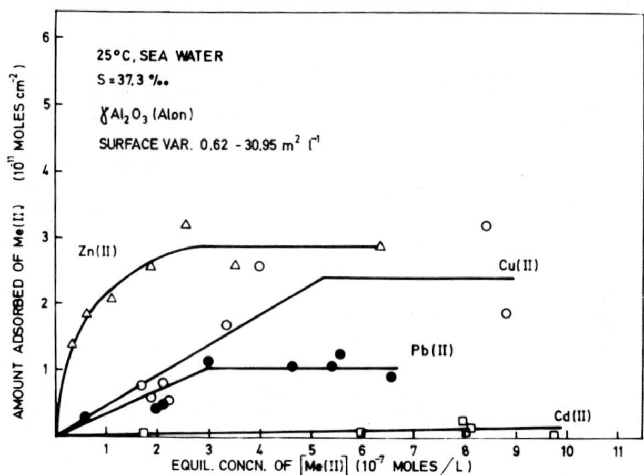

Fig. 6. Adsorption isotherms of Zn, Cu, Pb *and* Cd *on* γ Al_2O_3 *(Alon) in natural sea water. Data are plotted according to eq. (1), after being recalculated from experimental values presented in Table II.*

From adsorption curves in Fig. 6, the values Γ_∞ and $K \cdot \Gamma_\infty^{-1}$ were calculated, as well as the specific surface of γ Al_2O_3 (Alon) which is occupied by one molecule of adsorbate. Results are plotted in Table IV. The results show quantitatively the decreasing order in adsorption on Alon as follows: Zn > Cu > Pb >> Cd.

TABLE IV

Adsorption constants and specific surface of γ Al_2O_3 *(Alon) occupied by one metal ion of bivalent* Zn, Pb, Cu, Cd, *determined in natural sea water sample after equilibration (more than 5 hours) at 20°C.*

Metal	*Adsorbens*	Γ_∞ 10^{-12}	$K \cdot \Gamma_\infty^{-1}$ 10^4	*Specific Surface*
Zn	γ Al_2O_3 *(Alon)*	31	14	$(23\ \text{Å})^2$
Cu	"	24	51.6	$(26.3\text{Å})^2$
Pb	"	10	1.06	$(40.7\text{Å})^2$
Cd	"	*practically not adsorbed.*		

IV. DISCUSSION

The mechanisms of adsorption of heavy metal ions in sea water are very complex as it was shown by Parks (5). The charge on suspended solid particles can be either intrinsic to the structure and composition of the solid, arises from the interaction of the solid and water, or arises from specific adsorption of impurities.

For low charged inorganic species common in sea water (15) electrostatic contribution to free energy of adsorption, ΔG^o ads., is often smaller than contribution from covallent, hydrogen bonding or from solvation effect.

From the presented data in Figs. 2-6, it can be concluded that in natural sea water adsorption of trace metals investigated is not predominantly electrostatic and therefore does not obey the Gouy - Chapmen theory of the EDL. High selectivity observed indicates specific adsorption and Langmuir adsorption isotherm is used to fit the data. There is not universal agreement amongst authors as to the structure of the species formed, because there is very difficult to distinguish between surface coordination and hydrolysis models of adsorption as obvious from the following equations:

$$(2 \equiv AlOH) + Me^{2+} \rightleftharpoons (\equiv AlO)_2Me + 2\ H^+ \qquad (4)$$

$$(2 \equiv AlOH) + Me(OH)_2^o \rightleftharpoons (\equiv AlO)_2Me(OH)_2^{2-} + 2\ H^+ \qquad (5)$$

Both models will be discussed in connection with our experimental data. We shall try to explain, why is the second model according to our opinion more justified.

In the coordination model by Shindler (8) and Huang and Stumm (13) is adsorption of metal ions at oxide surfaces understood in terms of surface complex formation with deprotonated surfaces OH groups as ligands, according to equations:

$$(\equiv AlOH_2^+) \rightleftharpoons (\equiv AlOH) + H^+ \qquad (6)$$

$$K_{a_1}^s = \{\equiv AlOH\} \left[H^+\right] / \{\equiv AlOH_2^+\} \qquad (7)$$

$$(\equiv AlOH) \rightleftharpoons (\equiv AlO^-) + H^+ \qquad (8)$$

$$K_{a_2}^s = \{\equiv AlO^-\} \left[H^+\right] / \{\equiv AlOH\} \qquad (9)$$

$$(\equiv AlOH) + Me^{2+} \rightleftharpoons (\equiv AlOMe^+) + H^+ \qquad (10)$$

$$*\ K_1^s = \{\equiv AlOMe^+\} \left[H^+\right] / \{\equiv AlOH\} \left[Me^{2+}\right] \qquad (11)$$

$$(2 \equiv AlOH) + Me^{2+} \rightleftharpoons (\equiv AlO)_2Me + 2\ H^+ \qquad (12) = (4)$$

$$*\ \beta_2^s = \{(\equiv AlO)_2Me\} \left[H^+\right]^2 / \{\equiv AlOH\}^2 \left[Me^{2+}\right] \qquad (13)$$

The concentration condition is:

$$\left[Me^{2+}\right] + \{\equiv AlOMe^+\} + \{(\equiv AlO)_2Me\} = \left[Me\right]_{TOT} \qquad (14)$$

At the conditions of sea water is pH = 8.1. Equations (11), (13) and (14) can easily be solved with $\{\equiv AlOH\}$ as parameter. For simplicity the following new symbols are introduced:

$$\{\equiv AlOH\} = X \qquad (15)$$

Where X is concentration of surface species in mole per kg of the solid.

$$\left\{\equiv AlOMe^{+}\right\} + \left\{(\equiv AlO)_2Me\right\} / \left[Me\right]_{TOT} = y \tag{16}$$

Where y is the fraction of metal bound to surface complex and can have values $0 < y < 1$.

$$*K_1^s / 10^{-8.1} = a \tag{17}$$

$$*\beta_2^s / 10^{-16.2} = b \tag{18}$$

The solution of eqs. (11), (12) and (14), when symbols in eqs. (15 - 18) are introduced, present a third order curve:

$$y = (bx^2 + ax) / (bx^2 + ax + 1) \tag{19}$$

For assumed values of x and known values for a and b theoretical curve y, eq. (19), can be calculated.

It should approximately describe the behaviour of trace metals in sea water, if the model is justified.

Namely, equilibrium constants for surface complex formation have been determined only in 1 M $NaClO_4$ (8) and 0.1 M $NaClO_4$ (9) ionic strengths.

The complex formation constants determined in pure ionic medium at very high concentration of adsorbent and adsorbate cannot obviously predict the behaviour of trace metals in so complicated system as sea water.

We are inclined to believe that adsorption process can be rather described with eq. (5) or with:

$$(2 \equiv AlOH) + MeCO_3 \rightleftharpoons (\equiv AlOH)_2MeCO_3 \tag{20}$$

i.e. that adsorption bond is mainly hydrogen bonding.

The selectivity in adsorption on one adsorbent can be explained with different speciation of metal ions in sea water (15). Various selectivity order and amount of adsorption, which are found for different adsorbents suggest that speciation is not the only factor, but also the structural differences of surfaces play an important role.

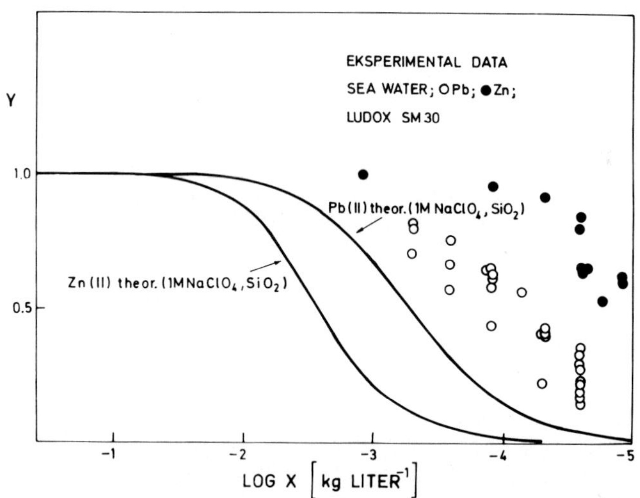

Fig. 7. Calculated theoretical curves for Pb(II) *and* Zn(II) *bound to* SiO_2*, according to surface coordination model (8). In this case is X =* {≡ SiOH} *. Experimental data from Table I are presented as points for comparison.*

The comparison of experimental points is shown in Fig.7 for Zn and Pb and calculated theoretical curves (8). Although the points seems to follow the shape of theoretical curves, the opposite selectivity order is found experimentally.

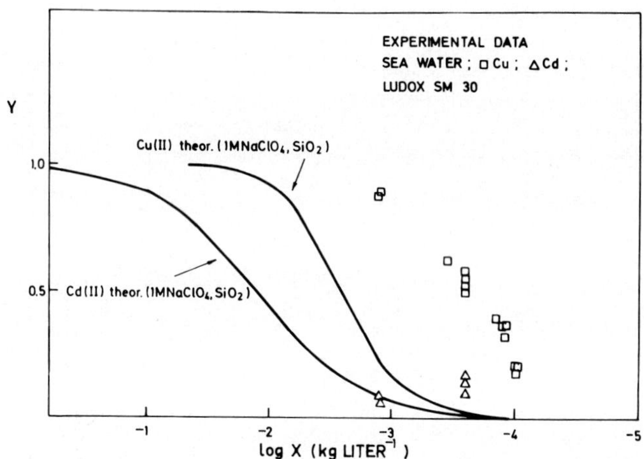

Fig. 8. Calculated theoretical curves for Cu(II) *and* Cd(II) *bound to* SiO_2, *according to surface co-ordination model (8). In this case is X =* $\{\equiv SiOH\}$. *Experimental data from Table I are plotted for comparison.*

In Fig. 8. is shown that experimentally determined adsorption of Cu(II) is much higher than predicted by theory (8). For $X < 10^{-3}$ kg l^{-1}, which are important for natural waters, cadmium is practically not adsorbed. It is also predicted by theory (8).

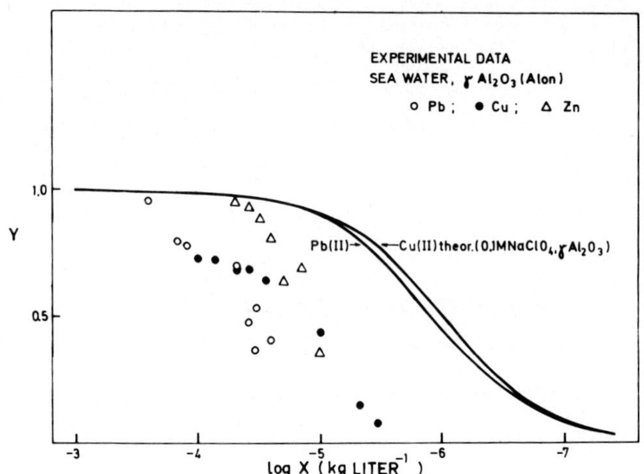

Fig. 9. Calculated theoretical curves for Pb(II) *and* Cu(II) *bound to* γ Al_2O_3 *(Alon), according to surface coordination model (8). In this case is* $X = \{\equiv Al_2O_3\}$. *Experimental data for* Zn, Cu, Pb *from Table II are plotted for comparison.*

The theoretical curves for Cu and Pb (see Fig.9) are compared with experimental data. Higher adsorption is predicted by theory than it was found experimentally. The values for a and b for Zn and Cd are not available for calculation.

It was found by Forbes, Posner and Quirk (10) for adsorption of $Hg(OH)_2^{o}$ and $HgCl_2^{o}$ on goethite at the same pH and ionic strength, that $HgCl_2$ is considerably less adsorbed, although both molecules are linear.

The same conclusion was obtained by Mc Naughton and James (16). Degens and Matheja (11) have suggested that HCO_3^- bonds to proteins with hydrogen bonds involving the amide hydrogen and carboxyl oxygen.

Some of the own results (6,7,12) indicate that adsorption of the species of the smallest charge (which can form hydrogen bonding), can satisfactorily explain the observed adsorption results.

Irreversible and very ready desorption of lead which we have found in 0.1 M KNO_3 (6) is a good indicator for weak bonding between metal ion and solid surface.

The value of Γ_∞ found for $PbCO_3^{o}$ on ludox is 1.75×10^{-12} (7), what is very near to value found in sea water, in Table III. Cadmium is present in sea water in the form of chlorocomplexes and therefore it cannot be adsorbed by hydrogen bonding.

We hope that, theoretical calculations (of N. Trinajstić and M. Marković, which are now in progress) will give the answer which type of bonding is energetically preferable.

Acknowledgement:

We thank to Prof. Werner Stumm for useful critical discussions. Also, we thank to Mr. Željko Kwokal, technical assistent, who has performed great part of experiments.

V. REFERENCES

1. Krauskopf, K.B., Geochim. Acta 9,1 (1956).
2. Riley, J.P. and Chester, R., "Introduction to Marine Chemistry", Academic Press, London and New York, 1971.
3. Horne, R.A., "Marine Chemistry", Wiley Interscience, 1969.
4. Neihof, R.A. and Loeb, G.J., Limnol. Oceanogr. 17,7 (1972).
5. Parks, G.A., "Adsorption in the Marine Environment", in Chemical Oceanography, editors Riley, J.P. and Skirrow, G., Vol 1, Academic Press London, New York, San Francisko, 1975.
6. Bilinski, H., Schindler, P., Stumm, W. and Zobrist., Vom Waser 43, 105 (1974).
7. Bilinski, H. and Stumm, W., EAWAG NEWS, Jan. (1973).
8. Schindler, P.W., Thalassia Jugosl, 11, (1/2), 101 (1975).
9. Stumm, W., Hohl, H. and Dalang, F., Croat. Chem. Acta (1976) in press.
10. Forbes, E.A., Posner, A.M. and Quirk, J.P., J. Colloid Interface Sci. 49, 403 (1974).
11. Degens, E.T. and Matheja, J., "Organic compounds in Aquatic Environments", editors Faust, S.J. and Hunter, J.V., Marcel Dekker, New York, 1971.
12. Bilinski, H., J. Inorg. Nucl. Chem. (1976) in press.
13. Huang, C.P. and Stumm, W., Surface Sci. 32, 287 (1972).
14. Stumm, W. and Morgan, J.J., "Aquatic Chemistry" p. 451 and 454, Wiley Interscience, New York, London, Sydney, Toronto, 1970.
15. The Nature of Seawater, ed. Goldberg, E.D., Dahlem Konferenzen, Berlin, March 10-15, p. 22 (1975).
16. Mc. Naughton, M.G. and James, R.O., J. Colloid Interface Sci., in press.

CHEMISORPTION OF SODIUM LAURYL SULPHATE ON γ-Al_2O_3

by

R. H. Yoon
DEPARTMENT OF ENERGY, MINES, AND RESOURCES

T. Salman
McGill University

For many oxide/ionic surfactant systems the adsorption depends on the surface charge of the solids, and the mechanisms are generally described by the coulombic attraction. The adsorption of sodium lauryl sulphate (SLS) on γ-Al_2O_3 is also charge dependent, but it may be explained by chemisorption. Infrared analyses of the adsorbed species show that the S-O vibrations correspond more closely to those of the aluminum lauryl sulphate than those characteristic for SLS or free LS^- ions. The adsorbed species cannot be desorbed by washing with water or various organic solvents even in an ultrasonic bath.

The heat of adsorption was measured in a microcalorimeter by mixing a 45 ml SLS solution with a 15 ml suspension containing 0.5 g γ-Al_2O_3. Between 0.2 and 0.4 surface coverages the molar integral heat of adsorption was found to be -1.25 Kcal/mole. This rather small enthalpy change for chemisorption is probably due to the endothermic heat effect involved in displacing water molecules from the hydrated surface.

A crystallographic model of epitactic adsorption of LS^- ions on the positively charged γ-Al_2O_3 surface is proposed. According to this model, a total of four water molecules in the first two adsorbed layers have to be displaced from the (001) cleavage plane for each LS^- ion adsorbed.

I. INTRODUCTION

The adsorption of many ionic surfactants on oxide

surfaces are charge dependent, i.e., the adsorption takes place on the oppositely charged oxide surfaces in an aqueous medium. When this is observed, the mechanism has been simply attributed to the coulombic attraction or physical adsorption(1,2). Chessick and Zettlemayer(3), however, stated that chemisorption of organic molecules on oxide surfaces is a much more extensive phenomenon than often realized. They showed that n-propyl alcohol chemisorbs at 25°C on rutile outgassed at 450°C. The chemisorption of alcohols on alumina was also shown by Hsing and Wade(4).

In the present work, the adsorption of sodium lauryl sulphate (SLS) has been studied in terms of surface charge, calorimetric heats of adsorption, infrared and desorption analyses. The results show that the adsorption is charge dependent, yet it is highly irreversible. Attempts have been made to explain the mechanism as a chemisorption process. A crystallographic model of the adsorbed species has been constructed to help understand the adsorption mechanism.

II. EXPERIMENT

A. Materials

The alumina sample used in the present work is the Linde Molecular Sieve Alumina Extrudate AB-72, which has 99.85% purity. Examination by x-ray diffraction verified that the sample is γ-Al_2O_3 with the cell edge, a_o=7.90Å. The BET specific surface as measured by nitiogen adsorption is 200 m^2/g. The alumina extrudate sample has the average pore diameter of 110 Å.

Specially pure sodium lauryl sulphate was purchased from BDH chemicals and used without further purification. Various organic solvents used for desorption tests and the NaOH and HCl used for pH adjustment were analytical grade. Double distilled water was used throughout the experiments.

B. Procedure

a. *Electrophoresis.* A Rank Brothers Particle Electrophenesis Apparatus, Mark II was used with a flat cell made of fuzed quartz. A dilute suspension of γ-Al_2O_3 of -3μ size was used for the electrophoretic measurements.

b. Calorimetric heat of adsorption measurement. A heat-flow type Calvet microcalorimeter[5] was used, which has chromel-constantan thermopiles as the sensing device. The rated sensitivity of the differential microcalorimeter was 60 μV/mw and the time constant was 400 seconds. The EMF was amplified by a Kethley Microvolt-ammeter Model 150B, which was connected to a strip chart recorder. The area underneath the curve representing the heat-flow was taken as the total heat output of a thermal reaction. At the recorder chart speed of 3mm/minute, the heat constant was 1.052×10^{-3}, 3.530×10^{-3}, and 1.0564×10^{-2} cal./cm^2 at 30, 100 and 300 μV, respectively. The calibration was made with a manganin heater immersed in paraffin oil.

The calorimetric cell was constructed from a 35 mm O.D. and 120 mm high Pyrex bottle with a 15 mm threaded mouth. Inside the cell a 14 mm O.D. glass tube was fuzed at the bottom. A 15 ml suspension containing 0.5g γ-Al_2O_3 sample was placed in the inner tube of the cell while a 45 ml SLS solution was added to the outer compartment. The pH of the suspension and that of the SLS solution were kept identical. The cell was sealed by a teflon stropper with a Viton o-ring which was held firmly in place by a Nylon threaded stopper. When thermal equilibrium is reached after placing the cell inside of the calorimeter, it was oscillated twice per minute through an arc of 180° to start the reaction. It was found that when using finely ground γ-Al_2O_3 extrudate sample (-400 mesh) the mixing was not adequate. Therefore, the samples used in the present work were coarsely crushed before use. In order to minimize the electronic noise caused by the agitation, a home-made low frequency filter was installed between the calorimeter and the amplifier. In calculating the heat of adsorption, corrections were made on the heat of dilution by following the method described by Skewis and Zettlemeyer[5]. Corrections due to the heat of mixing of γ-Al_2O_3 suspension were also made by running blank tests with distilled water replacing SLS solution.

c. Adsorption measurement. After the calorimetric measurements, the sample was centrifuzed and the clear solution was analyzed for SLS by a colorimetric method[7]. The concentration decrease was taken as the adsorption.

d. Infrared analyses and desorption tests. The centrifuzed γ-Al_2O_3 was filtered and half of it was dried under vacuum while the other half was dried after subjecting it to desorption tests by using water and various organic solvents. A one mg sample was suspended in a 220 mg KBr pellet, which was used as a window. A Perkin-Elmer Model 457 infrared spectrophotometer was used to take the spectrum.

III. RESULTS

a. Electrophoresis. From the electrophoretic measurements, the ζ-potential was calculate by using the Smoluchowski equation. As shown in Fig. 1, the point of zero charge (PZC) is found to be pH 8.1, which is close to the values reported by others[8].

b. Adsorption. Fig. 2 represents the results of adsorption tests obtained by using $5/3 \times 10^{-3}$ Moles/l SLS solution as a function of initial and final pH values. The adsorption takes place below the PZC demonstrating that it is charge dependent. The pH shift may indicate that OH^- ions are being displaced from the hydrated surface by the adsorbing LS^- ions. Fig. 3 shows that the heat of adsorption changes with pH, decreasing with increasing pH. Corrections were made for the heat of dilution of SLS based on Fig. 4.

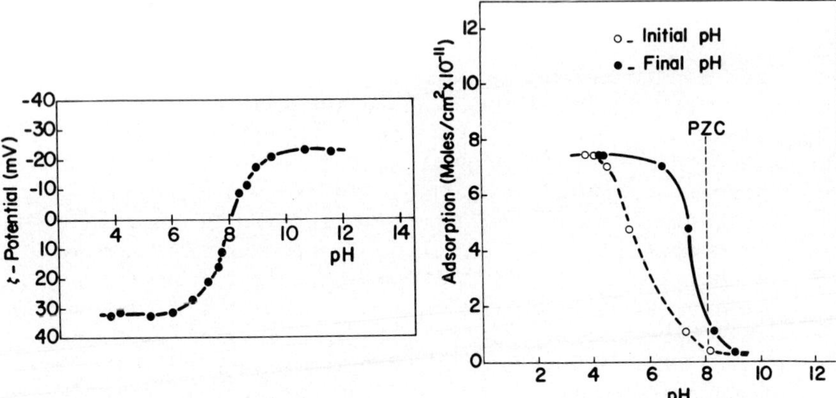

Fig. 1. ζ-potential at the γ-Al_2O_3/water inteface in the absence of supporting electrolyte as a function of pH.

Fig. 2. Adsorption of sodium lauryl sulphate on γ-Al_2O_3 as a function of pH.

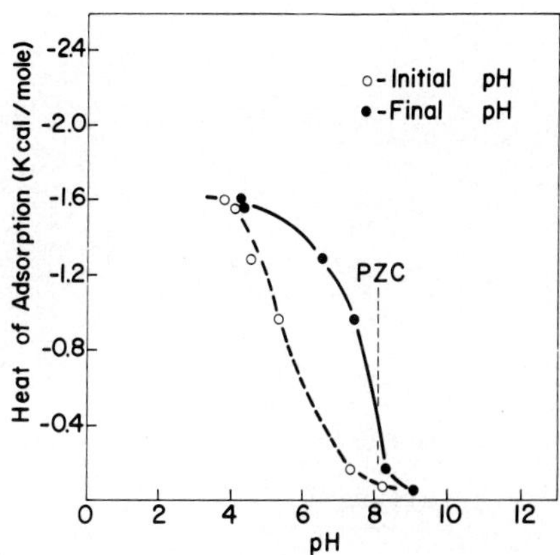

Fig. 3. Molar integral heat of adsorption of SLS on γ-Al_2O_3 as a function of pH.

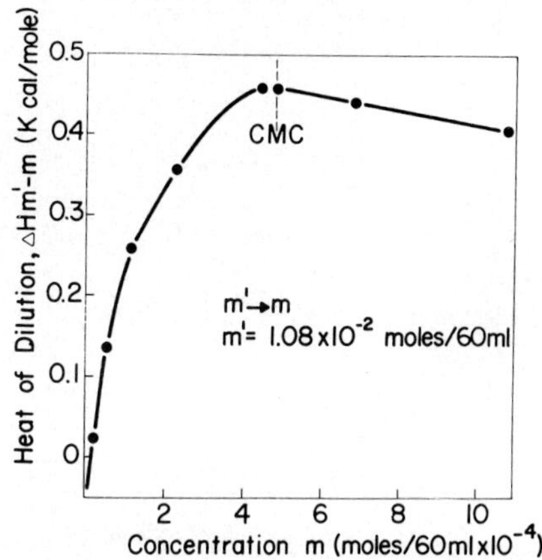

Fig. 4. Heat of dilution of sodium lauryl sulphate solutions.

The total integral heats of adsorption measured at the initial pH of 4.1±0.1 are shown in Fig. 5 as a function of surface coverage. The molar integral heat of adsorption is found to be approximately -1.25 Kcal/mole between 0.2 and 0.4 surface coverages. This value is close to the heat of adsorption of SLS on hematite (-1.36 Kcal/mole) reported by Shergold and Kitchener(9). Fig. 5 also shows that the molar integral heat of adsorption is reduced slightly when the equilibrium concentration becomes higher than the critical micelle concentration (CMC). The samples in this region are dispersed indicating micellar adsorption, while those treated at lower concentrations are well flocculated indicating a normal mode of adsorption. The fact that the CMC is reached before the monolayer coverage has been obtained is most likely due to the porosity of the sample. At very low surface coverages the molar integral heat of adsorption increases drastically with decreasing adsorption density as shown in Fig. 6.

c. Infrared analysis. Fig. 7A shows the IR spectra of SLS and related compounds. The SLS shows the degenerate stretching vibrations (ν_3) at 1250 and 1219 cm^{-1}. A doublet at 1015 and 996 cm^{-1} has been attributed by Shergold(10) to the stretching vibrations in the ˋO⁄C⁀Cˊ skeleton. The spectrum of LS^- ion obtained by using a BaF_2 cell shows two broad bands (ν_3) centered at 1245 and 1215 cm^{-1}, and ν_1 at 1062 cm^{-1}. The aluminium lauryl

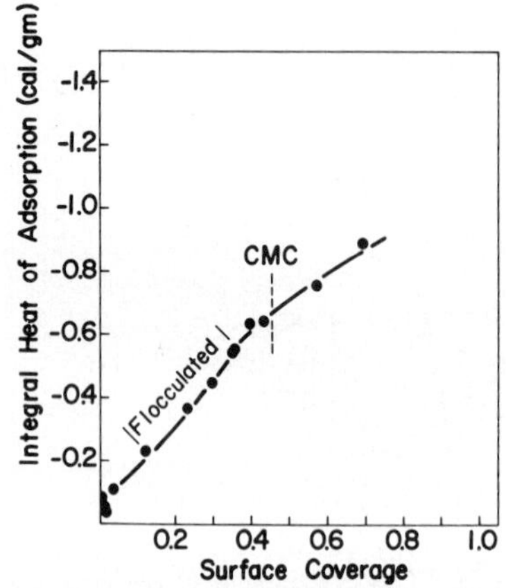

Fig. 5. Total integral heat of adsorption of SLS on γ- Al_2O_3 as a function of surface coverage at pH 4.1.

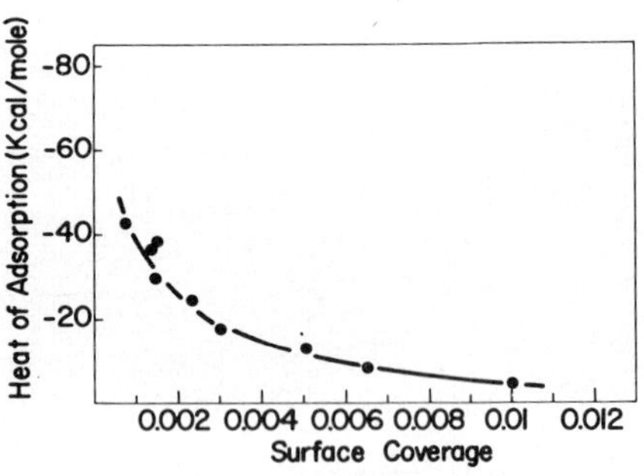

Fig. 6. Molar integral heat of adsorption of SLS on γ-Al_2O_3 as a function of surface coverage at pH 4.1.

sulphate (AlLS) shows a ν_3 triplet at 1262, 1229 and 1180 cm^{-1}, indicating C_{3v} symmetry due to coordination[11]. The ν_1 and O–C–C skeletral vibrations appear at 1068 and 970 cm^{-1}, respectively. The AlLS has been precipitated by mixing 3.3 x 10^{-3} Moles/l $Al_2(SO_4)_3 \cdot 18H_2O$ and 2×10^{-3} moles/l SLS solutions in equal volume at pH 3.6, where no precipitation of $Al(OH)_3$ occurs. It is soluble in ethyl-alcohol, but insoluble in benzene, acetone, or petroleum ether.

Fig. 7B shows the spectra of γ-Al_2O_3 samples treated with SLS solutions at various conditions. Apparently all the adsorbed species appear to have the ν_3 triplet at 1263, 1240, and 1200 cm^{-1} and ν_1 at 1065 cm^{-1}. It is difficult, however, to state whether it is a ν_3 triplet or doublet, since the γ-Al_2O_3 also shows a minor band at 1263 cm^{-1}. In any case, the adsorbed species is not identified by either SLS, LS^- ion or AlLS precipitate. Figs. 7B-b, -c, and -d confirm the results of the adsorption tests (see Fig. 2) by showing that the IR absorbance increases with increasing surface coverage. The sample conditioned at high SLS concentration (the equilibrium concentration is higher than the CMC) shows a broad band centered at 1235 cm^{-1} (Fig. 7B-d). However, after washing the sample in ethyl alcohol, it shows the same bands as the others of lower adsorption density. This indicates the micellar adsorption at high concentration, conforming to the results of the heat of

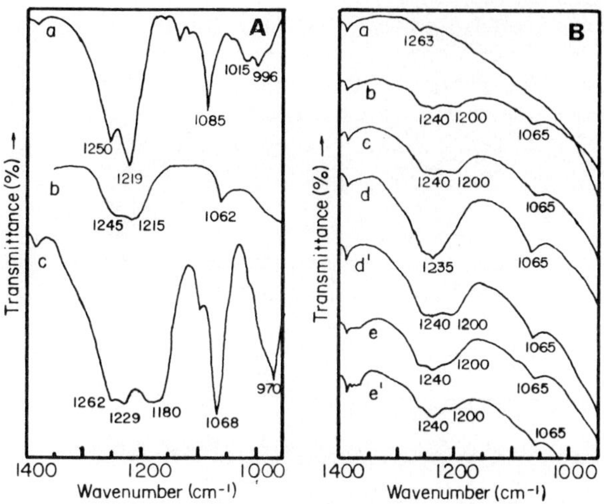

Fig. 7. (A) Infrared spectra of (a) sodium lauryl sulphate, (b) lauryl sulphate ion, and (c) aluminium lauryl sulphate. (B) Infrared spectra of (a) γ-Al_2O_3, and the samples treated with sodium lauryl sulphate solutions at the initial pH of 4.1 to give surface coverages of (b) 0.234, (c) 0.354, and (d) 0.693. d' shows the effect of washing sample d in ethyl alcohol. e and e' represent the sample treated with $5x10^{-2}$M solution of pH 4.5 (final pH 7.8) before and after washing in alcohol, respectively.

adsorption measurements (Fig. 5). The samples of lower adsorption densities, on the other hand, do not show any sign of desorption after repeated washing with water of various pH up to 11.5, alcohol, acetone, benzene, or petroleum ether. Desorption is not indicated even after the sample has been treated in an ultrasonic bath for an extended period. After drying, the samples float on the surface of water indicating that they are still hydrophobic. This clearly indicates that the SLS adsorbs strongly on γ-Al_2O_3 and the adsorption is irrevesible, i.e., chemisorption. The chemisorption appears to persist throughout the pH range of adsorption as indicated by Figs. 7B-e and e'.

IV DISCUSSION

The results obtained in the present investigation can easily be taken to suggest the physical adsorption mechanism in that the adsorption is charge dependent and the heat of adsorption is apparently small. The IR analysis reveals, however, contrary results: Firstly, the adsorbed species are not identified by either SLS, LS^- ion or lauryl sulphuric acid[10]. Secondly, the adsorption is highly irreversible as all desorption attempts have failed. Thirdly, if the spectrum of the adsorbed species could be considered as showing a ν_3 triplet at 1263, 1240 and 1200 cm^{-1}, this indicates the coordination of sulphate group[11].

Chemisorption was defined by Chessick and Zettlemoyer[3] as a process of creating new electronic states by either partial or whole electron transfer. In the γ-Al_2O_3/SLS system, the only electron transfer process would be between the surface aluminium ion and the adsorbing LS^- ion. In order to explain this possibility, the charging mechanism of γ-Al_2O_3 will be discussed with respect to the electronic state of the aluminium ion exposed on the surface.

a. Charging Mechanism of γ-Al_2O_3

Parks and deBruyn[12] have suggested that the PZC of an oxide should be at the pH of minimum solubility, at which the iso electric point of the solution (IEPaq) usually occurs. The solubility diagram of γ-Al_2O_3 (Fig. 8) has been constructed from the thermodynamic data compiled by Latimer[13], Stull[14], Sillen[15], and Hem[16]. The IEPaq in the γ-Al_2O_3/water system occurs at pH 7.85, which is close enough to the experimental PZC (pH 8.1) to substantiate the minimum solubility theory.

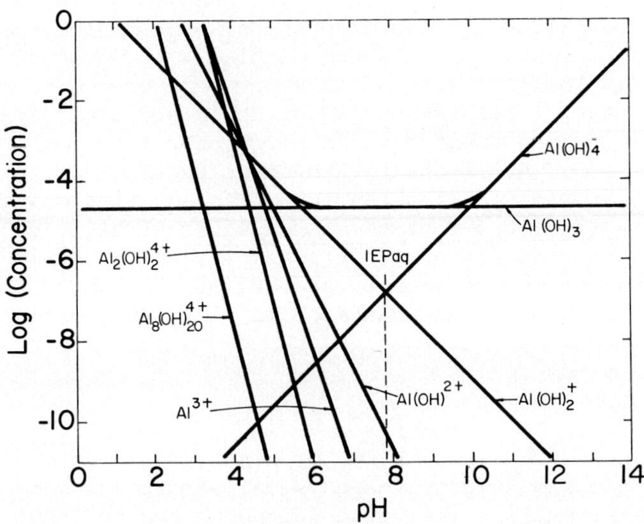

Fig. 8. Solubility Diagram of the γ-Al_2O_3/H_2O system.

The chief competitor of this mechanism is the adsorption of H^+ and OH^- ions on surface hydroxyl groups(8). For γ-Al_2O_3, the following equations may be written:

$$\underline{Al} - OH + H^+ \rightleftharpoons \underline{Al}^+ - OH_2 \quad (1)$$

$$\underline{Al} - OH + OH^- \rightleftharpoons \underline{Al} - (OH)_2^- \quad (2)$$

to indicate an equilibrium between the positive, neutral, and negative sites. The underscored symbols refer to the species that are included in the surface lattice. In Eq. (1) which represents the formation of a positive site, the $\underline{Al}$ species is shown to aquire the effective charge of +1. This is against the general convention that the positive sign is allocated to the proton of the water molecule(8).

b. Adsorption Mechanism

The heat of adsorption of SLS on γ-Al_2O_3, which has been found to be -1.25 Kcal/mole between 0.2 and 0.4 surface coverages, may be too small for a chemisorption despite the fact that the heat of adsorption increases drastically at lower surface coverages (Fig. 6). Shergold and Kitchener(9) have also reported a small heat of adsorption (-1.36 Kcal/mole) for the Fe_2O_3/SLS system. They have assumed that the predominant contribution toward the heat of adsorption comes from the hydrocarbon chain association. This is not likely the case, however, considering that the heat of precipitation of $Fe(LS)_3$ is only -1.65 Kcal/mole as reported by Mellgren, et. al.(17). The heat of precipitation of aluminium lauryl sulphate is also found to be in the same order of magnitude (-1.0 Kcal/mole of LS^- ion) as that of $Fe(LS)_3$. This small heat of precipitation is not likely to indicate the residual bonding between the $Al^{3+}.nH_2O$ and the LS^- ions, since the IR spectrum (Fig. 7A-c) shows a ν_3 triplet characteristic for coordination of the sulphate group(10). The small heat of precipitation is most likely due to the endothermic heat of dehydration involved in coordinating the $^-OSO_3$-R group with the hydrated aluminium.

The same reasoning could be used in explaining the small heat of chemisorption of LS^- ions on the positively charged γ-Al_2O_3 surface. Based on Eq. (1), the overall surface reaction may be written,

$$\underline{Al}^+ - OH_2 + LS^- \rightleftharpoons \underline{Al} - LS + H_2O \quad (3)$$

to indicate that the LS^- ion displaces H_2O from the positive site, which must be an endothermic reaction. Considering that the $Al^{3+}(aq)$ ion is one of the most strongly hydrated species, the small heat of chemisorption (-1.25 Kcal/mole) is not surprising. A higher heat of adsorption may, then, be expected on less strongly hydrated substrates. A substantially higher heat of adsorption has been measured for the $Cu(OH)_2$/SLS system as will be published elsewhere[18]. Sulphide minerals are known to be less strongly hydrated than most oxides: The heat of chemisorption of xanthate on galena has been reported to be approximately -10 Kcal/mole[19]. In general, heats of adsorption from the aqueous phase are lower than those from the gas phase or from non-polar organic solvents. This may be ascribed to the effect of hydration. According to Somonjai[20] the heats of chemisorption from the gas phase vary widely between -15 and -200 Kcal/mole. On account of the effect of hydration, the drastic increase in heat of adsorption at very low surface coverages may suggest that there are small fraction of sites that are not as strongly hydrated as the majority of the surface.

c. *Proposed Model of the Surface Compound.*

According to Balandin[21], there are two factors that determine the catalytic activity, namely the 'geometrical fit' of the reactant on the catalyst surface and the 'energetic fit' in terms of optimal bond strength. The latter has been given much more attention in discussing adsorption mechanisms. There are, nevertheless, a few investigations emphasizing the role of geometrical factors. Gaudin[22] has proposed a lattice fit of oleate on fluorite, and it has been later proved to be chemisorption by Peck[23]. Xanthat is well known to chemisorb on galena and Hagihara[24] has shown the geometrical fit. According to Groszek[25] the preferential adsorption of n-paraffin on graphite is also due to the geometrical fit.

It would be interesting to see if the geometrical fit theory can be applied to the γ-Al_2O_3/SLS system. The γ-Al_2O_3 may be represented by the formula $^{iv}Al^{3+}_{8}\ ^{vi}(Al^{3+}_{13\ 1/3}\square_{2\ 2/3})\ O_{32}$ showing the defect spinel structure. (The itallics denote the coordination number). When the crystal is immersed in water, the O^{2-} in the surface lattice will be protonated to become OH^- or H_2O species. The degree of protonation will determine the surface charge depending on the pH. At a relatively low H^+ concentration, protonation will be limited to those oxygens that form dangling bonds on the surface, in which case the positive site may be represented as $\underline{Al}^{v+} - OH_2$ following Pauling's electrostatic valency principle[26]. V is the ideal bond valence defined as the quotient of cationic charge by the coordination number[27]. For the 6- and 4- coordinated aluminium ions, v is $\frac{1}{2}$ and $\frac{3}{4}$, respectively. At relatively high H^+ concentration, protonation will proceed to other oxygens coordinated to the aluminum ions in the surface lattice. In this case, the positive site could aquire the effective charge of +1, as shown in Eq. (3). The protonation of more than one oxygen around a metal ion has been shown in Boehm's[28] model of hydrated anatase.

Fig. 9 represents the possible model of the surface compound formed by the adsorption of DS^- ions on the positively charged (001) plane of the γ-Al_2O_3. Most of the oxygens on the surface are assumed to have transformed into OH^- or H_2O species due to protonation, so that the aluminium ions in the positive sites aquire the effective charge of +1. The sulphate group is assumed to be a regular tetrahedron with the S-O distance 1.65Å [29]. Each $^-OSO_3R$ group is shown to have an edge-on orientation to 'fit' the atomic arrangement of the surface: Energetically, one of the two oxygens that touch the surface is coordinated to the $^{vi}\underline{Al}^+$-. The $^{iv}\underline{Al}$- species, which would have been at the corners and face centres of the unit cell, are considered to have gone into solution as hydrated aluminium ions. Geometrically, the oxygens of the $^-OSO_3R$ group are shown to occupy those crystallographic positions that would have been occupied by the oxygens of the crystal, if it would have continued. In this manner, the adsorbed species would take up the lowest potential energy sites of the surface. The area per adsorbed species is 31.6Å^2, which is larger than the cross sectional area (21Å). The present model is similar to the one that has been proposed for $Cu(OH)_2$/SLS system[31].

The model shows that each sulphate group displaces a total of four water molecules from the first two adsorbed layers, which gives rise to the endothermic heat effect discussed earlier. Most of this heat effect should be accounded for by the displacement of the water molecule in the positive site ($^{vi}\underline{Al}^+$ - OH_2), and the other three will contribute little as they are weakly bound to the surface. Some of the latter species that are displaced could be a combination of H_2O, H_3O^+, and OH^- depending on the pH. At relatively high pH toward the PZC, considerable amount of OH^- ions could be displaced and raise the pH during the course of adsorption as has been shown in Figs. 2 and 3. The pH drift could also be ascribed in part to the solubility of the γ-Al_2O_3. It may be stated, here, that the chemisorption of $^-OSO_3R$ species takes place only on the positive

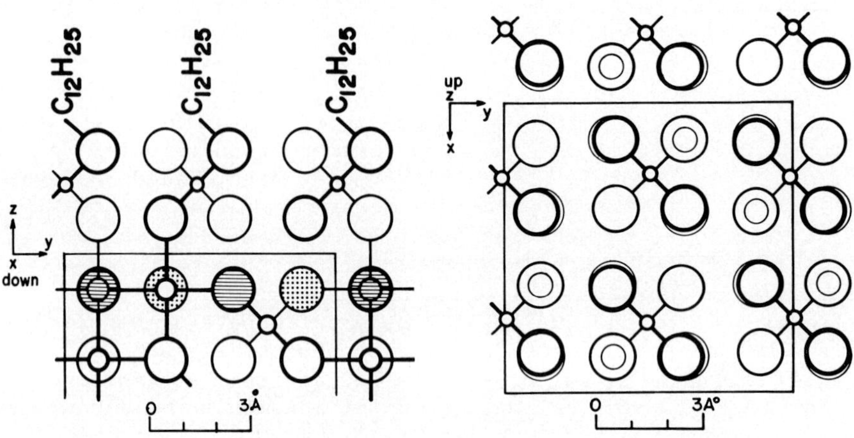

Fig. 9. Proposed model of the surface compound formed by the adsorption of sodium lauryl sulphate on γ-Al_2O_3: The largest circles represent oxygen atoms, shaded circles represent hydroxyl groups, and dotted circles represent water molecules. The smallest circles represent sulphur, the larger circles represent aluminium. The line thickness indicates the relative depth within the unit cell.

site ($^{vi}\underline{Al}^+$ - OH_2) but not on the neutral ($^{vi}\underline{Al}$ - OH) or negative sites ($^{vi}\underline{Al}$ - $(OH)_2^-$). This is most conceivably because the $^{vi}\underline{Al}$ - OSO_3R bond energy is larger than the energy required to rupture the $^{vi}\underline{Al}^+$ - OH_2 bond but smaller than that required to rupture the $^{vi}\underline{Al}$ - OH bond. However, the chemisorption on neutral or negative sites could be possible, when a stronger adsorbate such as a chelating agent is used.

It is difficult to explain the decrease of heat of adsorption with increasing pH (Fig. 3) on the basis of the chemisorption mechanism. A few explanations may, nevertheless, be made: Firstly, coadsorption around the chemisorbed species may be possible through hydrocarbon chain association. The slight sign of desorption of the sample conditioned at higher pH (Fig. 7B-e') may support this view. Secondly, the $^{vi}\underline{Al}$ - OSO_3R bond could be weakened due to the repulsion exserted by the oxygens that are not protonated at higher pH. Finally, nonstiochiometric adsorption of DS^- ions on the positive sites of fractional charges such as $^{vi}\underline{Al}^{\frac{1}{2}+}$ - OH_2 may be postulated. These sites could be present at higher pH as explained in a previous paragraph. According to Bérnard[30] the nonstiocheiometric surface compound is formed when adsorption takes place on localized sites. Chemisorption of this type may fit the definition of chemisorption describing the partial transfer of electrons(3).

V. SUMMARY AND CONCLUSIONS

Adsorption of SLS on γ-Al_2O_3 can be described as a chemisorption mechanism, despite the fact that the adsorption is charge-dependent and the heat of adsorption is only -1.25 Kcal/mole between 0.2 and 0.4 surface coverages. The apparently small heat of adsorption can be explained by the endothermic heat involved in displacing water molecules from the hydration sheath. The chemisorbed species could not be removed from the surface by washing with water or various organic solvents.

A proposed crystallographic model of the surface compound displays both the geometrical and energetic fit of the $^{-}OSO_3R$ group on the positively charged (001) plane of γ-Al_2O_3. The model is found useful in explaining the adsorption phenomena with respect to the surface hydration and the distribution of charge sites.

VI. ACKNOWLEDGEMENT

Authors wish to acknowledge discussions with Dr. F. Rinfrét, Dr. G. Donnay, Dr. A. A. Robertson, Dr. I. S. Butler, Dr. D. Patterson, and Mr. L. L. Sirois. Drawings were prepared by J. D. MacLeod. The present work is a part of RHY's PhD. thesis.

VII. REFERENCES

1. Iwasaki, I., Cooke, S.R.B., and Colombo, A.F., Report of Investigations 5593, U.S. Bureau of Mines, 1960.
2. Choi, H.S., and Whang, K.U., CIMM Bulletin, P. 466 (1963).
3. Chessick, J.J., and Zettlemoyer, A.C., "Proc. 3rd. Int. Congr. of Surface Activity", Vol. II, p. 174, 1960.
4. Hsing, H.H., and Wade, W.H., "Chemisorption of Alcohols on Alumina", presented at the 103rd AIME Annual Meeting, Dallas, Texas, 1974.
5. Gravelle, P.C., in "Advances in Catalysis", (D.D. Eley, H. Pines, and Paul B. Weisz, Eds.), Vol. 22, p. 191. Academic Press, New York, 1972.

6. Skewis, J.D., and Zettlemoyer, A.C., "Proc. Third Int. Congr. on Surface Activity", Vol. II, p. 401, 1960.
7. Krush, F., and Sonenberg, M., Aanalytial Chemistry 22, 175 (1950).
8. Parks, G.A., Chemical Review 65, 177 (1965), and in "Advances in Chemistry Series", 67, 121 (1967).
9. Shergold, H.L., and Kitchener, J.A., Int. J. of Mineral Proc. 2, (1975).
10. Shergold, H.L., Trans. IMM 81, C148 (1972).
11. Nakamoto, K., "Infrared Spectra of Inorganic and Coordination Compounds", 2nd Ed., p. 173, Wiley-Intercience, New York, 1970.
12. Parks, G.A., and deBruyn, P.L., J. Phys. Chem. 66, 967 (1962).
13. Latimer, W.M., "Oxidation Potentials", 2nd Ed., p. 282, Prentice-Hall, Englewood Cliffs, N.J. 1956.
14. Stull, D.R., "JANAF Thermochemical Tables", 1965.
15. Sillen, L.G., "Stability Constants of Metal-Ion Complexes", The Chemical Soc. London, 1964.
16. Hem, J.D. "Trace Inorganics in Water", Adv. in Chem. Ser. 73,'68.
17. Mellgren, O., Gochin, R.J., Shergold, H.L., and Kitchener, J.A., Proc. 10 the Int. Mineral Proc. Cong. p. 451 (1974).
18. Yoon, R.H., and Salman, T., "Heats of Adsorption of Ionic Sufactants on Metal Hydroxides", to be presented at the Symposium on Polymers at Interfaces, Montreal, Sept. 1976.
19. Mellgren, O., AIME Trans. 235, 46 (1966).
20. Somorjai, G.A., "Principle of Surface Chemistry", p. 244, Prentice-Hall, Englewood Cliffs, N.J., 1972.
21. Krylov, O.V., "Catalysis by Nonmetals", p. 93, Academic Press, N.Y. 1970.
22. Gaudin, A.M., Eng. and Mining J. 146 (12), 91 (1945).
23. Peck, A.S., Report of Investigations 6202, U.S. Bureau of Mines, 1962.
24. Hagihara, H., J. Phys. Chem. 56, 616 (1956).
25. Groszek, A.J., Proc. Roy. Soc. Lond. A. 314, 473 (1970).
26. Evans, R.C., "An Introduction to Crystal Chemistry", 2nd Ed., p. 180, Cambridge Univ. Press, 1965.
27. Donnay, G., and Allman, R., Am. Mineralogist 55, 1003, (1970).
28. Boehm, H.P., in "Advances in Catalysis", Vol. 16, p. 179, Academic Press, New York, 1966.
29. Wyckoff, W.G., "Crystal Structures", 2nd Ed. Vol. 3, p. 18, Interscience Pub., New York, 1965.
30. Bérnard, J., in "Problems of Nonstoichiometry", (A. Rabenau, Ed.), p. 209, Am. Elsevier, New York, 1970.
31. R.H. Yoon, and T. Salman, 103rd AIME Meeting, Dallas, 1974.

CONTINUOUS PHASE TRANSITIONS IN KRYPTON MONOLAYERS ON GRAPHITE AND BORON NITRIDE

Frederick A. Putnam and Tomlinson Fort, Jr.
Carnegie-Mellon University

ABSTRACT

Several laboratories have now confirmed by both LEED and vapour pressure experiments that there are two high-density phase transitions in krypton monolayers on graphite, and at least one on hexagonal boron nitride. Improved calculations of the thermodynamic functions in the neighborhood of the phase transitions are presented. The shapes of the singularities confirm that the transitions are continuous and not first order. It is argued that the transitions are actually manifestations of a single phenomenon, the tendency of the liquidlike monolayer to crystallize in registry with the adsorbate lattice. The classical model of Tsien and Halsey exhibits two phase transitions, but predicts them to be of first order. The phenomenological Landau theory of phase transitions also predicts first order transitions. Non-classical models which correctly predict the continuous nature of the transitions are discussed.

THE INTERACTION OF NON SPHERICAL MOLECULES WITH HOMOGENEOUS SURFACES

Robert A. Pierotti and Alvin C. Levy
Georgia Institute of Technology

ABSTRACT

High precision, high temperature data for the adsorption of nitrogen and carbon dioxide on the highly graphitized carbon black P-33(2700) are presented. Second and third gas-solid virial coefficients are determined from the data at a number of temperatures in the range 150 to 350°K. Two-dimensional virial coefficients are calculated from the gas-solid virial coefficients. Angle-averaged potential functions are used to obtain a potential of average force which is then used to evaluate the gas-solid interaction parameters for the systems studied. The effect of hindered rotation on gas-solid virial coefficients is investigated as well as are surface area determinations based upon high temperature adsorption techniques using non-spherical absorbates.

TEMPERATURE PROGRAMMED DEHYDRATION OF HYDROXYAPATITE

Frederick S. Baker, Ralph A. Beebe, Helga Furedi-Milhofer and Vladimir Hlady

Amherst College and Ruder-Boscovic Institute

The temperature programmed dehydration for a sample of near-stoichiometric hydroxyapatite has been investigated over the range from 20° to 700°C. Two experimental procedures have been employed: one (TPD) in which the water vapor is carried away from the sample in a dry helium stream and observed by a thermal conductivity detector and two (MTPD) in which the sample is subjected to a vacuum of the order of 10^{-6}-10^{-7} torr and water vapor is detected by a mass spectrometer. The hydroxyapatite was prepared under carefully controlled conditions by precipitation from aqueous medium. The hydroxyapatite has been characterized by various available techniques such as x-ray diffraction, electron microscope and infrared studies, as well as analysis for Ca/P atom ratio.

Two major peaks in both the TPD and MTPD spectra are observed: (1) near 90°C which is attributed to desorption of reversibly adsorbed water on the external surface of the apatite particles and (2) near 225°C which we attribute to the irreversible removal of water from ultrafine pores in the solid. The MTPD technique has been employed to demonstrate that the hydroxyapatite in the hydrated condition undergoes H_2O-D_2O exchange when soaked in heavy water. This applies to both the 90° and the 225° peaks.

RELATIVE INTERNAL AND TRANSLATIONAL ENERGY ACCOMMODATION OF MOLECULAR GASES ON SOLID SURFACES.[1]

Gerd M. Rosenblatt, Richard S. Lemons, and Clifton W. Draper
The Pennsylvania State University

I. ABSTRACT

A method has been developed to measure both translational and internal energy accommodation in gas-surface collisions from the surface temperature rise which occurs upon ultrasonic oscillation of a solid sample in a rarefied test gas. Thin, flat samples are vibrated normal to their exposed faces at rms velocities of ~3 x 10^4 mm sec^{-1} (1). About 35% more gas-surface collisions occur when the sample moves into the gas than when it recedes, so that, relative to the surface, incident gas molecules have more translational energy than do surface-equilibrated molecules. Gas-surface energy transfer causes the surface temperature to rise. A steady-state is attained due to thermal emission, conduction, and the fact that incident gas molecules are cooler internally than the heated surface (gas internal modes are independent of the surface motion). Analysis of the energy balance enables determination of both translational (γ_{trans}) and internal (γ_{int}) energy accommodation coefficients from measured steady-state temperature rises as a function of gas pressure. For the diatomics O_2 and N_2 on Cu, Pt, and Ni, $\gamma_{rot} \simeq 0.1\ \gamma_{trans}$. For the polyatomics CO_2, C_2H_6, and C_6H_6 on Cu, $\gamma_{int} \simeq (0.6-0.9)\ \gamma_{trans}$. Assuming, for the polyatomics, that $\gamma_{rot} \simeq \gamma_{trans}$ leads to values of γ_{vib} which agree with spectroscopic determinations (2).

II. REFERENCES

1. Lemons, R. S., and Rosenblatt, G. M., Surface Sci. 48, 432 (1975).
2. Black, G., Wise, H., Schechter, S., and Sharpless, R. L., J. Chem. Phys. 60, 3526 (1974)

---

1. Supported by the National Science Foundation

NMR STUDIES OF ADSORBED HYDROGEN ON SUPPORTED Pd AND Pt CATALYSTS*

L. E. Iton
Argonne National Laboratory

ABSTRACT

NMR measurements of hydrogen adsorbed on supported metal catalysts have been made for the first time by careful compensation for the support interference. Proton Knight shift (K, relative to methane) and cw linewidth measurements between 80°K and 340°K allow discrimination of high-temperature chemisorbed and low-temperature physisorbed (very weakly chemisorbed) species on both Pd and Pt.

The 10% Pd on carbon and 5% Pt on carbon catalysts had metal particles in the 15-60 Å diameter range. For 154 torr H_2 equilibrium pressure at 296°K, apparent activation energies for diffusion (ΔE_{diffn}) of the chemisorbed species in the 295-340°K range were estimated at $\sim$0.9 and $\sim$2.4 kcal/mole, respectively, for Pd and Pt. At T = 296°K, K $\sim$ + 9.5 ± 1.5 ppm, and - 50 ± 3 ppm, on Pd and Pt surfaces, respectively, reflecting the partial filling up of the d-band holes in Pd by dissolution of hydrogen. Adsorbed and absorbed (β-hydride) protons are discriminated for Pd by the temperature dependence of the linewidth; the observed K decreases above 296°K, but is temperature insensitive between 200-296°K. Reported K values for β-hydride protons in massive particles is much larger under comparable H_2 pressure.

Below 160°K, new adsorbed species appear. On Pd, the apparent ΔE_{diffn} $\sim$ 0.26 kcal/mole, and the observed K $\sim$ 3.2 ppm at 158°K, decreasing to a near zero value as T is lowered further. ΔE_{diffn} $\sim$ 0.4 kcal/mole for H_2 on the support in this range. On Pt (120 torr H_2) at 159°K, observed K $\sim$ - 21 ppm, but falls to a near-zero value at lower temperature.

*Based on work performed under the auspices of the U. S. Energy Research and Development Administration.

RAMAN SCATTERING FROM MOLECULES CHEMISORBED AT SILICA SURFACES

A. H. Hardin, *Energy Research Laboratories, CANMET*
and
B. A. Morrow, *University of Ottawa*

ABSTRACT

Raman scattering was observed for a series of inorganic halides and methyl halides chemisorbed on the surface of Cab-O-Sil HS5 silica. For Si-CH_3Cl_3 three bands were observed in the - CH_3 region at positions very near the valves observed by infrared methods. A very weak band was observed near 460 cm^{-1} on top of the broad SiO_2 envelope from the cell window and sample adsorbate. A similar result was obtained for $SiCl_4$. No distinct bands were found for BCl_3 or BF_3 after reaction with silica degassed and dehydrated in vacuum at 800°C. When $TiCl_4$ was reacted at 20°C, strong bands were observed at 390 cm^{-1} and 410 cm^{-1} (liquid $TiCl_4$, 386 cm^{-1}). The 390 cm^{-1} band decreased with sample evacuation at 20°C. The 410 cm^{-1} band increased in intensity when the reaction was carried out at higher temperatures. We attribute the band at 390 cm^{-1} to physisorbed $TiCl_4$ and the band at 410 cm^{-1} to chemisorbed - $Ticl_3$. The weakness or absence of bands for $SiCl_4$, BCl_3 and BF_3 may be due to strong coupling of the adsorbate modes with the bulk vibrations of the adsorbate.

ISOSTERIC HEAT OF ADSORPTION OF OXYGEN ON SILVER

Alvin W. Czanderna
Clarkson College of Technology

ABSTRACT

Adsorption isotherms have been measured on cleaned silver powder from 178° to 339° at oxygen pressures of 10^{-3} to 300 Torr using a vacuum ultramicrobalance. Adsorption equilibrium was found at all temperatures and pressures studied. The surface was prepared for the reproducible chemisorption studies and cleaned before the determination of each isotherm by an established method.[1] Seven isotherms were measured that spanned the fraction of the surface covered, θ, from 0.2 to 1.25. The isosteric heat of adsorption, q, was determined by application of the Clausius-Clapeyron equation at constant values of θ. From θ of 0.2 to 0.45, q decreases from 50 to 18 kcal/mole; it remains constant at 18 kcal/mole from 0.45 to about 0.85 and then decreases to zero at the highest coverages. The isosteric heat, the activation energy of adsorption,[2] and the activation energy of desorption[3] are used to support a model that dissociative chemisorption occurs on silver up to θ of 0.45 and that further dissociative and molecular chemisorption occurs above θ of 0.45. The nature of the chemisorbed states of oxygen on silver will be discussed in terms of models of mobile adsorbed species with lateral interactions and a two dimensional surface compound.

1. A. W. Czanderna, J. Phys. Chem. 70, 2120 (1966).
2. A. W. Czanderna, J. Phys. Chem. 68, 2765 (1964).
3. W. Kollen and A. W. Czanderna, J. Colloid Inter. Sci. 38, 152 (1972).

CHEMISTRY AT CATALYST SURFACES: THE OXIDATION OF OF SO_2 ON PLATINUM

C. Vayenas and H. Saltsburg
University of Rochester

ABSTRACT

One difficulty in elucidating reaction mechanisms of heterogeneous catalytic systems results from the failure to observe <u>*directly*</u> *the properties of surface species during the interaction with the catalyst.*

Utilizing a high temperature solid electrolyte (yttria-stabilized zirconia) one can construct an oxygen concentration cell. One electrode simultaneously functions as both electrode and catalyst and measurement of the cell EMF allows one to measure the oxygen activity of the catalyst. Using Pt to catalyse the oxidation of SO_2*, the oxygen activity,* a_{O_2}*, has been measured* <u>*on*</u> *the working catalyst while simultaneously observing the steady state chemical kinetics in a flow reactor. It has been found that the oxygen activity of the Pt is driven by gaseous* SO_2 *from the value characterized by equilibrium with gas phase oxygen to extremely low values at temperatures below 600°C. However, the product* $(P_{SO_2})_{gas}\,(a_{O_2})^{1/2}_{surface}$ *is a function only of temperature over the range 370-970°C. In the upper temperature range the gaseous oxygen activity is reached, the specific temperature limit being a function of the* SO_2 *pressure. This "equilibrium" together with the observed kinetics for the production of* SO_3*, (first order in* SO_2*), allows one to formulate a much less ambiguous reaction mechanism and sheds new light on the properties of the species involved in the catalytic reaction.*

THERMAL DECOMPOSITION OF METHANOL ADSORBED ON Al_2O_3 †

J. M. White and T. Matsushima
The University of Texas

The kinetic behavior of CH_3OH adsorbed on Al_2O_3 powder was studied by a thermal desorption method combined with a deuterium tracer technique in an ultra high vacuum system. CH_3OH is desorbed at low temperatures. Near 500°K and higher, H_2CO, H_2O, and CH_3OCH_3 become significant. Above 700°K, CO is predominant. H_2 is desorbed around the same temperature as H_2CO and CO. CO_2 is very small.

Ether desorbed from co-adsorption of CD_3OD and CH_3OH contains only CH_3OCH_3, CH_3OCD_3, and CD_3OCD_3 below 500°K, while the deuterium distribution in ether becomes random above this temperature. Hydrogen exchange between adsorbed methanols (methoxide) occurs above 500°K.

Thermal decomposition of preadsorbed CH_3OH in the presence of gas phase CD_3OD produces primarily CH_3OCH_3 from which we conclude that ether is formed through an interaction between two adsorbed methoxides. The reactivity of adsorbed methoxide will be discussed.

† Supported in part by the Office of Naval Research.

Abstract

THE EFFECT OF ADSORPTION TEMPERATURE ON FORMIC ACID FLASH DECOMPOSITION

John L. Falconer
University of Colorado

and

Robert J. Madix
Stanford University

Previous flash desorption studies of the decomposition of formic acid on clean Ni(110) following adsorption at 37°C indicated that following adsorption the formic acid intermediate decomposed autocatalytically upon heating the nickel. The present study used the techniques of flash desorption, isotopic labeling and Auger spectroscopy to study the decomposition of HCOOD and DCOOH following adsorption at -60°C on clean Ni(110).

The four reaction products from HCOOD decomposition were H_2, CO_2, CO and D_2O. The water product from decomposition was found, by comparison to the desorption of adsorbed water, to be reaction and not desorption limited. It was also found to be formed only from the acid hydrogen's of formic acid with a reaction order between zero and one.

The decomposition to form CO_2 and H_2 was not autocatalytic as it was for 37°C adsorption. The CO_2 and H_2 products were formed at a higher reaction rate with broader desorption peaks and a complex dependence on initial adsorbed coverage.

These experiments indicate the critical importance of adsorption temperature in flash decomposition studies.

KINETICS OF NH_3 DECOMPOSITION ON SINGLE CRYSTAL PLANES OF Pt

D. G. Loffler and L. D. Schmidt
University of Minnesota

ABSTRACT

The rates of NH_3 decomposition on (111), (100, (110), and (210) planes of Pt have been measured between 600 and 1300°K and 10^{-2} and 1 torr in a flow system with mass spectrometric gas analysis. Surfaces were in the form of thin discs which were heated by a focussed light beam to achieve uniform temperature. The (210) plane has the same activity as polycrystalline wires or foils. The rates decrease in the order (210) > (110) > (111) > (100) with the rate on (100) being at least 10 times less than (210). Temperature and pressure dependences agree quite well with a Langmuir Hinshelwood expression from which reaction and adsorption activation energies on each plane are obtained. For (100) and (111) the rates are so low that reaction may be occuring only on the edges of the crystals.

THE CATALYTIC BEHAVIOR OF A Cu/Ni SURFACE ALLOY

R. J. Madix, *Stanford University*
T. J. Dickinson, *Washington State University*
and
D. Ying, *Stanford University*

ABSTRACT

The use of modern techniques of surface physics allows the study of the relationship between the structure and composition of solid surfaces and their catalytic activity and selectivity. In this work the decomposition of formic acid on a (110) oriented single crystal alloy with the bulk composition 90% nickel and 10% copper was studied. The surface composition was measured by Auger spectroscopy to be 35% nickel-65% copper, in accord with previous work of Helms, Spicer and Yu. The decomposition reaction was studied by flash desorption spectroscopy following adsorption of the formic acid at 200K.

The decomposition products observed were H_2, CO_2, H_2O and CO. Formic acid itself also desorbed at a low temperature as the surface was heated, indicating a rather weak binding energy. The products were observed to occur in three distinct temperature regions; namely, 200-300K, 300-450K and 500-600K. The dominant reaction products, H_2 and CO_2, were formed simultaneously in the middle range of temperature by a single rate-limiting step with a first-order rate constant, $k = 2.10^9 \exp. \{-19 \text{ kcal/gmole}/RT\}$. The CO_2/CO selectivity was nearly 10:1.

The characteristics of the decomposition on the alloy surface showed no resemblance to the decomposition on <u>clean</u> Ni (110) even though the bulk composition was 90% nickel. Qualitatively, the mechanism of decomposition, the activity and selectivity, resembled the decomposition on a Ni(110)(4x5)C surface (surface carbide) studied previously in our laboratory. These results indicate that (1) the catalytic behavior of metals depends primarily on the composition of the outermost layer of atoms and only secondarily on bulk composition and (2) that metallic alloys formed with group 1B elements may show catalytic properties similar to metal carbides.

FRACTURING OF A POROUS CATALYST PARTICLE FOLLOWING WETTING BY A REACTIVE LIQUID

Arthur S. Kesten, Pierre J. Marteney and
Joseph J. Sangiovanni
United Technologies Research Center

ABSTRACT

The long term effectiveness of packed bed catalytic reactors can be limited by degradation of the catalyst particles due to particle breakup. Breakup caused by large pressure gradients has been illustrated under conditions in which a reactive liquid (hydrazine) wetting the outside surface of a catalyst particle blocks the escape of gaseous decomposition products. Wetting by the very reactive hydrazine results in exothermic decomposition and buildup of gaseous decomposition products near the gas-liquid interface. Pressure buildup is very rapid, while dissipation of pressure throughout the particle is slow because of the very small pore sizes and correspondingly high pressure drops. Pressure can be alleviated if (a) gas pushes liquid out of the pores, (b) gas escapes from pores which have not been wet, or (c) the particle fractures. The greater the tendency of the particle to wet and the higher the fraction of the particle which is wet, the more likely the particle is to fracture.

In order to permit visual investigation of the wetting and the subsequent breakup of particles, an experiment was developed using high speed motion pictures to examine breakup in a flow reactor in which the catalyst particle can be isolated from and then exposed to the hydrazine flow stream. Single particle tests have been performed as a function of liquid temperature and flow rate and catalyst particle size and history. Effects of these variables on the frequency and mechanism of breakup are discussed.

POLYMER COLLOIDS AS HETEROGENEOUS CATALYSTS

R. M. Fitch and C. Gajria
University of Connecticut, Storrs

ABSTRACT

The surface chemistry of synthetic polymer colloids is being investigated with particular emphasis on their potential ability to serve as "tailor-made" heterogeneous catalysts. In this study the kinetics of the "self-destruction" of surface polymer, catalyzed by strong acid surface groups, have been observed.

Aqueous latexes of poly(methyl methacrylate) (PMMA) and poly(methyl acrylate) (PMA) were prepared by polymerization of the corresponding monomers using a redox initiator. The strong acid end-groups derived from the initiator migrate to the surface of the colloidal particles during the polymerization. This surface concentration was measured by conductometric titration of the latex in the acid form.

The kinetics of the hydrolysis of the surface acrylate ester repeating units was studied as a function of surface concentration of end-groups and the nature of the counter-ions. The rate of hydrolysis was followed by measuring the concentration of the carboxyl groups formed by conductometric titration as a function of time. Pseudo-zero order kinetics were observed for a given latex. The rate depended on the surface concentration of H^+ *strong acid counter-ions.*

Differences between PMMA and PMA behavior, energies of activation and practical applications also will be discussed.

CONFIGURATIONS OF TERRACE-LEDGE-KINK ARRAYS: LEED AUGER STUDIES OF UO_2(100) VICINAL SURFACES*

W. P. Ellis and T. N. Taylor
University of California
Los Alamos Scientific Laboratory

ABSTRACT

The crystallography of UO_2(∿100) vicinal surfaces has been examined by LEED for correlation with previous UO_2(∿111) studies.[1] *Auger electron spectroscopy was utilized to monitor surface cleanliness, and optical transforms were employed as an indication of the types of diffraction displays expected. Correlations between LEED patterns and optical simulations were especially useful when characterizing surfaces with high kink and ledge densities. For the (∿111) surfaces under dynamic conditions of surface transport, elevated temperatures were seen to produce microfacets and rotated ledges. For vicinal (∿100) surfaces reported here for the first time, the {711} for example is a terrace-ledge array with ledges parallel to the <110>. The simplest configuration consisting of equally spaced steps of minimum height was stabilized by prolonged exposure to 10^{-4} Pa O_2 at 530°C. A total of five (∿100) surfaces have been examined under dynamic conditions and will be discussed. These studies are preliminary to an examination of the catalytic properties of UO_2 vicinal surfaces in the manner used to examine stepped Pt surfaces.*[2]

1. *W. P. Ellis, Surface Sci. 45 (1974) 569.*
2. *R. W. Joyner, B. Lang and G. A. Somorjai, J. Catalysis 27 (1972) 405.*

Work performed under the auspices of the United States Energy Research and Development Administration.

RATE OF PENETRATION OF A FLUID INTO A POROUS BODY III. GLASS FIBRES AND TiO_2 POWDER

Robert J. Good and N.-J. Lin
State University of New York at Buffalo

ABSTRACT

Data are reported which confirm the empirical significance of the modification of the Washburn equation, previously derived by the senior author. The rate of uptake of liquid by a porous body, which has been equilibrated with the vapor of the liquid, is less than that of the same system which has been outgassed, by as much as 15%, depending on the liquid and on the thickness of the adsorbed film. At coverage less than one monolayer, water and n-butanol are more effective than hydrocarbons and methanol in reducing the rate of penetration. The results are discussed in terms of the film pressure as computed from adsorption isotherms.

I. INTRODUCTION

A theory has recently been derived [1,2,3], which showed that at least part of the surface free energy of an outgassed, porous solid can be converted into kinetic energy of a penetrating liquid. This amounted to a modification of the Washburn theory of pore penetration [4,5]. The prediction was made [1,2], that liquids should penetrate an outgassed, porous solid at an appreciably faster rate than an equilibrated solid.

---

[1] Good, R. J., Chem. Ind. (1971), 600.

[2] Good, R. J., J. Coll. and Interface Sci., 42, 473 (1973).

[3] Good, R. J., and Lin, N.-J., J. Coll. and Interface Sci., 54, 52 (1975).

[4] Washburn, E. D., Phys. Rev., 17, 374 (1921).

[5] Rideal, E. K., Phil. Mag., 44, 1152 (1922).

A confirmation of that prediction was reported in 1974 [3], using data obtained with glass capillaries. The immediate objective of the present work was to complete the test of that prediction, by extending the study to packed powder beds and sintered bundles of fibres.

With the materials employed in this work, the hydraulic radius was small enough that the gravitational force on the liquid was negligible compared to the capillary force. So the rate of penetration was independent of the orientation of the sample.

II. THEORY

The modified Washburn equation, for zero contact angle, is [2,3]

$$\frac{d(x^2)}{dt} \leqq \frac{r(\gamma_{\ell v} + \gamma_{s(t=0)} - \gamma_{sv})}{2\eta} \tag{1}$$

where x is the distance penetrated, r the pore radius, η the viscosity, $\gamma_{\ell v}$ the liquid-vapor interfacial tension, γ_{sv} the surface free energy at the saturated solid-vapor interface, and $\gamma_{s(t=0)}$ the surface free energy of the solid as it exists at zero time in the experiment. In using Eq. (1), we assume that the effect of gravity on flow can be neglected. If π_o is the film pressure of the adsorbed gas at the solid-gas interface at zero time, and π_e is that for the surface equilibrated with the saturated vapor, then the excess driving force term is

$$\gamma_{s(t=0)} - \gamma_{sv} = \pi_e - \pi_o \tag{2}$$

If dissipative processes, other than the friction of the liquid behind the meniscus, take place during the advance of the liquid (for example, evaporation from the liquid surface, diffusion down the pore, and adsorption on the solid) then all the excess driving force may not be used in accelerating the liquid. An efficiency factor, α, may be employed in Eq. (1) to convert it to an equality; and we may write, combining (1) and (2),

$$\frac{d(x^2)}{dt} = \frac{r[\gamma_{\ell v} + \alpha(\pi_e - \pi_o)]}{2\eta} \tag{3}$$

It was shown in [3] that if there is a film, of thickness δ which is comparable to r, on the solid, then the right side of (3) must be multiplied by $r/(r - \delta)$. This may be an important correction in the case of the saturated solid, if the hydraulic radius is of the order of 10^3Å or less. Applying this modification to (3) and integrating,

$$x = r\sqrt{\frac{[\gamma_{\ell v} + \alpha(\pi_e - \pi_o)]t}{2\eta(r-\delta)}} \tag{4}$$

The volume of liquid in n pores, filled to distance x, is

$$v = \pi r^2 nx$$

$$v(t) = \pi n r^3 \sqrt{\frac{[\gamma_{\ell v} + \alpha(\pi_e - \pi_o)]t}{2\eta(r-\delta)}} \tag{5a}$$

$$= K_v t^{1/2} \tag{5b}$$

In order to interpret the results of treatment in terms of effects on π and α, it is better to compare runs in terms of K^2 rather than K.

$$K_v^2 = \frac{\pi^2 n^2 r^6 [\gamma_{\ell v} + \alpha(\pi_e - \pi_o)]}{2\eta(r-\delta)} \tag{6}$$

The percent change in K_v^2, based on the rate constant for the baked-out solid, is

$$\Delta K_v^2\% = \frac{100\ \alpha(\pi_e - \pi_o)}{\gamma_{\ell v}} \tag{7}$$

if δ is small. This method of handling data has the advantage that the deviation of the real, porous solid from the model of a bundle of a parallel capillaries, expressed as a multiplicative tortuosity coefficient, cancels out.

To treat gravimetric data on penetration, we note that

$$K_w = \rho K_v \tag{8}$$

so that the time dependence of the weight w is

$$w(t) = K_w t^{1/2} \tag{9}$$

provided that flow into the pores is the only process that is going on.

III. EXPERIMENTAL

A. Equipment

For the gravimetric method, braided bundles of glass fibres were employed. Each bundle was sintered, to maintain dimensional stability during wetting. The fibres were suspended from a Cahn electrobalance, above a beaker of the test liquid. The support for the beaker was raised with a servomotor drive (see Fig. 1) arranged so that the drive cut

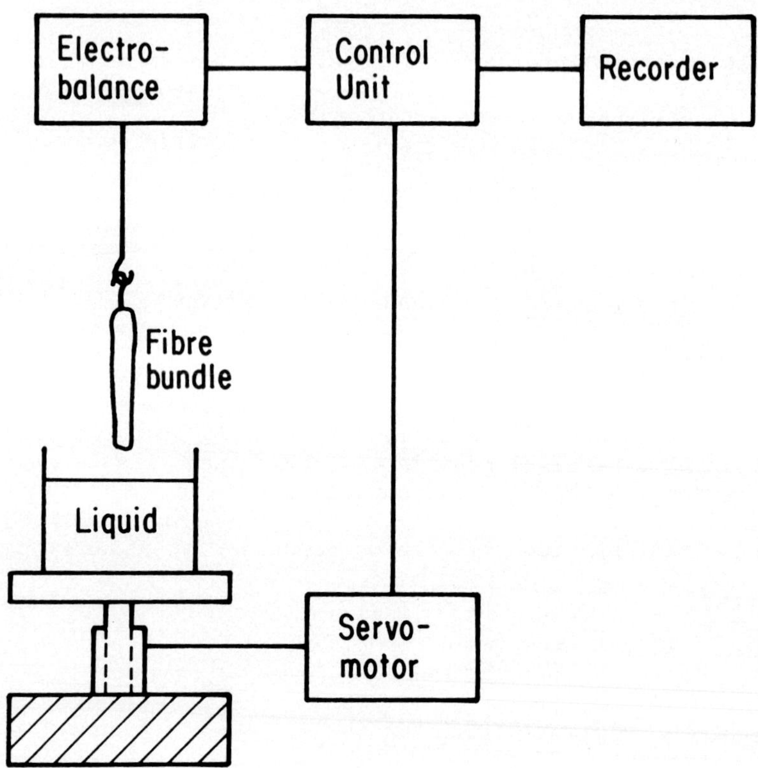

Fig. 1. Gravimetric Apparatus for Sintered Braid of Fibres.

off when the balance registered a sudden increase in weight. (This automatic control was only marginally better than manual control.) The change in weight, from the time when contact was made, was recorded; it was graphed as w vs. $t^{1/2}$, to determine K_w, see Fig. 2. The initial transient, which

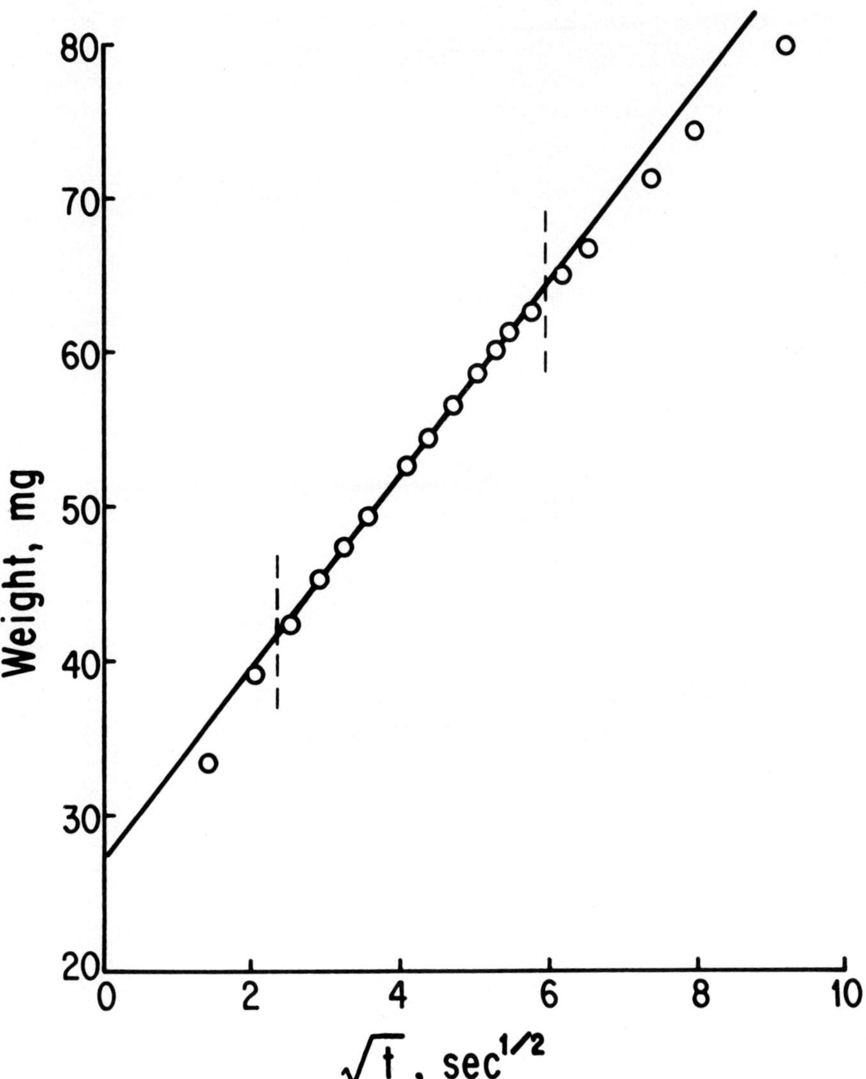

Fig. 2. Dodecane Uptake by Glass Fibre Bundle.

corresponded to bulk flow up the outside of the bundle to form a curved meniscus, is not shown. The fact that points in the first section of the curve lie below the straight line was attributed to flow of a very thin film, creeping up the outside of the bundle. The downward curvature in the last section was attributed to evaporation from the surface of the bundle, which became significant when the external surface area that was saturated with liquid became large. Because of this evaporation, low-boiling liquids could not be used in the gravimetric studies. A least squares analysis was made, of the data in the central portion of each run, for w vs. $t^{1/2}$.

Fig. 3 shows, schematically, the volumetric apparatus.

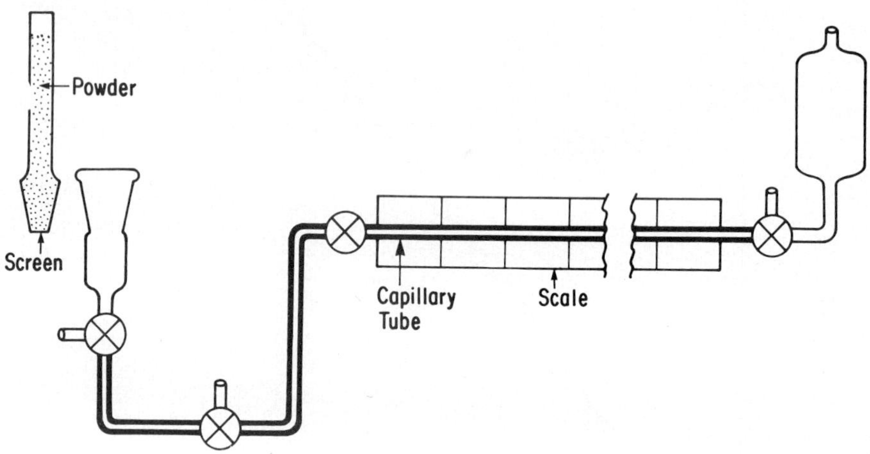

Fig. 3. Volumetric Apparatus for Powder.

The powder sample was contained in a 14/35 ground glass joint, with a stainless steel screen at the bottom for support. The volume-measuring part of the equipment was exactly horizontal. After the liquid was allowed to make contact with the powder, the stopcock at the far right was opened to the air, so that a column of air was drawn in, and the end of the column of liquid could be observed as it moved past the marks on the scale.

Volumetric runs were made in alternation, and analysed in sets of three: A pre-saturated run and the baked-out runs that preceded and followed it. This was to minimize the effects of changes in packing with run number. There was a trend towards decreasing K_V for baked samples, over a sequence of up to 20 sets of runs on a sample.

B. Materials

The TiO_2 powder was rutile, kindly furnished by Dr. G. D. Parfitt, of Tioxide International, Ltd. Its nitrogen surface area was 8.6 m^2/gm. It was packed into the sample tube, with gentle tapping to settle it. The average porosity was 0.758 ± 0.020, corresponding to a hydraulic radius of 853 ± 22 Å (95% confidence level). The samples were outgassed at 200°C and about 0.5 torr, for 10 hours, and cooled to room temperature under vacuum. Adsorbed films were laid down by flowing nitrogen, which was within 1% of saturation with the vapors of the test liquid, over the sample for various lengths of time. The saturation of the gas was accomplished with a thermostatted train of bottles, through which the gas was bubbled. The amount adsorbed was determined by weight.

The glass fibres were from National Filter Media Co., style 132-009-01. Their specific surface area was about 0.2 m^2/gm. They were formed into small braids. From the apparent density, the hydraulic radius was estimated to be about 40,000 Å. The braids were cleaned by boiling for 1 hour in 70% HNO_3, washed, and sintered at 675°C for 24 hours. After cooling and exposure to air, and between penetration runs, they were baked out for 10 hours at 200°C and 0.5 torr. The saturation of the samples was done with the same equipment used for the TiO_2. The surface area was too small for accurate weighing of the amount adsorbed, at low coverage, so a constant saturation time (2 hours) at constant gas flow rate was employed, and it was assumed that film thickness was constant.

The pore size was small enough, for both the glass and the rutile systems, that the gravitational force on the penetrating liquids was small compared to the capillary force.

The liquids used were: distilled water; Fisher Certified grades of benzene, toluene, n-hexane, methanol, 1-butanol, 1-octanol, Fisher Spectranalyzed n-heptane and iso-octane, and Eastman n-decanol.

C. Results

Table I shows the average excess rate of uptake of liquids by the glass fibre braids, expressed as $\Delta K_w^2\%$.

Table I

Liquid	Average ΔK_w^2%	Confidence limits for 99.5% confidence level
n-dodecane	3.8	± 2.1
1-octanol	6.9	± 2.8
1-decanol	9.9	± 3.5

These results establish that outgassing causes a statistically significant acceleration of the flow of liquid into a porous, silicate glass body, and confirm the preliminary results [3], which were obtained with single glass capillaries.

Table II shows the results for rutile powder, for multilayer films of various adsorbates.

Table II

TiO_2 powder with multilayer films, vs. outgassed powder, ΔK_v^2%.

Liquid	Range of film thickness, δ, Å*	ΔK_v^2%	Confidence limits	At level, %
Benzene	40 - 50	11.3	± 0.04	99
Toluene	20 - 25	4.8	± 2.5	95
n-Hexane	40 - 70	2.2	± 1.9	90
n-Heptane	60 - 80	12.3	± 3.7	99
iso-Octane	50 - 65	11.1	± 2.0	99
Water	7.5 - 12.5	5.3	± 3.7	95
Methanol	30 - 50	4.34	± 2.9	99
1-Butanol	6.5 - 12.5	5.9	± 1.1	95

*δ based on weight adsorbed, and assumption that density of adsorbed film equals density of bulk liquid.

The ranges of average film thickness, δ, were chosen from the graphs of the data, to represent the upper regions of coverage attained. (No great effort was made to achieve uniform ranges of δ for this purpose. Layers of water and butanol thicker than 12.5 Å were not obtained under the conditions and saturation times employed.) As with the glass fibres, it is clear that there is a statistically significant effect on rate of penetration, due to outgassing vs. pre-saturation.

The increase of ΔK_V with film thickness was notable: see Figs. 4 and 5. In Fig. 4, for n-heptane, the results fit the

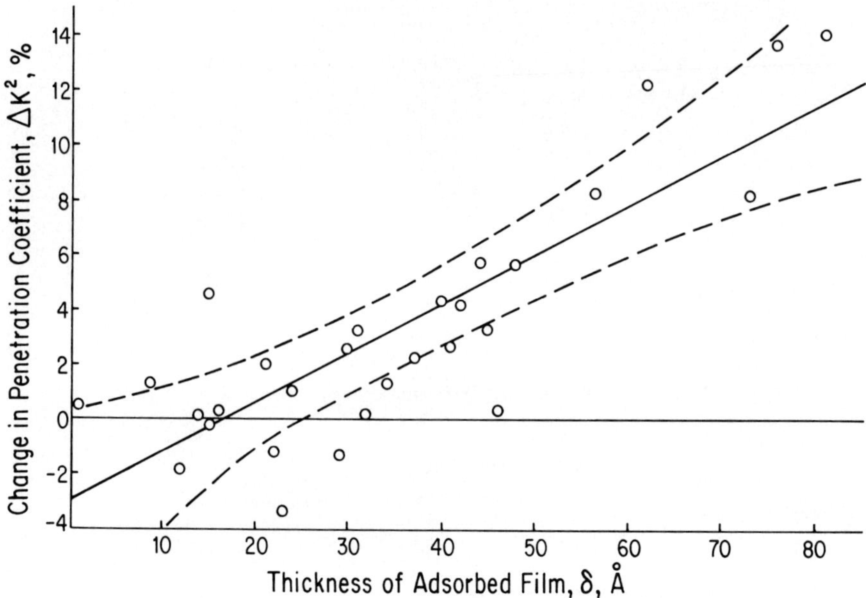

Fig. 4. Effect of film thickness, δ, on change in penetration coefficient, $\Delta K_v^2\%$, for n-heptane into rutile powder bed.

equation,

$$\Delta K_v^2\% = -2.9 + 0.18\delta$$

At the 95% level, the intercept is not significantly different from zero. The slope is greater than zero, at the 99.5% confidence level. It is interesting that there is no significant indication of ΔK_v^2 levelling off at high coverage, even out to over 60 Å. The same was true of all the other liquids. Table III shows the results of linear regression analysis of the data.

Table III

Regression analysis of TiO_2 data. $\Delta K_v^2\% = a + b\delta$.

Liquid	a, %	b_δ %/Å	Confidence level for conclusion b >0	b ≃ 0	a not >0	a > 0
Benzene	4.1	0.15	95%		95	
Toluene	-0.9	0.31	99.5		99.5	
n-Hexane	-2.4	0.08	97.5		99.5	
n-Heptane	-2.9	0.18	99.5		99.5	
iso-Octane	3.2	0.14	99.5			99.5
Water	3.4	0.24		90		99.5
Methanol	1.4	0.10	95		90	
n-Butanol	4.2	0.30	90			95

Fig. 5 shows the data for water. It would be expected

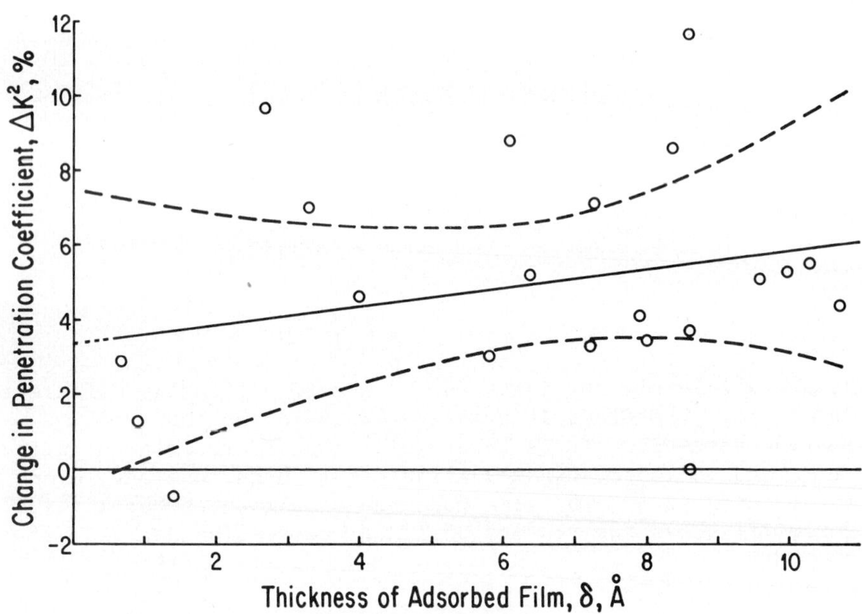

Fig. 5. Effect of film thickness on change in penetration coefficient, $\Delta K_v^2\%$, for water into rutile powder bed.

that water would be more strongly adsorbed than alkanes, on high-energy sites, and would consequently be more effective in reducing the surface free energy from its maximum value, γ_s, at low coverage. This would lead to a "shoulder" in the curve of ΔK^2 vs. δ, and to a non-zero intercept of the straight line extrapolation to zero thickness. In spite of the scatter, we can assert the presence of a non-zero intercept at the 99% confidence level. The data for n-butanol, (with which relatively few runs were made) resemble water, in this regard. Methanol, however, resembled the hydrocarbons, in having an intercept that was not distinguishable from zero. It is notable that, for water and n-butanol, values of ΔK_v^2% which were significantly greater than zero were obtained in the sub-monolayer region of coverage. With the hydrocarbons, only iso-octane behaved in this way; the average value of ΔK_v^2 was about 4%, for δ between zero and 10 Å. For the rest of the hydrocarbons, ΔK_v^2 was approximately zero, for δ less than monolayer thickness.

IV. DISCUSSION

The scatter of data is appreciable, in this study, particularly in the volumetric measurements. We will not, at this time, estimate the extent to which differences between liquids on a solid are intrinsic, vs. being due to experimental error. The major part of the scatter for any single liquid was probably caused by fluctuations in the packing density. (Graphs of K_v^2 vs. run number, for baked samples, showed a scatter about a smooth curve which corresponded roughly to the distribution of values of ΔK_v^2% about a straight line.) Another possible cause of scatter is imperfect attainment of uniform distribution of adsorbate over the whole powder sample.

The scatter of the data was not so serious, however, as to interfere with our drawing the following conclusion: The modified version of the Washburn equation [1,2] is confirmed.

We may generalize, that when accuracy of better than, perhaps, ± 10% is required, explicit consideration should be given to the effect of adsorbed films on the driving force for penetration of porous bodies by liquids. Our unexpected finding of the dependence of rate on film thickness in the high multilayer region for hydrocarbons, and on the presence of sub-monolayer quantities of water or n-butanol, show that this may be quite a serious matter.

Table IV gives estimates of $\alpha(\pi_e - \pi_o)$ for the systems studied, together with two systems from Ref. 3.

Table IV

Apparent excess driving force for penetration, $\alpha(\gamma_S - \gamma_{SV})$, for baked out surfaces vs. nearly saturated surfaces.

Liquid		$\alpha(\gamma_S - \gamma_{SV})$
(a) In glass systems		
Benzene	capillary*	0.4
Dodecane	fibre braid	0.8
n-Octanol	capillary*	3.1
n-Octanol	fibre braid	1.9
n-Decanol	fibre braid	2.8
(b) Rutile powder, large film thickness δ		
Benzene		2.7
Toluene		1.4
n-Hexane		0.6
n-Heptane		1.1
Iso-Octane		1.5
Water		5.7
Methanol		0.8
n-Butanol		1.4

*Ref. 3

Estimates of π for various adsorbates on rutile, using data of Every et al. [6], indicate that for water, π reaches a value of over 100 ergs/cm^2 with coverage $\delta \leq 3$ Å; for methanol, π reaches 60 ergs/cm^2 with $\delta < 4$ Å; and for n-hexane, π reaches 30 ergs/cm^2 with $\delta < 15$ Å. Comparing these values of π with Table IV, we may conclude that α is less than 0.05. Since the experimental values of π were obtained on a different sample of rutile, we must consider this value to be only a rough estimate. (Gas adsorption data on our TiO_2 will be reported in a separate communication.) But qualitatively, it is clear that α is small, and probably varies from liquid to liquid on the same solid.

The dependence reported above, of rate of penetration on film thickness in the region of $\delta = 50$ to 80 Å for an alkane, lead us to speculate on some details of the meniscus structure for an advancing liquid front in a pore. If we treat the flow as being driven by the component of liquid

[6] Every, R. L., Wade, W. H., and Hackerman, N., J. Phys. Chem., 65, 937 (1961).

surface tension parallel to the flow direction, then the surface tension in the region of tangency and immediately beyond it must be greater than that of the bulk liquid. This amounts to an interpretation of the excess rate of flow, as being related to the Marangoni effect.

A question remains, which appears to have troubled some surface chemists: How to reconcile the excess penetration rate with the theory based on the Laplace equation. By the derivation which starts with the Laplace pressure across a meniscus, the driving pressure for flow in a circular capillary of radius r is $2\gamma_{\ell v}/r$. The condition of the solid ahead of the advancing meniscus should not affect this local pressure. The pardox, i.e. that the condition of the solid does influence rate of advance of the meniscus, can be resolved by allowing the curvature at the center of the capillary to be greater than 1/r. The system is in a flowing state, and so there is no *a priori* requirement that the meniscus be in its equilibrium shape. We may hypothesize that the meniscus deviates from spherical shape, possibly in the direction of being paraboloid. We have not attempted, in this work, to determine what the actual shape will be, however.

Acknowledgement. This work was supported by the Petroleum Research Fund. We thank Dr. D. A. Cadenhead and Mr. J. Stetter, Department of Chemistry, State University of New York at Buffalo, for carrying out the surface area measurement.

ELECTRICAL DISPERSION OF LIQUIDS FROM CAPILLARY TIPS - A REVIEW

Raghupathy Bollini [1] and Steven B. Sample [2]
1 *Southern Illinois University at Edwardsville*
2 *University of Nebraska, Lincoln, Nebraska*

Numerous studies have been conducted on the atomization or dispersion of liquids from capillary tips with the aid of applied electric fields. The process makes possible the production of charged droplets having charge-to-mass ratios that vary over a very wide range. The process lends itself to a broad spectrum of applications, and as a result, investigations of this process have been carried out by researchers from many different areas. Applications of electrical dispersion include preparations of thin films of radioactive salts for nuclear cross sectional studies, printing, painting, and deep space thrustors.

This paper will summarize experimental and theoretical results in the field of electrical dispersion, and will attempt to establish a number of general principles and simple models. The paper will also discuss the effects of various physical parameters on the electrical dispersion process in terms of current experimental data.

I. INTRODUCTION

This paper summarizes experimental observations and conclusions reached to date on the process of atomization or dispersion of liquids from capillary tips with the aid of electric fields. The diameters of the charged liquid particles generated by electrical spraying vary from 1 to 1000 μ. Several different modes of spraying (1-4) have been observed, some of which are erratic and some of which are periodic in nature. Under certain conditions the emitted drops are uniform in size (monodisperse) and are uniformly charged.

The electrical atomization process depends on the following parameters:

1. the surface tension constant of the liquid
2. the conductivity of the liquid
3. the viscosity of the liquid
4. the type of capillary (metal or glass)
5. size of the capillary
6. the magnitude of the electric field at the liquid meniscus as the latter forms at the tip of the capillary
7. whether the liquid meniscus is positively charged or negatively charged
8. the pressure within or mass flow rate of the liquid
9. the atmosphere in which the spraying process is taking place (5)
10. the symmetry of the spraying configuration.

This review paper will attempt to give the present understanding of the effects of the above parameters on the spraying process.

Theoretical analysis of the process is made difficult by the dynamic nature of the problem, e.g. the variation of the size of the liquid meniscus, the motion of the liquid meniscus, and the temporary shielding of the meniscus by the emitted drops and ions. The electric field also varies as a result of the above mentioned effects at the meniscus, and is therefore difficult to calculate. However, the approximate electric field at the meniscus in the static case has been calculated (6), and some qualitative analyses have been made to predict the charge-to-mass ratios of the emitted drops.

It should be pointed out here that this paper is concerned only with those cases in which the mass flow rate (without the electric field applied) is either zero or very low, so as to cause the liquid to drip at the rate of 1 to 5 drops a second. This paper does not deal with the case in which the mass flow rate is high enough to form a jet. There is a large body of literature concerned with the formation and break-up of a charged liquid jet (7-10) [see (8) for a more complete list of references for this phenomenon].

II. EXPERIMENTAL APPARATUS

The basic spraying apparatus used in air is shown in Fig. 1. If a glass capillary is used then the high voltage lead has to be inserted into the liquid reservoir as shown in Fig. 1 of Ref (2). It is very important to keep the tip of the needle as flat as possible without any burrs or defects, since any defects present at the tip cause the liquid meniscus to climb the sides of the capillary and the results

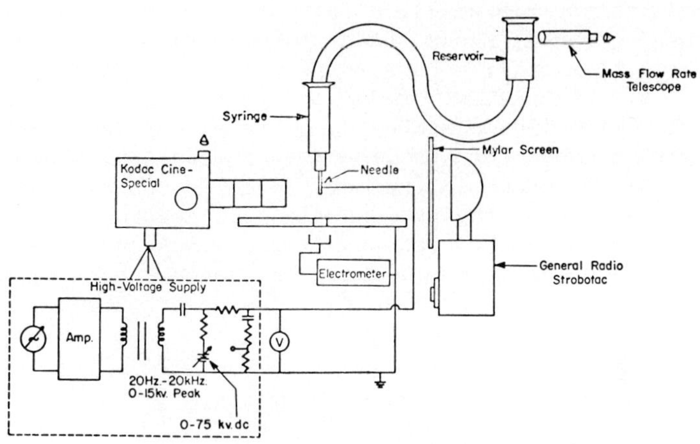

Fig. 1. Experimental apparatus for studying the electrical spraying process.

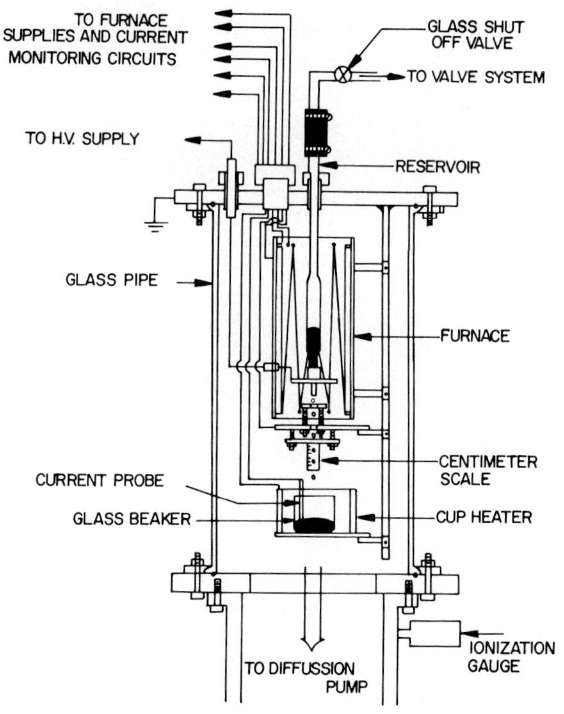

Fig. 2. Schematic of the experimental apparatus used for the electrical spraying of molten metals in vacuum.

obtained are not reproducible (3). For spraying materials which are not in a liquid state at room temperature, a vacuum environment is necessary to maintain the purity of the material (ie. to prevent oxidation), and to maintain the material in the liquid state. A schematic of the experimental apparatus used for spraying molten metals is shown in Fig. 2. A microscope is generally used to observe the meniscus when the spraying is periodic. A stroboscope may be used to slow the process for direct observation and for making slow-motion movies. When the process produces distinct droplets, the charge on the drops can be measured by two different methods. One method involves collecting the drops in an insulated metal cup grounded through an electrometer, and then dividing the electrometer current by the droplet emission frequency. The second method involves the use of a Faraday Cage, as described in (11).

III. EXPERIMENTAL OBSERVATIONS

The experimental observations vary depending on the parameters of the liquid, the size of the capillary, and the electric potential applied to the liquid.

A. Conducting Liquid with Positive and Negative Potentials and Small Needles

Small needles in this context means needles having an outer diameter (o.d.) of less then 0.8 mm. The mass-flow rate is such as to cause the liquid to drip at the rate of 1-5 drops per second without an applied potential. The size of the drops is determined by the surface tension constant of the liquid, density of the liquid, and by either the o.d. of the needle if the liquid wets needle, or by the inner diameter (i.d.) of the needle if the liquid does not wet the needle. The dripping mode is described in detail in (12). The size of the drop decreases and the frequency of emission increases as the potential is increased.

The transition from dripping to spraying takes place at about 5000 volts (with 2 cm. spacing between the needle tip and the plate). This transition is characterized by a sharp increase in frequency of emission, and a sharp decrease in mass-flow rate and drop size. In fact, without the strobe to slow the process the individual drops cannot be detected and a continuous jet seems to emanate from the meniscus. The process is periodic and the frequency of emission of the drops can be controlled over a small bandwidth by the addition of an ac voltage (3). The same kind of periodic mode was observed with molten metals in vacuum (13), but the droplet emission frequency was much lower than that in the case of water. For room temperature liquids, Ref (3) reports two stable regions as a function of voltage (5.0 to 6.5 kV and 7.0 to 11.0 kV). In (4) only one stable region (4.5 kV - 9.0 kV) was observed.

Above 11 kV the spraying becomes very erratic, in some cases no meniscus at all forms at the tip of the needle, and occasionally luminosity can be observed. If the meniscus wets the sides of the needle, the spraying becomes some what unsymmetrical and modes similar to those described in Fig. 3 of (1) frames j and k can be observed. In case of molten metals (13) the stable spraying is observed over a very narrow range of voltages (6.0 - 7.5 kV). Results with negative applied voltages seem to be similar to those obtained with positive voltages (4).

B. Non-Viscous Low Conductivity Liquids with Positive Potential and Large Needles

In the case of large needles, (o.d. greater than 0.8 mm) the phenomenon occuring at the meniscus is described in (4). The experiments seem to have been carried out at zero initial mass-flow-rate. The voltage at which spraying begins is higher than with conducting liquids. The periodic mode mentioned for smaller needles and conducting liquids was not observed. At different voltages different types of corona were observed. The main meniscus [called main drop in (4)] was stable with no ejection of water between 7.5 kV and 9 kV. A very fine spray of droplets was observed between 6.8 kV and 7.5 kV. At around 10.5 kV the main meniscus violently disrupted, in a manner similar to that which was observed with smaller needles and conducting liquids.

With negative applied voltages between 6.7 kV and 8 kV, fine sprays of droplets accompanied by different types of corona were observed. The stable region observed with positive voltages was not detected, and the violent disruption of the meniscus began at 8 kV.

C. Low-Conductivity Liquids with Positive Potential and Small Glass Capillaries

With low conductivity liquids like distilled water, alcohol, etc. and small glass capillaries one can observe a periodic mode similar to that described in Section A for conducting liquids (2). In addition to this periodic mode, it is possible to observe a "smoke" of fine particles coming off of the capillary tip (2,5,14). The size of the drops in this smoke or cloud are estimated to have a diameter on the order of 1 micron, and appear to be uniform in size for a given spraying configuration. The spraying process depends on the composition of the atmosphere in which it is taking place (5). The smoke of particles cannot be obtained at pressures below 600 Torr with normal oxygen nitrogen composition as in atmosphere present in the chamber. If the percentage of oxygen is increased, the smoke can be produced at lower pressures than those noted above [see Ref (5)].

D. Low-Conductivity and High Viscosity Liquids with Positive Potentials

Liquids with high-viscosity are observed to behave differently (6,15) than those with lower viscosity. Glycerin, for example, does not yield a common spraying mode at all. Initially, with increasing potential, the dripping frequency increases as in the case of low-viscosity liquids. However, when the potential reaches about 5 kV (for a plate-to-needle distance of 2 cms) the dripping completely stops and the meniscus becomes stable and no spraying is observed. At higher potentials, glycerin produces a fine jet which has a diameter on the order of 10 to 20 microns. The jet can have a stable length of a few centimeters before it breaks up into droplets. Glycerin doped with sodium chloride exhibits periodic spraying over a very narrow voltage range, but quickly reverts to the jet mode described above at higher potentials.

IV. DISCUSSION

The two competing forces at the meniscus are: first, the surface tension force and, second the electrostatic force on the charges induced on the meniscus. At low potentials the surface tension force is higher than the electrostatic force, and hence the electrostatic force simply acts with the gravitational force to increase the dripping frequency (3,12). The transition potential can then be anticipated to occur at the potential where the electrostatic force equals the surface tension force. Using this criteria, minimum spraying potentials were calculated (3,6), and the computed and experimental values agree well. This confirms the fact that the spraying process is a field-dependent phenomenon.

For periodic spraying to exist the liquid must be able to replenish the charges at the surface at the rate demanded by the dynamic equilibrium conditions of droplet formation. In other words, the conductivity of the liquid must be high enough so that the electrical relaxation time of the liquid is short compared to the inverse of the droplet emission frequency (16). At potentials higher than the minimum spraying potential, the meniscus and the spraying process remain stable due to shielding of the meniscus by emitted ions (4). The actual field at the meniscus may in fact decrease with increases in voltage (due to ion shielding), which would then account for the increase in the size of both the meniscus and the emitted drops after the initial dramatic onset of periodic spraying (3).

Taylor (6,17) has concluded from his calculations and experimental results that the conical shape attained by a meniscus at the tip of a capillary raised to a high potential, is similar to that attained by isolated drops that have been stressed by applied electric fields nearly to the point of rupture. His calculations indicate that this conical end should then be drawn out into a filament, as is the actual case observed with non-conducting and viscous liquids. Drozin (18) reaches a somewhat similar conclusion, since he also compares the phenomenon occuring at the ends of an electrically-stressed isolated drop to that occuring at a capillary tip raised to a high potential. Drozin predicts that the field strength will increase at points of local fluctuations on the surface of the meniscus, thereby giving rise to threads of liquid shooting out at high velocities.

A somewhat different explanation for the spraying process in the case of non-conducting liquids can be given in terms of the effects of nonuniform fields on dielectrics as studied by Pohl (19). Pohl's experiments have shown that dielectric liquids become polarized and move into regions of highest field strength, an effect he calls dielectrophoresis. Pohl in (19) has an illustration of an experiment in which liquid drops in the form of a fountain are thrown up from a flat liquid surface under the influence of nonuniform fields.

The fan mode, in which a smoke of droplets in the one micron range are emitted, has been explained by Zawidzki et. al. (5) as a chemical process. In Ref (5), the authors think that ionization of oxygen molecules will occur in the capillary corona, and that the resulting negative ions will discharge onto the water-gas interface. Zawidzki et. al. conclude that in the absence of suitable substances with which to react in solution, (at low conductivities therefore) the released oxygen ions will recombine in the liquid and "boil off". This process will contribute to the mechanical rupturing of the surface, which in turn will result in the formation of very small droplets.

The charge-to-mass ratio of the individual emitted droplets in most of the experiments seems to be roughly one half of the maximum stable limit calculated by Rayleigh (20). Pfeifer and Hendricks (21) extended Vonnegut and Neubauer's energy-minimization analysis of the disintegration of a charged droplet, and found that the charge-to-mass ratios of the resultant particles should indeed equal one-half of the Rayleigh limit.

As mentioned in Section III - A, the mass flow rate decreases after the transition from the dripping mode to the periodic spraying mode. Two possible mechanisms have been mentioned for this decrease by Sample and Bollini (3). One of these is the reduction in the average meniscus size, which gives rise to a greater back pressure that would tend to reduce the massflow rate. The second mechanism involves a possible increase in surface tension as the frequency of droplet emission increases sharply. In the periodic spraying mode new surfaces are formed in less than 10 msec; therefore a dynamic value for surface tension might be appropriate (22). In the case of water, the surface tension reaches a maximum dynamic value of 110 dynes/cm, as opposed to the static value of 73 dynes/cm. This increase in the effective surface tension could lead to a significant reduction in mass flow rate.

V. CONCLUSIONS

It is clear from this brief discussion that the fundamental processes underlying electrical dispersion of liquids from capillaries is not fully understood. For example, it is not yet clearly understood why some spraying configurations give rise to sharply periodic modes, while other configurations yield only random emissions. Then too, several different theories exist to explain the conical jet or fountain mode observed with dielectric liquids at high potentials.

In spite of these unanswered questions, electrical dispersion of liquids is beginning to find an increasing number of applications including preparation of thin films of radioactive salts for nuclear cross sectional studies (23), printing (24), painting (25,26), and deep space thrustors (27).

VI. REFERENCES

1. Zeleny, J., Proc, Cambridge Phil. Soc. 18, 71, (1915).
2. Vonnegut, B., and Neubauer, R.L., J. Colloid Sci. 7, 616 (1952).
3. Sample, S.B., and Bollini, R., J. Colliod Interface Sci. 41, 185 (1972).
4. English, W.N., Phys. Rev. 74, 179 (1948).
5. Zawidzki, T.D., Petriconi, G.L., and Papee, H.M., Z. Angew Mathematik Physik 14, 441 (1963).
6. Taylor, G.I., Proc. Roy Soc. Lond. A. 313,453 (1969).
7. Lindblad, N.R., and Schneider, J.N., J. Sci. Instr. 42, 635 (1965)
8. Dabora, E.K., Rev. Sci. Instr. 38, 502 (1967).
9. Melcher, J.R., and Warren, E.P., J. Fluid Mech. 47, 127 (1971).
10. Mason, B.J., Jayaratne, P.W., and Woods, J.D., J. Sci. Instr. 40, 247 (1963).
11. Hendricks, C.D., J. Colloid Sci. 17, 249 (1962).
12. Raghupathy, B., and Sample, S.B., Rev. Sci. Instr. 41, 645 (1970).
13. Bollini, R., Sample, S.B., Seigal, S.D., and Boarman, J.W., J. Colliod Interface Sci. 51, 272 (1975).
14. Neubauer, R., and Vonnegut, B., J. Colloid 8, 552 (1953).
15. Zeleny, J., Phys. Rev. 10, 1 (1917).
16. Hayt, W.H., "Engineering Electromagnetics", p. 154 McGraw Hill, New York, 1974.
17. Taylor, G.I., Proc. Roy. Soc. Lond. A 280, 383 (1964).
18. Drozin, V.G., J. Colloid Sci. 10, 158 (1955).
19. Pohl, H.A., J. Appl. Phys. 29, 1182, (1958).
20. Lord Rayleigh, Phil. Mag. 14, 184 (1882).
21. Pfeifer, R.J., and Hendricks, C.D., Phys. Fluids 10, 2149 (1967).
22. Netzel, D.A., Hoch, Geraldine, and Marx, T.I., J. Colloid Sci. 19, 774 (1964).
23. Lauer, K.F., and Verdingh, B., Nucl. Instr. Methods 21, 161 (1963).
24. Winston , C.R., U.S. Patent 3060 429 (1962).
25. Hines, R.L., J. Appl. Phys. 37, 2730 (1966).
26. Miller, E.P., and Spiller, L.L., Paint and Varnish Production, June,July (1964).
27. Cohen, E., Rept. ARL-63-88, Space Technology Laboratories, Redondo Beach, California 1963.

TEMPERATURE DEPENDENCE OF CONTACT ANGLES ON POLYETHYLENE TEREPHTHALATE*

A.W. Neumann, Y. Harnoy, D. Stanga, A.V. Rapacchietta
University of Toronto

This study is a further step in the investigation of conformational and structural changes of materials by means of surface techniques. Contact angles on polyethylene terephthalate (PET) were obtained as a function of temperature using the capillary rise technique. The measurements were carried out in the temperature range from 20°C to 90°C.

Results show, for the two types of PET investigated, commercial grade and research grade, that in addition to the well known glass transition at about 75°C there is a transition at 30°C.

In order to check if these results were not only surface effects, differential scanning calorimetry (DSC) measurements, as a typical means of the investigation of bulk effects, were performed. Results from DSC were found to be in agreement with those obtained by the surface technique.

I. INTRODUCTION

A. The Use of Surface Techniques for the Detection of Transitions

Temperature induced phase changes in substances are usually accompanied by more or less discontinuous changes in the physical properties of the substance.

It is therefore reasonable to expect also a change in interfacial tension or surface free energy of a solid substance due to structural changes.

---

* This paper represents in part, the M.A.Sc. thesis of Y. Harnoy (University of Toronto, Toronto, Canada, 1975).

A discontinuity in the interfacial tension should be connected, in view of Young's equation (1),

$$\gamma_{SV} - \gamma_{SL} = \gamma_{LV} \cos \theta_e \qquad [1],$$

with a discontinuity or at least a change of slope in the temperature dependence of the equilibrium contact angle θ_e.

Advancing contact angles θ_a are measured in this investigation; it was shown (2,3) that the advancing contact angle on a smooth heterogeneous solid surface is the equilibrium contact angle θ_e for the lower surface free energy portions of the surface.

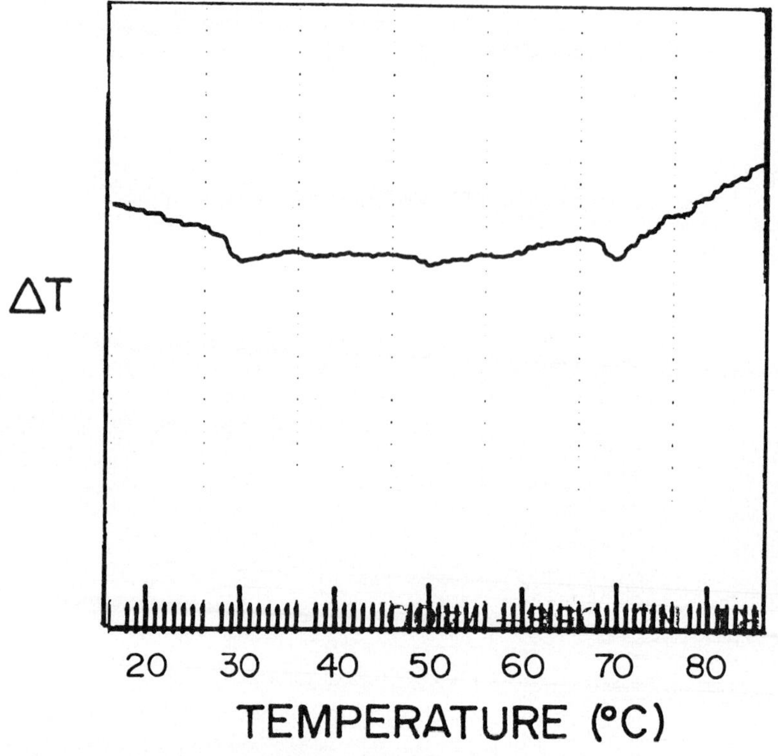

Fig. 1. DSC scan for polyethylene terephthalate.

Contact angle measurements were used previously for detection of phase transitions in hexatriacontane (5) and in cholesteryl acetate (6) as well as some polymers: polytetrafluoroethylene (Teflon) (7) a vinyl chlorid vinyl acetate copolymer and a chlorinated rubber (8). These transitions were also found by means of bulk techniques (e.g. by differential scanning calorimetry).

The thermal effects in conformational changes studied so far were typically large; indeed, the DSC run with Teflon reproduced in reference (7) was performed at a relatively low sensitivity setting to prevent the pen from running off scale. In Fig. 1, a DSC scan of polyethylene terephthalate (PET) is reproduced; in spite of using the highest practicable sensitivity, a relatively large sample of 21 mg and a relatively high scanning rate of 10°/minute, the glass transition temperature T_g just above 70°C shows up only as a small peak and a moderate change in slope. In addition to the T_g which has been reported previously, there is also a small peak near 30°C (and possibly a smaller one still at 50°C). Although all the features of the DSC scan in Fig. 1 were not very pronounced, they were perfectly reproducible.

It is the aim of this investigation to study the temperature dependence of contact angles in the range from 20°C to 90°C and to detect, if possible, the glass transition as well as the as yet unknown transition near 30°C.

II. EXPERIMENTAL

A. The Capillary Rise at a Vertical Plate Technique

The technique used for measurements of the temperature dependence of contact angles is the technique of the capillary rise at a vertical plate (4-8). The capillary rise on a sufficiently wide (2-3 cm) vertical smooth solid plate immersed into a liquid is, for the central 1 - 1.5 cm of the plate, virtually independent of edge effects and the three-phase line in this central part is straight.

The measurement of the capillary rise h can be performed very accurately (to 2×10^{-4} cm) with a cathetometer.

The contact angle θ can be calculated from the capillary rise h using

$$\sin \theta = 1 - \frac{(\rho_L - \rho_V) g h^2}{2\gamma_{LV}} \qquad [2]$$

obtained by integration of the Laplace equation of capillarity.

The values of γ_{LV}, ρ_L, ρ_V for the various liquids used were taken or calculated from I.C.T. (9) and the Handbook of Chemistry and Physics (10).

With the accuracy of γ_{LV}, ρ_L, ρ_V and g as available from the above sources and the experimental error in the determination of h, the resulting relative error in θ is usually a small fraction of a degree (0.2° or less).

The design and construction of the apparatus for the measurements of temperature dependence of contact angles largely follows the description in (7) with a number of modifications and improvements to suit the needs of our particular investigation.

B. Sample Preparation

The mounting of the solid sample at which the capillary rise is measured can be done in various ways according to the form in which the sample is available. When the solid is available in stiff sheets, the sample can be cut and simply mounted for immersion. If the sample is in the form of flexible thin sheets (foil) they can be stretched on a small special mounting frame made for this purpose to ensure a flat surface.

The PET of commercial grade (about 0.16 mm thick sheets, obtained from CELANESE CANADA LTD., Toronto) was stiff enough and no stretching or support was needed. It had, however, a relatively rough surface; and though the three-phase line looked straight through the cathetometer, the roughness could be observed through a microscope.

For our purposes a clean and smooth solid surface is desired and therefore the PET samples (of commercial grade) were prepared in the following way:

1) Cleaned by immersion into warm detergent (ALCONOX) for 10 minutes, then rinsed through with hot water.

2) Rinsed with distilled water; dipped in methanol for 5-10 minutes and washed with acetone.

3) Dry sample inserted between two gold coated microscopic glass slides and pressed at about 270°C for 45 minutes and cooled.

4) The PET was separated from the glass slides by immersion in acetone. The gold remained on the PET film and was peeled off subsequently.

The surfaces of PET samples treated in this way showed remarkable improvements in smoothness when observed under the microscope.

PET of research grade (about 0.05 cm thick films obtained from Imperial Chemical Industries Ltd., England) had a sufficiently smooth surface and only the cleaning steps 1 and 2 were applied. These samples had to be mounted on the special frame to keep them flat. Reagent grade glycerol and glycol were used for the contact angle measurements without further treatment.

C. Measurement Procedure

A sketch of the apparatus is shown in Fig. 2. After

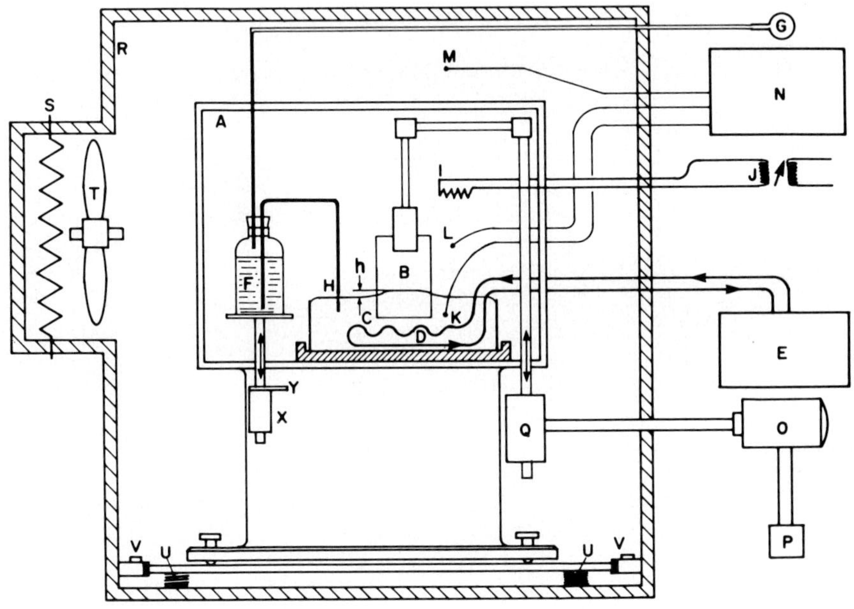

Fig. 2 Apparatus. A: inner enclosure; B: vertical measuring plate; C: measuring liquid in a rectangular glass cell; D: heating coil for the liquid; E: thermostat; F: bottle with liquid to regulate the level in C; G: pneumatic control for F; H: tip of liquid supply tube which also facilitates location of liquid level between it and its mirror image; I: electric heating element for inner enclosure; J: variable transformer to control I; K: thermocouple for temperature measurement of liquid in C; L: thermocouple for temperature measurement of the inner enclosure A; M: thermocouple for temperature measurement of the outer enclosure R; N: temperature-time recorder; O: electric motor to

raise and lower the measuring plate B; P: speed control for O; Q: transmission and speed reduction system; R: outer enclosure; S: electric heating element for the outer enclosure R; T: fan; U: springs and foam pads; V: rubber vibration damping device; X: screw for raising or lowering the bottle F; Y: lever to rotate the bottle F.

cleaning the glass cell C and the capillary glass tube H (with detergent, distilled water, methanol and acetone), the cell was filled with liquid to a level slightly above the brim. It was maintained this way throughout the experiment. (In the case of capillary depression, not used in this study, liquid is filled to below the cell's level).

The cathetometer alignment needs special attention. Because of its high magnification the object field is very small and therefore it is necessary to rotate the cathetometer about its vertical axis between reading of the three-phase line and reading of the tip and its mirror image. The tip and the three-phase line have to be located at the same distance from the axis of the cathetometer since any focus adjustments between readings of the two might change the reading slightly. The tip's location is therefore adjusted before the experiment by the screw X and the lever Y. It is most important to make sure that the cathetometer is aligned horizontally and remains so when rotated about its axis during the experiment.

When measurements begin, the door of the outer enclosure is closed and all controls are operated from the outside.

The first measurements start at room temperature; there were no obvious provisions in this apparatus for measurements below room temperature. Subsequent measurements were made by increasing the temperature of the three independent systems in steps of about 2-3°C.

To avoid evaporation and condensation of liquid in the inner enclosure, its temperature should be kept always slightly higher (about 0.2°C) than the temperature of the liquid, and the temperature inside the outer enclosure should be maintained somewhat higher still. The fan T was shut off at the moments when cathetometer readings were taken, to minimize the risk of vibration at the liquid surface.

Four separate readings were taken for each temperature and results are presented with error limits representing 95% confidence limits assuming a student t distribution.

III. EXPERIMENTAL RESULTS

A. Temperature Dependence of Capillary Rise and Contact Angles

Contact angles were studied on PET by measuring the capillary rise of glycerol and glycol, both with commercial as well as research grade PET.

The values of the contact angles were calculated from the capillary rise data using Eq. [2].

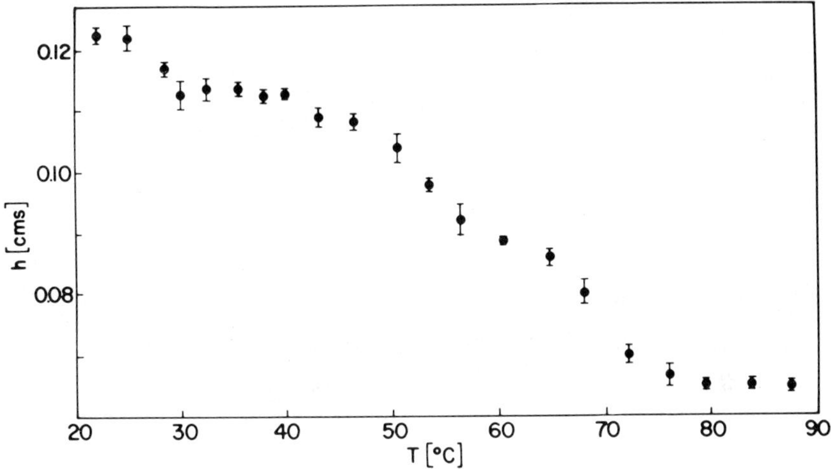

Fig. 3. Temperature dependence of capillary rise of glycerol at polyethylene terephthalate (research grade).

Figs. 3 and 4 show the results of the capillary rise measured with glycerol on two different specimens of research grade PET. The glycerol curves demonstrate a prominent change of slope at 30°C and at about 70°C.

It therefore appears that the two conformational changes indicated in the DSC scan of Fig. 1 are indeed also detected in the temperature dependence of the contact angles. Fig. 5 represents the actual contact angles calculated from the capillary rise data in Figs. 3 and 4; in addition, it also contains the data from a run with the commercial PET, up to a temperature of about 45°C. We note that there does not seem to be a major difference between the two kinds of PET specimens used in this paper.

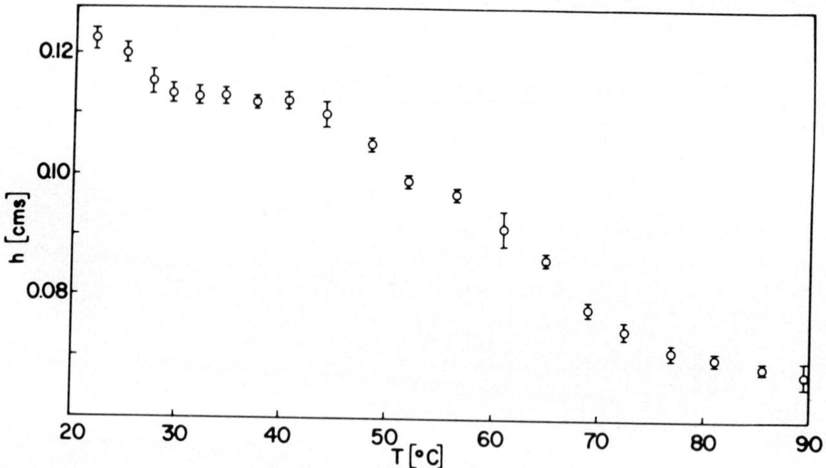

Fig. 4. Temperature dependence of capillary rise of glycerol at polyethylene terephthalate (research grade, different specimen).

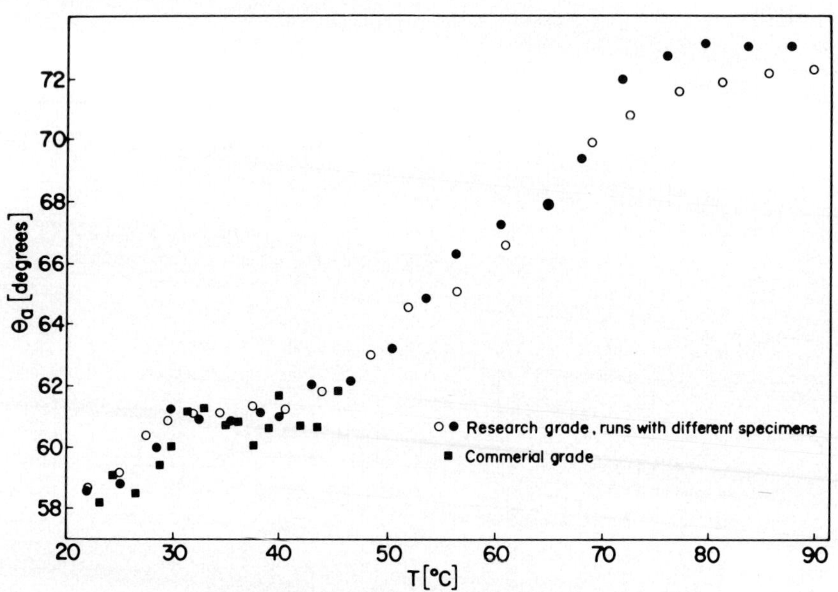

Fig. 5. Temperature dependence of contact angles for polyethylene terephthalate and glycerol.

Fig. 6 gives the results of the capillary rise measured

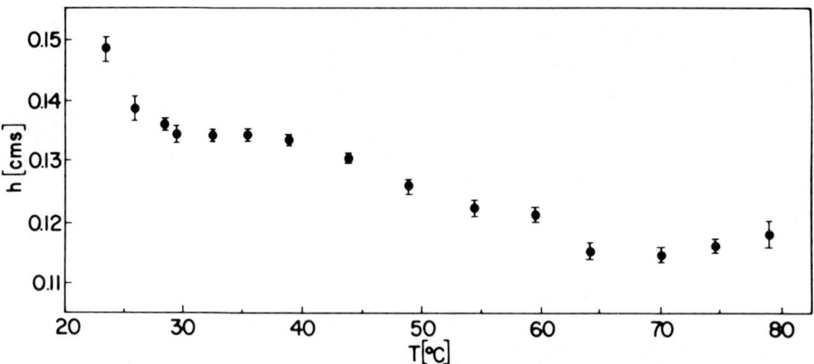

Fig. 6. Temperature dependence of capillary rise of glycol at polyethylene terephthalate (research grade).

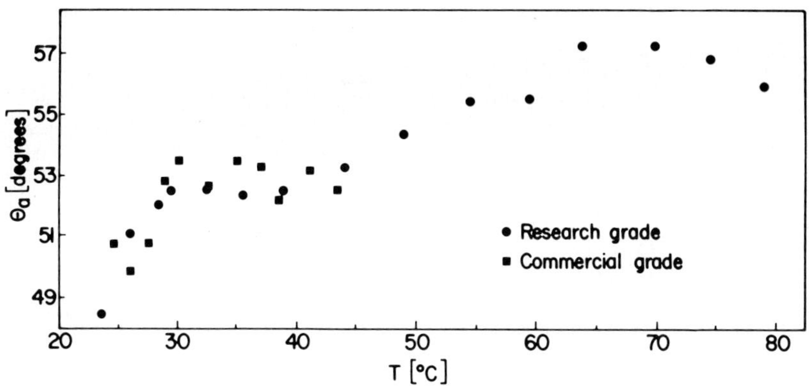

Fig. 7. Temperature dependence of contact angles for Polyethylene terephthalate and glycol.

with glycol on research grade PET. Fig. 7 shows the temperature dependence of contact angles for the glycol run shown in Fig. 6; in addition the data for a run with the commercial PET up to 45°C is also included. We note that the main features observed in the runs with glycerol (Figs. 3-5) are also present in Figs. 6 and 7.

B. Discussion of Results

The capillary rise technique for measuring contact angles and the DSC technique used in this study for the investigation of PET show the same results. That is, transitions take place at about 30°C and 70°C.

The reproducibility of the results is generally quite good. There are, however, some deviations, for example shifts in the temperature of the glass transition T_g, even for the same specimen or when investigated with different liquids.

These deviations are presumably caused by the thermal history of the specimen which affects the degree of crystallinity, and by sorption of liquid into the surface. Reproducibility can also be affected by excessive stretching on the stretching frame.

The above factors had been shown to influence the transitions of PET (11-20).

According to a recent investigation on the structural changes in glassy PET by Siegmann and Turi (13), glassy PET is not amorphous in the usually accepted sense of the term and glassy PET is not a completely frozen system; the structure can be altered and more order can be gained at temperatures below T_g. Their investigation demonstrated that annealing of PET about 30°C below T_g results in a PET with different structural behaviour. We are inclined to attribute the 30°C conformational change tentatively to such processes.

C. Conclusions

The results of this study reaffirm that the capillary rise technique is a powerful tool for detection of transitions in polymers. The measuring equipment used in this study was specifically designed for the 20°C - 90°C tempereture region but the technique can be used below 20°C or above 90°C with a suitable apparatus. The virtues of the technique are apparent; it provides us with a high precision measurement of contact angles over a considerable range of temperatures.

The results obtained confirm the well-known glass transition of PET near 75°C.

The transition of 30°C not reported previously was found and confirmed by means of DSC.

IV. ACKNOWLEDGEMENTS

This investigation was supported, in part, by NRC grant No. A8278. The research grade PET was obtained courtesy of E.L. Zichy, Polymer Science Division, I.C.I. England.

V. REFERENCES

1. Young, T., in "Miscellaneous Works" (G. Peacock, Ed), Vol. 1, J. Murray, London, 1855.
2. Neumann, A.W., Adv. Colloid Interface Sci. 4, 105 (1974).
3. Neumann, A.W. and Good, R.J., J. Colloid Interface Sci. 38, 341 (1972).
4. Neumann, A.W., Z. Phys. Chem. (Frankfurt) 41, 339 (1964).
5. Hellwig, G.E.H. and Neumann, A.W., in "Int. Congr. Surface Activity", Section B, p. 687, Barcelona, 1968.
6. Hellwig, G.E.H. and Neumann, A.W., Kolloid - Z. Z. Polymere 229, 40 (1969).
7. Neumann, A.W. and Tanner, W., J. Colloid Interface Sci. 34, 1 (1970).
8. Funke, W., Hellwig, G.E.H. and Neumann, A.W., Angew. Makromol. Chem. 8, 185 (1969).
9. "International Critical Tables", Vols. 3 and 4, McGraw-Hill, New York, 1928.
10. "Handbook of Chemistry and Physics", 53rd edition, The Chemical Rubber Co., Cleveland.
11. Holdsworth, P.J. and Turner-Jones, A., Polymer 12, 195 (1971).
12. Roberts, R.C., Polymer 10, 117 (1969).
13. Siegmann, A. and Turi, E., J. Macromol. Sci.-Phys. B10, 689 (1974).
14. Bair, H.E., Salovey, R. and Huseby, T.W., Polymer 8, 9 (1967).
15. Sweet, G.E. and Bell, J.P., J. Polymer Sci. A-2 10, 1273 (1972).
16. Coppola, G., Fabbri, P., Pallesi, B., Alfonso, G.C., Dondero, G. and Pedemonte, E., Macromol. Chem. 176, 767 (1975).

17. Nealy, D.L., Davis, T.G. and Kibler, C.J., J. Polymer Sci. A-2 8, 2141 (1970).

18. Thompson, A.B. and Woods, D.W., Trans. Faraday Soc. 52, 1383 (1956).

19. Illers, K.H. and Breuer, H., J. Colloid Sci. 18, 1 (1963).

20. Reddish, W., Trans. Faraday Soc. 46, 459 (1950).

This paper represents in part, the M.A.Sc. thesis of Y. Harnoy (University of Toronto, Toronto, Canada, 1975).

WETTING PROPERTIES OF LIVE CLEAN SKIN

Allan H. Rosenberg, George L. Cohen
Bristol-Myers Products

A variety of surface active agents were applied to live clean skin. Water contact angles on skin so treated indicate that wetting is enhanced in the general order of cationics > anionics > non-ionics. Although cationic and to a lesser extent anionic compounds enhance wetting of live clean skin by polar liquids little or no effect on wetting of non-polar liquids is observed. This result is consistent with our previous finding that clean live skin exhibits a hydrophobic surface. Quantitative determination of cetyl pyridinium chloride (CPC) adsorption on live clean skin was accomplished. Comparing total amount of CPC sorbed with water wetting properties of skin treated with CPC we find no correlation between total CPC adsorption and enhanced wetting properties.

I. INTRODUCTION

Contact angle measurements on human skin have been reported by several researchers (1-5). Although variations in contact angles are noted there is general agreement that human skin is hydrophobic (i.e., a low energy surface) and is poorly wetted by polar liquids. The variation in reported contact angles, especially for water, is likely due to differences in the amount of lipid material removed from the skin surface during sample preparation (5-6).

In addition to wetting properties of clean skin it is important to characterize the effect of deposition of surface active agents on wetting since enhanced wetting of skin is a prime factor in developing effective cosmetic and health care products that are topically applied. Surfactant effects on skin wetting have been reported by Ginn and Jungerman (7). From contact angle measurements on skin pretreated with surfactant solution they concluded that wetting by water is enhanced in the general order of cationic > anionic > nonionic.

The intent of this work is to further characterize wetting properties of skin treated with surface active material and also to investigate the relationship between quantitative absorption of a surface active agent and wetting properties of skin treated with the same surface active agent.

II. EXPERIMENTAL PROCEDURES

A. Contact Angle Measurements

Contact angles were measured with a rame-hart contact angle goniometer equipped with a specially designed finger holder. The dorsal area of the ring or forefinger was used. All measurements were made by the sessile drop method using advancing angles at 25°C. For each contact angle determination, the average value of four droplets (reading both sides of the drop) was designated as the contact angle. Angles obtained in this manner were reproducible to within ± 2° for each individual liquid on a specific subject.

B. Preparations of Skin Treated with Surfactant

The subject's hands were thoroughly washed with Ivory soap, rinsed with ethyl alcohol and distilled water and then air dried. Contact angles of water on skin cleaned in this manner were always greater than 100°. The fingers were then soaked in a solution of the surfactant under study for 10 minutes. After soaking in surfactant solution the fingers were rinsed by dipping in distilled water four times. The fingers were then air dried and contact angles were determined for water and other liquids. In some cases further rinses were carried out and contact angles remeasured. The change of contact angle as a function of the number of rinses gives a qualitative indication of substantivity of the material sorbed on the skin surface.

C. Adsorption of Cetylpyridinium Chloride (CPC) by Skin

CPC adsorption onto skin was determined by a continuous flow UV analysis technique using the Beckman Kintrac VII Spectrophotometer. An aqueous CPC solution has an absorption maximum at 258 nm which follows Beer's law in the region of interest (.01% CPC). No measurable adsorption of CPC was observed on glass or plastic tubing leading to the optical cells of the instrument.

The subject's clean right hand was placed in a thermally regulated beaker containing 200 ml of a 0.01% CPC solution maintained at 37°C. Decreasing UV absorption due to CPC depositing on the hand is recorded. Since water extracts materials which absorb at 258 nm from clean skin, the subject's left hand was placed in a beaker containing 200 ml of distilled water at 37°C and the increase in absorption was used as a blank correction. (Both the right and left hands were found to give off equal amounts of material absorbing at 258 nm). Liquid from both beakers was continually pumped thru optical flow cells. The amount of CPC adsorbed on the skin is calculated from loss of absorbance in the right hand beaker corrected for absorbance due to extracted material from skin determined from the left hand beaker.

III. RESULTS & DISCUSSION

A. Effect of Surface Active Agents on Water Contact Angles

Water contact angles were measured on skin treated with several surface active materials. Additionally, a qualitative indication of substantivity was obtained by observing the increase in contact angle with repeated rinsing. Table 1 summarizes these studies.

TABLE 1

Water Contact Angles on Human Skin Treated with Surface Active Agents

Materials (Wt.% in H_2O)	*# of Rinses*	*Subject & Angles (°)*			
		1	*2*	*3*	*4*
.2% Cetyltrimethyl-ammonium Bromide (CTAB)	4	0	0	0	33
	8	0	0	0	50
	12	0	52	0	61
	16	0	55	0	69
.2% Tetradecyltri-methylammonium Bromide (TDTAB)	4	0	0	0	48
	8	0	0	0	55
	12	0	56	0	66
	16	51	62	0	78
.2% Sodium Lauryl Sulphate (SLS)	4	64	65	67	67
	8	71	73	74	82
	12	77	76	81	87
	16	-	83	-	94
.2% Sodium Myristate	4	0	0	0	0
	8	74	74	70	83
	12	90	81	75	88
	16	93	-	84	87

Materials (Wt. % in H_2O)	*# of Rinses*	*Subject & Angles (°)*			
		1	*2*	*3*	*4*
.2% Brij 99 [a]	4	82	88		
	8	84	94		
	12	90	102		
	16	94	-		
.2% Tween 80 [b]	4	83	92		
	8	90	97		
	12	92	-		
	16	95	-		

[a] *Polyoxyethylene (20) Oleyl Ether*

[b] *Polyoxyethylene (20) Sorbitan Monooleate*

Skin treated with cationic surfactants, cetyltrimethylammonium bromide (CTAB) and tetradecyltrimethylammonium bromide (TDTAB), show the greatest enhancement of water wetting as noted by the large number of zero contact angles (i.e. spreading of the droplet). Skin treated with the anionic agents sodium lauryl sulfate (SLS) and sodium myristate also improve water wetting properties of the skin but not to the same extent as the cationic materials. We see that water angles of 0° and 65° are observed for skin treated with sodium myristate and SLS respectively after 4 rinses. However, water angles of 70-80° are noted after eight rinses for these compounds. The nonionic materials Brij 99 and Tween 80 have very little effect on water wetting as seen by the small reduction in observed contact angles for treated skin compared to clean untreated skin. From the above results it appears that wetting of skin by water is enhanced in the order of cationic > anionic > nonionics or that substantivity increases in this same order. This finding is consistent with the results reported by Ginn and co-workers (7).

Table 1 shows that the cationic materials are the most difficult to remove since the smallest increase in contact angle is observed after 16 rinses. Anionic materials are easier to remove as noted by the increase in contact angles after rinsing. Nonionics are the easiest compounds to remove as seen from relatively large contact angles after only a few rinses. Thus, as seen above, the order of substantivity appears to be cationic > anionic > nonionic assuming that change in contact angle represents substantivity.

B. Effect of CTAB and SLS on Wetting of Skin with Various Liquids

The effect of CTAB and SLS on contact angles of various liquids on skin is shown in Table 2.

TABLE 2

Contact Angles of Various Liquids on Human Skin Treated with 0.2% CTAB and SLS Solutions[a]

Liquid	γ_L	γ_{LD}	γ_{LP} [b]	Average Contact Angle		
				SLS	*CTAB*	*Clean Skin*
Water	72.0	22.0	50.0	0,56±1[c]	0	107±2
Formamide	58.2	32.0	26.2	0	0	79±7
Ethylene Glycol	47.6	27.1	20.5	0	0	70+5
Diiodo-methane	50.6	50.6	0	54±3	47±7	55±7
Tetrabromo-ethane	49.7	49.7	0	44±5	33±3	42±5
Tetrachloro-ethane	36.1	36.1	0	0	0	0
Mineral Oil	31.9	31.9	0	0	0	0

[a] *Results are quoted ±SD of the parameter; Five subjects used for SLS & CTAB runs, ten subjects used for clean skin study.*

[b] *values from reference (4)*

[c] *3 subjects show average values of 56; 2 subjects show zero contact angle.*

The table includes values for surface tension γ_L, dispersion force component of surface tension γ_{LD} and polar force component of surface tension γ_{LP} where $\gamma_L = \gamma_{LD} + \gamma_{LP}$ for all liquids employed. It is seen that wetting of skin treated with CTAB or SLS by liquids having large polar force components such as water, formamide and ethylene glycol is greatly improved (i.e., zero or low contact angles are observed) compared to clean untreated skin. In contrast, contact angles of non-polar liquids (i.e., liquids having little if any γ_{LP} component) are little affected by the presence of surface active agents on the skin surface.

It is clear that when a clean hydrophobic skin surface is treated with CTAB or SLS, the new surface exhibits different wetting properties. The new surface may be a CTAB or SLS film or a composite CTAB-skin or SLS-skin surface. Wetting properties of treated skin surface will depend upon the critical surface energy of the surface with respect to polar and non-polar interaction of the impinging liquid droplet.

C. Adsorption of Cetylpyridinium Chloride (CPC) on Skin and Wetting Properties of Skin Treated with CPC

Tables 3 and 4 summarize the effect of CPC on wetting properties of skin and the extent of adsorption of CPC onto skin from a 0.01% solution at 37°C.

TABLE 3

Adsorption of Cetylpyridinium Chloride (CPC) on Human Skin From 0.01% Solution at 37°C

	Amount Sorbed (mg)[a]		
Time (Minutes)	*Subject 1*	*Subject 2*	*Subject 3*
1	0.3	0.3	0.6
2	0.8	0.8	1.6
3	1.3	1.2	2.0
4	1.6	1.4	2.2
5	1.8	1.7	2.4
6	2.0	1.9	2.7
7	2.1	2.2	2.8
8	2.2	2.4	2.9
9	2.3	2.4	3.0
10	2.4	2.6	3.2

[a]*Average of two or more runs; ±10% variation*

TABLE 4

Water Contact Angles on Human Skin Treated with CPC

	Contact Angles 0.2% CPC			*Contact Angles 0.01% CPC*		
	Subjects			*Subjects*		
# of Rinses	*1*	*2*	*3*	*1*	*2*	*3*
4	0	0	0	0	0	51
8	0	0	38	53	0	64
12	0	0	49	68	0	79
16	38	0	54	76	53	88

These investigations were carried out in order to determine if wetting properties of skin treated with ionic surfactants, such as CPC, are directly related to the amount of CPC adsorbed on the skin surface or whether other factors must be considered.

When comparing wetting and absorption data, it is always implied that easily adsorbed materials will show good substantivity (i.e. easily adsorbed materials will be difficult to remove by rinsing). This assumption is valid for most adsorption phenomena. Notice in Table 4 the subject-to-subject variation in wetting properties of skin after CPC treatment. Table 3 shows that CPC adsorption appears to approach a plateau at 10 minutes contact time. The amount of CPC after 10 minutes varies between 2.4 - 3.2 mg depending upon the subject. Comparing the extent of adsorption of CPC onto skin to wetting properties of skin treated with CPC for subjects 1, 2 and 3, it is seen that improved wetting of skin is not related to the amount of CPC adsorbed on the skin surface. This suggests that subject-to-subject variation in skin protein composition and orientation leads to differences in CPC - skin interactions which in some manner (as yet unknown) dictate the wetting properties of the skin surface containing adsorbed CPC.

REFERENCES

1. Adamson, A.W., Kunichika, K., Shirley, F and Orem, M.J., J. Chem. Educ. 45, 702 (1968).
2. Ginn, M.E., Noyes, C.M. and Jungermann, E., J. Colloid Interface Sci. 26, 146 (1968).
3. Schott, H., J. Pharm. Sci. 60, 1893 (1971).
4. Rosenberg, A., Williams, R., Cohen, G., J. Pharm. Sci. 62, 920 (1973).
5. Shimi, A.E., Goddard, E.D., J. Colloid Interface Sci. 48, 242 (1974).
6. Zatz, J., J. Pharm. Sci. 64, 1080 (1975).
7. Ginn, M.E., Dunn, S.C., Jungermann, E., J. Amer. Oil Chem. Soc. 47, 83 (1969).

HYDROGEL-WATER INTERFACE

Frank J. Holly and Miguel F. Refojo
Eye Research Institute of Retina Foundation

The surface of acrylic hydrogels demonstrates only partial wettability by water even when in a fully hydrated state. The high contact angle hysteresis observed with water may indicate that the gel surface is capable of changing its polarity through conformational and orientational changes of its surface polymeric chains depending on the polarity of the liquid phase adjacent to it. However, possible changes in surface morphology of the hydrogels due to the vertical component of water surface tension at the periphery of the sessile droplet cannot be ruled out as a potential cause of contact angle hysteresis.

Chemically uncrosslinked PHEMA gels have water wettabilities similar to those of covalently crosslinked PHEMA gels. The wettability by water of thin, uncrosslinked PHEMA coatings is different from that of PHEMA gels; the receding contact angle of water is significantly smaller on the PHEMA coatings regardless of the type of underlying substrate.

The adsorption of macromolecules, such as the uncrosslinked gel matrix polymers, and biopolymers, takes place with positive or negative changes in the gel-water interfacial tension. Even when the hydrogel-water interfacial tension increases as a result of polymer adsorption, a considerable increase in water wettability of the polymer-coated gel can be observed.

A comparison of the surface properties of hydrogels with that of biointerfaces such as cell and tissue boundaries indicates that certain types of hydrogels may serve as realistic models for biointerfaces.

Hydrogels play an increasingly important role as contact lenses and biomaterials. Since biocompatibility is usually determined by the extent or rather the lack of interaction of the hydrogel boundaries with dissolved biopolymers and cellular elements, the basic properties of the hydrogel-water and hydrogel-biopolymer solution interfaces are of some importance.

Gels in general consist of macromolecules which are crosslinked to form a three-dimensional network or gel matrix. If the repeating unit of the macromolecule contains hydrophilic groups, the gel matrix will spontaneously imbibe water until the water content reaches an equilibrium value. The magnitude of the equilibrium hydration will depend on the relative hydrophilicity of the polymer network and on the degree of crosslinking.

We have been studying the interfacial properties of acrylic hydrogels such as poly(2-hydroxyethyl methacrylate), PHEMA, poly-

(glyceryl methacrylate), PGMA, poly(2-hydroxyethyl acrylate), PHEA, and poly(acrylamide), PAA (Table 1).

Table 1
Acrylic Hydrogels Studied

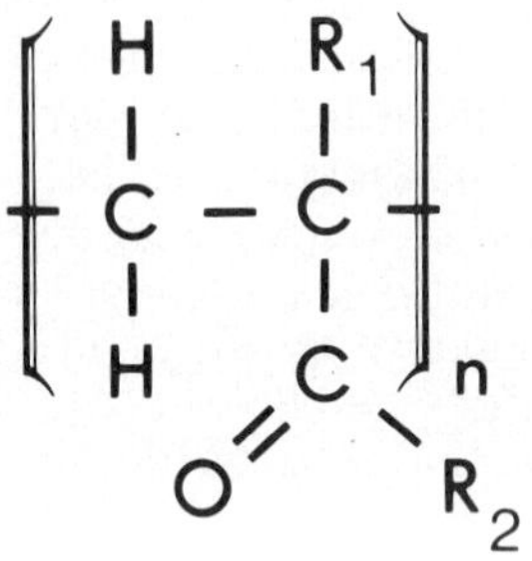

$$\left[\begin{array}{cc} H & R_1 \\ | & | \\ C & - \; C \\ | & | \\ H & C \\ & O{=}\;\; \backslash R_2 \end{array}\right]_n$$

GEL OR SOLID		R_1-	R_2-
1.	PMMA	$-CH_3$	$-OCH_3$
2.	PHEMA	$-CH_3$	$-OCH_2CH_2OH$
3.	PGMA	$-CH_3$	$-OCHOHCH_2OH$
4.	PHEA	$-H$	$-OCH_2CH_2OH$
5.	PAA	$-H$	$-NH_2$

I. SURFACE HYDROPHILICITY

It has been a matter of some controversy whether the surface of these hydrogels is hydrophilic. As we have recently pointed out (1) the answer depends on the criterion for surface hydrophilicity that is used.

When wetting is considered as a contact angle phenomenon, a hydrophilic surface may be defined as one that is completely wetted by water. This condition implies that the advancing contact angle of water on the surface is zero. (In practice, however, many solids with a small but finite receding contact angle of water are also considered to be hydrophilic.) In such a case, the water spontaneously spreads on the surface because the solid-vapor interfacial tension is equal to or greater than the sum of the liquid surface tension and the solid-liquid interfacial tension.

When wetting is considered as a capillary phenomenon, the inner wall of the capillary may be defined as hydrophilic if the capillary rise of water is positive. This condition implies that the water contact angle is less than 90° on the material of the capillary wall; thus this criterion of hydrophilicity is much less stringent than the first. In such a case, the solid-vapor interfacial tension is greater than the solid-liquid interfacial tension.

The third criterion for hydrophilicity requires that the solid surface exhibit a preference toward water in the presence of a condensed, nonpolar phase such as n-octane. This condition implies that the water-octane contact angle on the solid should be less than 90°, i.e. that the interfacial tension at the solid-water interface is less than that at the solid-octane interface.

Table 2
Water Wettability of Hydrogels and Solids

Gel or Solid	*Water Content (%)*	*Contact Angle or Water in Air (^{o})*	*Contact Angle of Water in Octane (^{o})*
PE	0	94.0 ± 1.0	153.2 ± 2.2
PMMA	1.5	72.6 ± 1.5	120.0 ± 1.1
PHEMA	38.9	59.5 ± 2.3	88.0 ± 1.3
PHEA	71.9	44.0 ± 2.5	79.3 ± 2.1
PGMA	73.3	41.3 ± 1.3	63.3 ± 1.6
PAA	77.7	10.1 ± 0.9	11.5 ± 1.5

Table 2 contains the water-in-air and the water-in-octane advancing contact angles for polyethylene, PE, and poly(methyl methacrylate), PMMA, and for the hydrogels. According to these data, when the first criterion of hydrophilicity is employed, none of the materials investigated is hydrophilic to the degree that it is completely wetted by water. By the second criterion, only PE is hydrophobic; PMMA and the hydrogels are hydrophilic. The third criterion yields the most realistic answer instinctively expected on the basis of the relatively large bulk water content in hydrogels. According to this criterion, all hydrogels are hydrophilic to various degrees, while both PMMA and PE are hydrophobic.

II. CONTACT ANGLE HYSTERESIS

One of the most striking things about hydrogel wettability is the large difference observed between the advancing and the receding contact angle values. In other words, the contact angle hysteresis for water on hydrogels is unusually large. Table 3 contains the receding contact angle of water and the relative contact angle hysteresis value, defined as

/1/ $$H_R = (\theta_A - \theta_R)/\theta_A$$

for the two polymeric solids and for the hydrogels. All the hydrogels exhibit contact angle hysteresis about twice as large as that on PMMA, while the hysteresis on PE is almost zero.

Table 3
Contact Angle Hysteresis

Gel or Solid	*Water Content*	*Receding Angle*	*Relative Contact Angle Hysteresis*
PE	0	$92 \pm 2°$	0.02
PMMA	1.5	$49 \pm 2°$	0.33
PHEMA	39.	$17 \pm 1°$	0.72
PHEA	72.	$18 \pm 2°$	0.59
PGMA	73.	$10 \pm 2°$	0.76
PAA	78.	$4 \pm 1°$	0.60

It is of interest to note that while no correlation was found between the gel water content and gel wettability for any given gel type, the relative contact angle hysteresis appears to increase with decreasing wettability for each type of gel investigated (2).

We have considered the possible causes of contact angle hysteresis (Table 4) described by Adam (3).

Table 4
Known Causes of Contact Angle Hysteresis

1. Surface roughness or heterogeneity
2. Physical interaction* (hydration, dissolution, soluble contaminants)
3. Chemical interaction* (hydrolysis, decomposition, etc.)
4. Stereo-chemical interaction* (conformational or orientational changes at the interface)

**between the solid and water*

By producing optically smooth hydrogel surfaces, which are in physical equilibrium with water and also are chemically inert to water, we may assume that the first three possible causes of contact angle hysteresis have been eliminated. This leaves the fourth possibility, given in Table 4, which is also the most plausible. Thus, it is likely that through orientational and conformational changes depending on the interfacial molecular force field, the surface chain segments take on a hydrophobic configuration when adjacent to a nonpolar or a gaseous phase, and a hydrophilic configuration when exposed to liquid water or any other condensed hydrogen-bonding phase.

It should be mentioned here that there is another possible cause of hysteresis not included in Table 4: the hydrogels are not rigid; they are rather elastic solids and as such are presumably subject to surface deformation under the stress induced by the vertical component of the liquid surface tension at the periphery of the sessile droplet. If such a surface irregularity did, in fact, exist, it could result in a considerable contact angle hysteresis. We suspect that such a gradual change in surface morphology may be responsible for the time dependence of the water contact angle on hydrogels. A slight decrease of the advancing and a slight increase of the receding contact angle of water with time observed (4).

III. EFFECT OF CROSSLINKING ON WATER WETTABILITY

Among the dissolved matrix polymers investigated, uncrosslinked PHEMA is not soluble in water. This fact accounts for the formation of PHEMA gels that are not chemically crosslinked and yet can be swollen to equilibrium in water. The polymer matrix of such uncrosslinked PHEMA in water is held together by secondary bonds so effectively that even its equilibrium water content is similar to that of chemically crosslinked PHEMA gels (~40%).

Table 5 contains the contact angle data obtained with chemically crosslinked and uncrosslinked PHEMA gels by the sessile droplet and the captive bubble techniques (4). Essentially, there is no difference between the water wettabilities of these two gel types. The only difference, which may not be real, is related to methodology. Thus, the crosslinked PHEMA appears to be less wettable when examined by the captive bubble technique, as opposed to the sessile drop method (cf. ref. 4), while the uncrosslinked PHEMA is more wettable by the captive bubble technique.

Table 5
Effect of Chemical Crosslinking on Phema Gel Wettability

Phema Gel Type	*Equilibrium Water Content*	*Contact Angle*			
		Water-in-Air		*Air-in-Water*	
		Advancing	*Receding*	*Advancing*	*Receding*
crosslinked	40%	$78\pm1^\circ$	$20\pm1^\circ$	$81\pm1^\circ$	$26\pm2^\circ$
uncrosslinked	39%	$72\pm3^\circ$	$21\pm2^\circ$	$66\pm3^\circ$	$21\pm4^\circ$

PHEMA hydrogel has been employed as a coating material for various prostheses (5), so the wettability of thin PHEMA coatings was also investigated. Three different substrates--glass, PMMA, and PE--were coated with PHEMA dissolved in methanol. After the methanol was evaporated completely from the PHEMA coating, the coated solids were stored in distilled water for varying lengths of time. After each storage time interval, the water wettability was determined. Some of the results obtained can be seen in Table 6.

Table 6
Wettability of Chemically Uncrosslinked Phema Coatings
(cast from absolute methanol and stored in distilled water)

Material Coated	*Advancing Contact Angle*			*Receding Contact Angle*		
	1 hour	*1 day*	*long time**	*1 hour*	*1 day*	*long time**
Polyethylene	45 ± 2°	51 ± 5°	68 ± 1^o(24)	10 ± 1°	12 ± 1°	9 ± 1^o(24)
Poly(methyl methacrylate)	47 ± 1°	47 ± 1°	63 ± 1^o(26)	9 ± 2°	12 ± 2°	8 ± 1^o(26)
Glass	44 ± 1°	48 ± 1°	65 ± 1^o(32)	7 ± 2°	10 ± 2°	8 ± 1° (32)

* *time of storage in distilled water; the number of days is indicated in parentheses*

The advancing contact angles are initially low on the coatings but they increase with increasing water storage time. After being exposed to water for 20 hours, the water wettability of the PHEMA-coated solids becomes indistinguishable from that of PHEMA gels. The receding contact angle on the PHEMA coatings, however, shows no such increase with storage time, so its value remains about one-half of the water receding angle on PHEMA gels. When the thickness of the PHEMA gel coatings was increased about fourfold by repeated coating procedures, the initial wettability by water was higher and the storage time dependence was greatly diminished. The receding contact angle values, however, remained about the same as on thin coatings.

IV. POLYMER ADSORPTION AT HYDROGEL-WATER INTERFACES

We have investigated the effect of some water-soluble polymers on the gel-liquid interface by comparing the adhesion tension of water and of the polymer solutions on three hydrogels: PHEMA, PGMA, and PAA. Polyethylene was included as a reference material. The test solutions contained one of the polymers also used in the gel matrix--uncrosslinked PHEMA, PGMA, and PAA--or one of three biopolymers: bovine serum albumin (BSA), egg lysozyme (LYZ), and bovine submaxillary mucin (BSM), a surface active, highly soluble glycoprotein. All these polymers were dissolved in water at 1% concentration. The exception was uncrosslinked PHEMA, which, being insoluble in water, was dissolved in a urea solution (0.04%) at 0.2% concentration. When PHEMA dissolved in urea was used in the sessile drop over a gel, the gel had also been equilibrated in the same urea solution prior to the measurement.

The surface tension of the solutions, and the advancing contact angle of the solutions on each substrate were determined. The change in interfacial tension at the solution-substrate boundary resulting from polymer adsorption, i.e. the film pressure of the polymers adsorbed at the interface, π_i, was calculated using the Fowkes-Harkins equation (6):

/2/ $$\pi_i = \gamma_s \cos\theta_s - \gamma_w \cos\theta_w$$

where subscripts s and w represent solution and water, respectively. The implicit assumption in the above equation is that the solid-vapor interfacial tension is the same for both the solution and the water.

The π_i values are shown as a function of the polymer film pressure at the solution-air interface (π_o) in Figure 1. In this graph, every vertical line (π_o = constant) corresponds to a given solution, while the various symbols represent the substrates, hydrogels, and polyethylene.

Due to the lack of hydrogen-bonding across the interface, the water-polyethylene interfacial tension is rather high. Thus, the film pressure of the polymers adsorbed at the surface of the solution and at the solution-polyethylene interface are comparable for all the solutions, i.e. the values fall near the straight line $\pi_i = \pi_o$. The π_i values obtained with PAA gel show just the opposite effect; the results fall on the line $\pi_i = \pi_o$. This behavior results from the relatively high water wettability of the PAA gels. For small contact angles:

/3/ $$\cos\theta_s \simeq \cos\theta_w \simeq 1$$

so the Fowkes-Harkins equation /2/ becomes

$$\pi_i \simeq \gamma_s - \gamma_w = -\pi_o$$

The peculiar behavior of the polymer solutions on the PGMA and PHEMA gels can be explained on the basis of the different hydrophilic characteristics of the matrix polymers in these gels. When the dissolved polymer is more hydrophilic than the matrix polymer (uncrosslinked PGMA and PAA on crosslinked PHEMA), the

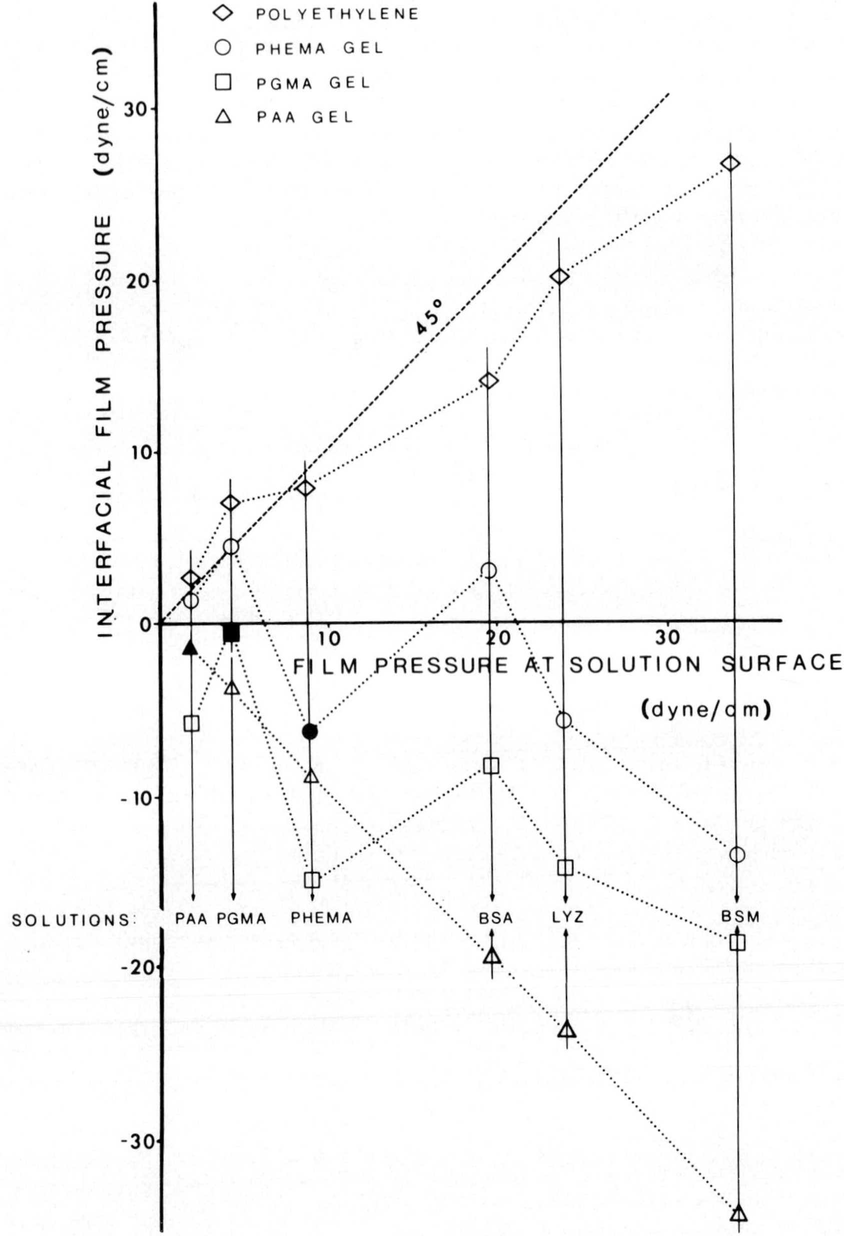

Fig. 1. Interfacial film pressure as a function of surface tension depression of polymers adsorbed on hydrogels and polyethylene. Solid symbols indicate homopolymer adsorption (the adsorbing polymer is the same as the polymer constituting the gel matrix).

interfacial film pressure is positive. When the polymer dissolved in the droplet is equally or more hydrophobic than the matrix polymer (uncrosslinked PHEMA on crosslinked PHEMA, PGMA, or PAA), then the interfacial film pressure becomes negative.

It is important to note that the increase in interfacial tension ($\pi_l < 0$) does not necessarily indicate a negative Gibbs surface excess concentration of the soluble polymer at the interface, since the Gibbs adsorption equation is not valid for macromolecular solutes. Furthermore, negative solute film pressure at the interface does not necessarily indicate a net increase in free energy. Endothermic macromolecular adsorption processes are driven entropically (7) while other related processes in the system (desorption of water, conformational changes in the macromolecules, changes in water structure, etc.) all contribute to the free energy change so that the adsorption process is accompanied by an overall decrease in free energy.

Table 7
Comparison of Hydrogel and Tissue Boundaries

Property	*Hydrogel*	*Tissue*
Pliability	Yes	Yes
High water content	Yes	Yes
Macromolecular surface composition	Yes	Yes
High mobility of surface structure	Yes	Yes
Negative charge density	DMC*	Yes
Low interfacial tension against water	DMC*	Yes
Large contact angle hysteresis	Yes	IPL**
Biopolymer adsorption and denaturation	DMC*(?)	No
Platelet adhesion	DMC*(?)	No

* *depends on matrix composition*

** *in the presence of lipids*

In a qualitative way, we have attempted to compare the known properties of hydrogels with that of biosurfaces such as cellular monolayers and tissue boundaries (Table 7). There are many similarities, possibly due to the biased selection of the properties. We believe, however, that many of these properties play a role in cellular interaction; thus in properly designed hydrogels they may favorably contribute to biocompatibility. It appears that the chemistry and structure of the gel matrix rather than the magnitude of the equilibrium water content *per se* determine the properties of the gel boundaries. Further research into the intricacies of the hydrogel interfaces may provide additional clues as to the design and formulation of more biocompatible materials.

REFERENCES

1. Holly, F.J., and Refojo, M.F., in "Proceedings of the International Conference on Colloidal Surface Science" (E. Wolfram and T. Szekrényesi, Eds.), Vol. 2, p. 159, Akadémiai Kiadó, Budapest, Hungary, 1976.
2. Holly, F.J., and Refojo, M.F., in "Hydrogels for Medical and Related Applications" (J.D. Andrade, Ed.), ACS Symposium Series, Washington, D.C., 1976 (in press).
3. Adam, N.K., "Physics and Chemistry of Surfaces," p. 79. Clarendon Press, Oxford, 1930.
4. Holly, F.J., and Refojo, M.F., J. Biomed. Mater. Res. 9, 315 (1975)
5. Hoffman, A.S., and Ratner, B.D., in "Hydrogels for Medical and Related Applications" (J.D. Andrade, Ed.), ACS Symposium Series, Washington, D.C., 1976 (in press).
6. Harkins, W.D., in "Recent Advances in Surface Chemistry and Chemical Physics" (F.R. Moulton, Ed.), p. 19, The Science Press, Lancaster, PA, 1939.
7. Koral, I., Ullman, R. and Eirich, F., J. Phys. Chem. 62, 541 (1958).

ACKNOWLEDGMENT

This work was supported by PHS Grant No. EY-00208 and EY-00327 from the National Eye Institute, National Institutes of Health. The technical assistance of Ms. Fee-Lai Leong, and the editorial assistance of Ms. S. Flavia Blackwell are hereby gratefully acknowledged.

EFFECT OF PRESSURE ON THE SURFACE TENSION OF WATER: ADSORPTION OF PROPANE AND DIMETHYL ETHER ON WATER AT TEMPERATURES BETWEEN 0 AND 50°C[1]

R. Massoudi[†] and A. D. King, Jr.
University of Georgia

ABSTRACT:

The variation of surface tension with pressure has been determined for water in equilibrium with compressed propane and dimethyl ether at temperatures between 0 and 50°C using the capillary rise method. It is found that dimethyl ether suppresses the surface tension of water to a remarkable extent at low to moderate pressures. The surface tension of water is an order of magnitude less sensitive to propane.

Adsorption isotherms and the thermodynamic parameters characterizing these isotherms are evaluated for these systems. Dimethyl ether exhibits type I isotherms[2] at all temperatures. The corresponding heat of adsorption is greater than the heat of vaporization of dimethyl ether and is found to decrease with increasing surface coverage. With propane the opposite is found in that adsorption of propane follows isotherms having a type III shape and the heat of adsorption is smaller than the heat of vaporization. The data are taken to indicate the existence of hydrogen bonds between water molecules at the surface and adsorbed ether molecules.

INTRODUCTION:

While the bulk properties of water in equilibrium with dense gases have been the subject of considerable research over the years, comparatively little attention has been paid to the surface properties of such systems. References 3-10 constitute a reasonably complete bibliography on this subject.

The results of these studies, all of which entail measuring changes in surface tension with pressure, indicate that in general the degree of adsorption of the gaseous component increases with molecular polarizability of the gaseous species as would be expected if physical forces of the van der Waals' type were the principal mode of interaction. Furthermore, the data correlates well with those obtained with higher molecular weight vapors at subambient pressures[10, 11, 12]. In several instances, namely CO_2, N_2O, and ethylene[10], the degree of adsorption is found to be somewhat greater than expected suggesting that additional weak interactions, either electrostatic or chemical in nature, must exist between these gases and water molecules at the surface. These perturbations are small however and the picture that emerges from these studies is one in which the aqueous interface constitutes a low energy surface at which gas phase molecules concentrate primarily as a result of ordinary dispersion interactions.

There are two instances, both involving ethers, which appear to contradict this model. Experimental studies involving water with diethyl ether[13] and n-propyl ether[14] indicate that vapors of these compounds exhibit an exaggerated affinity towards the aqueous interface. Hartkopf and Karger[12] have examined the thermodynamic parameters that characterize the adsorption of n-propyl ether at low coverages and suggest that hydrogen bonding between water molecules at the surface and the ether oxygen atom is primarily responsible for the unusual surface activity of this compound.

This paper reports results obtained in measuring the changes in surface tension of water induced by a third member of this class of compounds, namely dimethyl ether. Measurements involving the homomorphic compound propane are included for comparison. Dimethyl ether seemed particularly appropriate for such an investigation since it is not only the simplest member of the ether family but also closely approximates the monomer unit present in hydrophilic portions of polyoxyethylene nonionic surfactants.

EXPERIMENTAL SECTION:

The changes in surface tension with pressure under compressed dimethyl ether and propane were determined using the capillary rise technique. The experiments were carried out using a thermostatted bomb fitted with 3.5 x 0.75 inch viewing port which utilizes a 0.5 inch thick glass window. A more detailed description of this bomb and the experimental procedures can be found in refs. 10 and 15. The capillary rise was measured to 0.03 mm with a cathetometer located externally; and pressures were determined to better than 1% using Bourdon gauges which were periodically calibrated against a dead weight tester. The temperature in the bomb was controlled to better than $\pm 0.2^{\circ}$C as determined by a thermocouple located in the central cavity of the bomb. A single capillary having a 0.5 mm diameter bore was used in all measurements. The capillary is suspended from a moveable stainless steel rod passing through an "O" ring seal in the top of the bomb to a micrometer drive mounted outside, thus permitting a series of measurements to be made using the same cross section of capillary for the upper meniscus regardless of the pressure. All data were taken with a receding meniscus.

Equilibrium between bulk gas and the surface is established very rapidly with these systems. In the case of propane, the capillary rise assumed a constant value within minutes following a change in pressure. Considerably different behavior was observed with the more soluble dimethyl ether[16]. Upon increasing the pressure of this gas by some increment, the capillary rise would immediately drop to some initial value then rise slowly as the gas subsequently dissolved into the bulk aqueous phase causing the pressure to decrease. The reverse behavior was observed whenever the pressure was decreased incrementally and evolution of dissolved gas followed. For most measurements, time spans in excess of 30 minutes were required in order for equilibrium to be established between the unstirred bulk phases. In several instances the pressure was deliberately changed by large increments so that the capillary rise could be monitored throughout the period of equilibration. The values thus obtained were found to exhibit the same pressure dependence within experimental error as those measured under true equilibrium conditions. These observations again illustrate the rapidity with which a quasi-equilibrium is established between the surface, the gas phase, and the upper layers of the bulk liquid.

Surface tensions were calculated using the relation[17]

$$\gamma = \tfrac{1}{2}\, r\, g\, (h + r/3)\,(\rho_l - \rho_g) \qquad (1)$$

where γ is the surface tension, g is the acceleration of gravity, and h is the capillary rise observed in the capillary whose radius at the upper meniscus is r. The symbol ρ_l and ρ_g designate the densities of the bulk aqueous and gas phases respectively. Experimentally, h was determined at a series of pressures for each gas at a given temperature. The product $(h + r/3)(\rho_l - \rho_g)$ was calculated at each pressure assuming the liquid density to be independent of pressure[18]. The concentration of water vapor in the bulk gas was assumed to be negligible to the extent that P V T isotherms for the pure gases[19] were satisfactory for obtaining accurate values of ρ_g at the various pressures. In each experiment, the capillary rise was measured prior to the introduction of the gas and the value thus obtained was combined with the literature value for the surface tension of water against air at that temperature[20] to establish the magnitude of the quantity rg/2 appropriate to that experiment. The micrometer drive mechanism was used to adjust the elevation of the capillary suspended in the bomb so that the upper meniscus was positioned at the same place along the capillary bore at each subsequent measurement.

In calculating values for the surface tension using eq.(1) it is assumed that the contact angle between water and glass is zero and independent of pressure. While this is undoubtedly a good approximation with propane, it was felt that this assumption should be checked in the case of dimethyl ether considering its solubility and the remarkable depression in capillary rise induced by this gas. Accordingly, the depression in surface tension by dimethyl ether was determined independently using a maximum bubble pressure apparatus of the Sugden design[17] with nitrogen and dimethyl ether as the bubble forming gases. This method, which is independent of contact angle, yielded a value of 21.3 ± 0.3 dyne/cm for the surface pressure of dimethyl ether over water at 0.98 atm. and 25°C. This agrees well with the values calculated from the capillary rise measurements using equation (1)(Fig.1).

The propane and dimethyl ether used in this work were C P grade or the equivalent, purchased from Linde division of Union Carbide, having quoted purities of 99.0 %. Laboratory distilled water was used without further purification.

RESULTS AND DISCUSSION:

The experimental results obtained for the pressure induced changes in surface tension are shown in Figs. 1 and 2. The data are represented as surface pressure, $\pi = \gamma_0 - \gamma$, with γ_0 and γ denoting the surface tensions of water in equilibrium with pure saturated vapor and dense gas respectively. The experimental error based on repetitive measurements is ± 0.1 dyne/cm for propane and ± 0.2 dyne/cm in the case of dimethyl ether. Also a small directional error may be present in the dimethyl ether data due to uncompensated changes in the density of the liquid phase arising from dissolved ether at the higher pressures.

The temperature span over which the data could be collected was limited by problems with condensation of water vapor on the cell window at temperatures above 50°C and in the case of propane, by solid hydrate formation at lower temperatures. The upper hydrate quadruple point above which no hydrate can form is 5.69°C for the propane-water system.[21] At temperatures below this, crystals of the solid hydrate were observed to deposit on the inner surface of the glass container holding the water whenever the pressure exceeded the decomposition pressure. This in itself did not prevent measurements of capillary rise since the lower meniscus could still be observed. However, Herrick and Gaines[9] report that with H_2S and water, the deposition of hydrate along the capillary bore alters the contact angle hence capillary rise to a significant extent. Therefore, in order to avoid any ambiguities that might arise from such effects, no data was recorded with propane below 10°C.

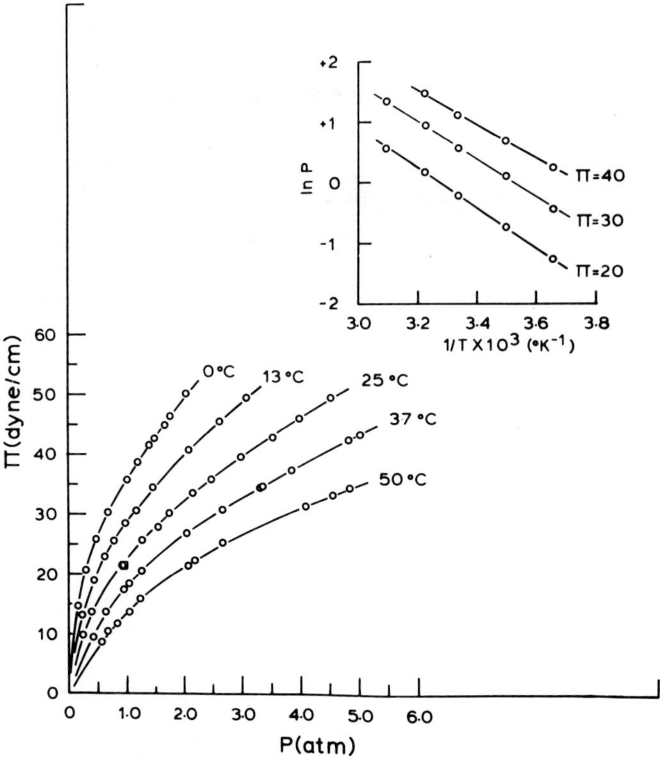

Fig. 1. Surface pressures of dimethyl ether on water.
□: Value obtained with maximum bubble pressure method.

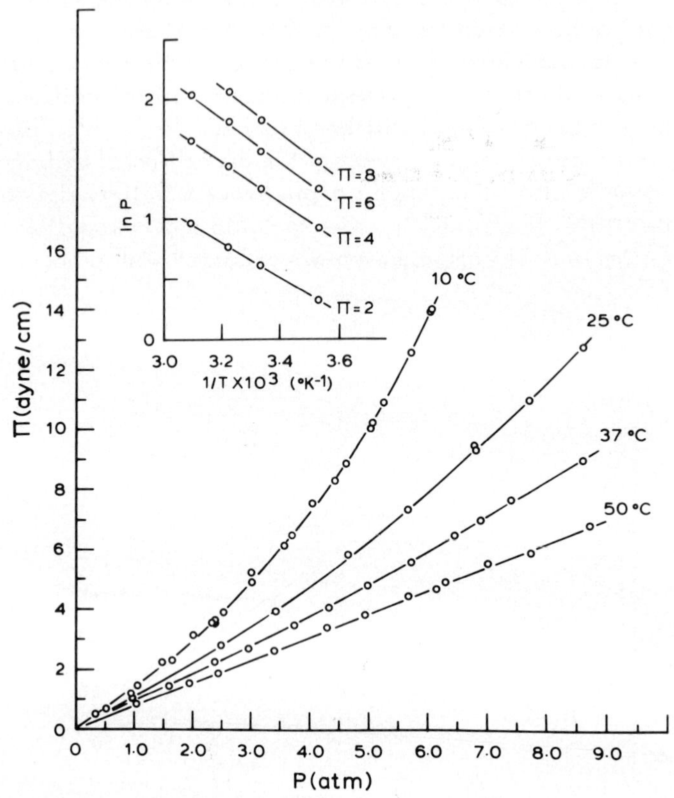

Fig. 2. Surface pressures of propane on water.

A similar hydrate has been reported in the case of dimethyl ether. Although the lattice constant for the solid hydrate of this gas has been reported[22], a careful search of the literature has failed to reveal any information regarding the phase behavior of this system. Therefore, the surfaces exposed to this gas were continuously monitored with the cathetometer telescope during the low temperature measurements. However, no evidence was found to indicate hydrate formation under the conditions encountered in these experiments. Therefore, we believe the low temperature data of Fig. 1 are accurate within limits of the experiment.

The most obvious feature of Figs. 1 and 2 is the remarkable extent to which dimethyl ether suppresses the surface tension of water. When compared to the effect produced by its homomorph propane, it is seen that the surface tension is an order of magnitude more sensitive to the ether. It is clear that dimethyl ether has a much greater affinity for the aqueous surface. It is also seen that there is a notable difference in the curvature of the data shown in these two figures.

The corresponding surface excess concentrations for the gases, $\Gamma_2^{(1)}$, have been calculated according to the common convention which places the Gibbs dividing plane such that the surface excess of water equals zero using the Gibbs equation.

$$(\partial \gamma / \partial \ln P) = -\Gamma_2^{(1)} \, Zk\,T \tag{2}$$

Here Z represents the compressibility factor of the pure gas at pressure P and temperature T while k designates the Boltzmann constant. The derivatives in equation (2) were obtained by fitting the data of Figs. 1 and 2 to a quadratic in ln P and differentiating the resulting expansions. The values thus obtained are shown plotted as a function of gas partial pressure in Figs. 3 and 4. The vapor pressures of liquid propane and dimethyl ether are known[23–25] so that one can compare the surface coverages of Figs. 3 and 4 on a reduced basis as shown in Fig. 5.

It is obvious from the data shown in these figures that the substitution of an ether oxygen for a methylene carbon profoundly alters the adsorption characteristics of these molecules. In the case of dimethyl ether, the initial portion of each isotherm exhibits a rapid rise with pressure. Each isotherm is concave to the pressure axis and the curvature of the individual isotherm is seen to increase with temperature. The molecular area of dimethyl ether, taken to be $\bar{V}^{2/3}$ where $\bar{V}$ denotes the molecular volume of the liquified ether, is estimated to be 24 $Å^2$ using densities found in ref. 23. It is clear, therefore, that inspite of the downward curvature in Fig. 3, nothing resembling saturation at monolayer coverages ($\sim 4 \times 10^{14}\,cm^{-2}$) results. With propane on the other hand, the isotherms exhibit a distinctly anti-Langmuir shape and the curvature is seen to decrease with increasing temperature. The molecular area of propane, as estimated from liquid densities[24], is 28 $Å^2$/-molecule so that a completed monolayer would be expected to correspond to a surface coverage of about $3.6 \times 10^{14}\,cm^{-2}$. Thus, in Fig. 4 one sees that with propane the low temperature isotherms also pass smoothly into multilayer coverages.

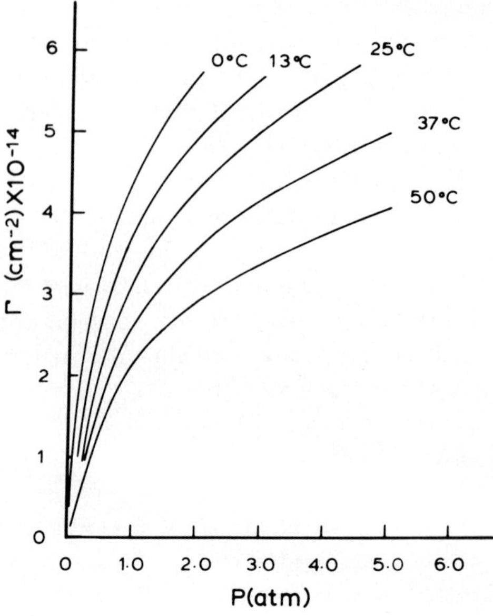

Fig. 3. Surface excess concentrations of dimethyl ether on water.

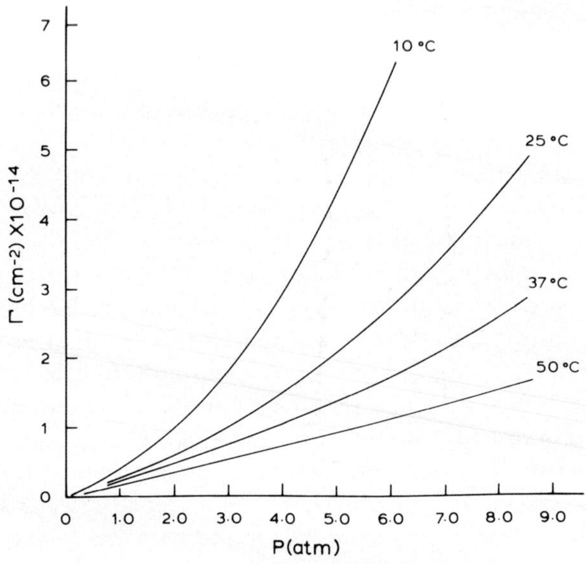

Fig. 4. Surface excess concentrations of propane on water.

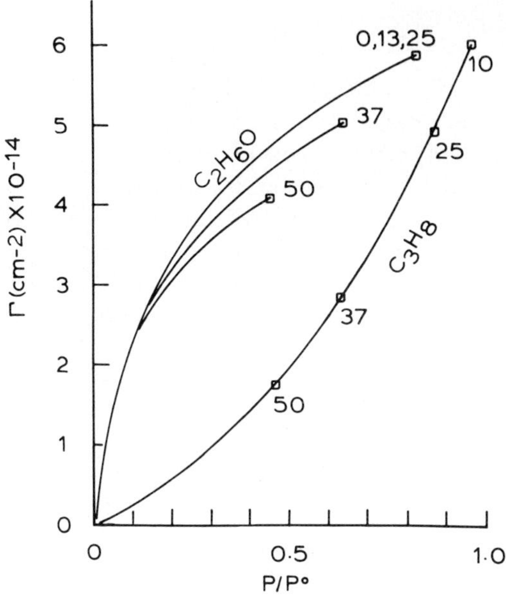

Fig. 5. Surface excess concentrations of dimethyl ether and propane plotted as a function of reduced pressure.

An interesting feature emerges when the isotherms of the two gases are compared on a reduced basis (Fig.5); namely that the low temperature data, which span nearly the whole range, appear to approach nearly identical coverages at saturation pressures, P°. This suggests that in contrast to the situation at lower coverages, the factors influencing the degree of adsorption in the multilayer region are quite similar for the two gases. The fact that the individual isotherms for propane all fall on one curve indicates that the isosteric heat of adsorption for propane on water is about the same as its heat of vaporization. The deviation between the isotherms for dimethyl ether on the other hand, suggests that the heat of adsorption for the ether exceeds its heat of vaporization.

Surface pressures are directly measured quantities in these experiments and thus are known to a higher degree of accuracy than the surface excess concentrations derived from them. Therefore, it is advantageous to interpret the adsorption characteristics of these gases in terms of equilibrium rather than isosteric heats of adsorption. The equilibrium heat of adsorption, ΔH, is defined by the equation[26].

$$(dP/dT)|_{\pi} = (-\Delta H/T)/(V_g - V_s). \qquad (3)$$

The quantities V_g and V_s represent the molar volumes of the adsorbate in the gas phase and adsorbed state respectively. If the thermodynamic properties of the adsorbent are assumed to be unaffected by the presence of the adsorbate molecules, then the equilibrium heat of adsorption can be thought of as resembling a heat of liquification for molecules of the adsorbate at a surface pressure π. The gases studied here, particularly propane, deviate significantly from ideality at the higher pressures. Consequently the molar volume in the gas phase is smaller than that for an ideal gas by a factor Z; i.e. $V_g = ZRT/P$. Therefore, after following the usual approximation of neglecting V_s and rearranging terms, eq. 3 becomes:

$$d \ln P/d(1/T)|_{\pi} = \Delta H/ZR. \qquad (4)$$

Since equilibrium exists between the gaseous and adsorbed phases, the corresponding entropy of adsorption, ΔS, is simply:

$$\Delta H = T\Delta S. \qquad (5)$$

Table I lists values for equilibrium heats and entropies of adsorption which have been calculated at several surface pressures from the data shown in the insets of Figs. 1 and 2. In the case of dimethyl ether, the heats of adsorption decrease with increasing surface coverage. The values are considerably larger than the heat of liquification of dimethyl ether which is -4469 cal/mole at 25°.[23] With propane on the other hand, the heats of adsorption are constant within experimental error and are somewhat smaller than the heat of liquification for this gas which is -3606 cal/mole at 25°.[24]

The surface pressure data for propane in Fig. 2 are reasonably linear at low pressures to that fairly accurate estimates can be made for the thermodynamic parameters that characterize adsorption of this gas in the Henry's law regime. If the Kemball-Rideal convention for standard states is adopted,[27] i.e. P = 1 atm for the gas phase and $\pi = 0.0608$ dyne/cm for the surface, then the free energy of adsorption in the zero pressure limit becomes:

$$\Delta G_A = -RT \ln\{[\lim_{P\to 0}(\frac{\partial \pi}{\partial P})_T] / 0.0608\}. \qquad (5)$$

Table I. Thermodynamic Functions for Adsorption at 25°C.

Adsorbate: Dimethyl Ether

π (dyne/cm)	$-R \frac{\partial \ln P}{\partial (1/T)}\Big\vert_{\pi}$ (cal/mole)	$P_{298,\pi}$ (atm)	$Z_{298,\pi}$	$-\Delta H_{298,\pi}$ (cal/mole)[a]	$-\Delta S_{298,\pi}$ (cal/deg mole)[b]	$-\Delta S_{calc}$ (cal/deg mole)
20	6500	0.8	0.99	6400	21	
30	6100	1.75	0.97	5900	20	
40	5500	3.0	0.95	5200	18	
22.5		1.0	0.98	(6300)[c]	21	19

Adsorbate: Propane

π (dyne/cm)	$-R \frac{\partial \ln P}{\partial (1/T)}\Big\vert_{\pi}$ (cal/mole)	$P_{298,\pi}$ (atm)	$Z_{298,\pi}$	$-\Delta H_{298,\pi}$ (cal/mole)[a]	$-\Delta S_{298,\pi}$ (cal/deg mole)[b]	$-\Delta S_{calc}$ (cal/deg mole)
2	3000	1.85	0.98	2900	10	
4	3200	3.45	0.93	3000	10	
6	3600	4.85	0.90	3200	11	
8	3800	6.1	0.88	3300	11	
1		1.0	0.98	(3100)[d]	10	14
0.0608				2800	4	8

a. Estimated error: ± 200 cal/mole

b. Estimated error: ± 1 cal/deg mole

c. Interpolated assuming ΔH varies linearly with pressure

d. average value

The free energies of adsorption for propane calculated according to this convention are -1800, -1700, -1650 and -1600 cal/mole at 10, 25, 37 and 50° respectively. The corresponding heat of adsorption calculated according to the Gibbs-Helmholz relation is: $\Delta H_A = -2800 \pm 300$ cal/mole. This agrees well with the heats of adsorption at finite coverages and reinforces the conclusion that with propane the heat of adsorption does not change significantly with surface coverage.

The last column of Table I contains values calculated for the entropy changes expected for adsorption from the gas phase at 1 atm assuming that one translational degree of freedom is lost in the process:[28]

$$\Delta S = -R \ln \left[M^{1/2} T^{3/2} \Gamma \right] + 68.1 \;. \qquad (6)$$

A comparison with the experimental value for dimethyl ether indicates that the ether molecules are adsorbed strongly enough to have lost the equivalent of a whole translational degree of freedom. With propane however, the less strongly adsorbed molecules retain a residual increment of entropy amounting to about 4 cal/deg mole.

In general the shapes of the isotherms for propane in Figs. 4 and 5 closely resemble those obtained for higher molecular weight hydrocarbons in the early investigations by Cutting and Jones[11]. However, a comparison reveals that the low temperature isotherms of propane more closely resemble true type III isotherms, i.e. become asymtotic to P°, than do those for the higher alkanes. This is reasonable in view of the similarity between the heat of adsorption and heat of liquification of propane since the distinction between adsorption and true condensation would be expected to be slight at multilayer coverages.

In a more recent series of investigations,[12] Karger has directly determined the adsorption isotherms of various alkanes on water at low temperatures using gas chromatographic techniques. He finds that the free energies of adsorption at 12.5° for normal alkanes in the zero coverage limit differ one from the next by 0.45 kcal/mole per methylene group in accordance with Traube's rule. His data, when extrapolated to C_3H_8 predicts a free energy change $\Delta G_A = -1.9$ kcal/mole for the adsorption of propane from 1 atm to the Kemball-Rideal standard state. This agrees well with the experimental value of -1.8 kcal/mole at 10° found here. Correspondence with Traube's rule is generally interpreted as indicating that the adsorbate molecules have a preferred configuration parallel to the adsorbent surface which in this case has the effect of equalizing contributions of the skeletal carbon atoms to $\triangle G_A$. Karger also notes that experimental entropies of adsorption for the higher members of the alkane series exceed that expected for the loss of one translational degree of freedom. He attributes this clustering effect which hinders the vibrational and rotational freedom of the larger adsorbed alkanes. His data show that this descrepancy diminishes with carbon number; and in the case of n-pentane the calculated and observed entropies are about equal. The data here suggests that this trend continues into the smaller members of the alkane series with the adsorbed molecules now retaining an excess entropy, presumedly due to low frequency vibrations normal to the surface and relatively unhindered rotational modes of motion. Thus, one sees that the propane data here fits into a more general picture in which the aqueous surface behaves like an inert low energy surface upon which the vapors of the saturated hydrocarbons can adsorb.

There is little data available with which the results found here for dimethyl ether can be compared; although, as noted earlier, Karger[12] has observed that n-propyl ether adsorbs quite strongly at the aqueous surface; a fact that he attributes to hydrogen bonding. One can however, compare the data for dimethyl ether with that of its homomorph propane. The structures and optical polarizabilities[29] of these two molecules are quite similar so that one expects the dispersion forces between these molecules and the aqueous surface to be nearly identical. Consequently the difference in adsorption behavior observed for these two molecules can be largely attributed to additional forces of attraction specific to the ether linkage and water, the most obvious being hydrogen bond formation.

Several features of interest emerge when the data for these gases are compared in this context. To begin with the isotherms of Figs.3-5 indicate that dimethyl ether concentrates at the aqueous surface to a much greater extent than propane at low coverages as would be expected if the ether formed hydrogen bonds with water molecules at the surface. However, as noted earlier, the isotherms for the ether begin to approach those for propane at coverages in excess of a monolayer, with the surface excess concentrations becoming nearly equal at the respective saturation pressures. Such behavior is expected here since the build up of second and successive layers should be determined largely by van der Waals forces between the gas molecules which are about equal for these two gases. The fact that the heats of adsorption for dimethyl ether decrease with increasing coverage is in accord with this picture since each value represents a weighted mean of the enthalpy contributions of the first and subsequent layers of dimethyl ether at the aqueous surface.

Secondly, in contrast to the results for propane, the entropy of adsorption for the ether is seen to be somewhat greater than that expected for the loss of one translational degree of freedom. It is clear that the adsorbed ether molecules are much more firmly bound to the aqueous surface than are molecules of propane.

Finally, as was mentioned above, it is reasonable to expect dispersion force contributions to the heat of adsorption to be approximately equal for these two gases. Furthermore, various lines of evidence, both experimental[12,30] and theoretical[31], point to the fact that water molecules in the outermost layer of liquid water are preferentially oriented with hydrogens directed towards the gas phase. Therefore, it is reasonable to assume that the formation of a hydrogen bond between an adsorbed ether molecule and the aqueous surface is not accompanied by any major reorganization of the water molecules at the surface. Thus, one might expect the difference between the heat of adsorption of dimethyl ether and that for propane to be a reasonably accurate reflection of strength of the bond formed. The difference in enthalpies listed in Table I for these gases at low coverages is seen to be 3.5 kcal/mole which is certainly the magnitude expected for such a hydrogen bond.

CONCLUSION:

Experiments measuring pressure induced changes in the surface tension of water reveal that dimethyl ether adsorbs on the aqueous surface to a much greater extent than its homomorph, propane. The results indicate that the aqueous interface represents an inert low energy surface with respect to the adsorption of propane. The strong adsorption of dimethyl ether is attributed to hydrogen bond formation between ether molecules and molecules of water in the surface region.

ACKNOWLEDGMENT:

The authors are grateful for the support provided by the National Science Foundation (NSF Grant No. GP-38386).

SUPPLEMENTARY MATERIAL AVAILABLE:

Tabulated values for the experimentally determined surface tensions have not been included here for the sake of brevity. They are available from one of these authors (ADK) upon request.

References and Notes

† Current address: Faculty of Sciences, The University of Isfahan, Isfahan, Iran.

1. This work was supported by a grant from the National Science Foundation (Grant No. GP-38386)

2. S. Brunauer, "The Adsorption of Gases and Vapors" Vol. 1, Princeton University Press, Princeton, N.J. 1945
3. E.W. Hough, M.J. Rzasa, and B.B. Wood, Jr. Trans. AIME, 192, 57 (1951)
4. E.W. Hough, B.B. Wood, Jr., and M.J. Rzasa, J.Phys. Chem. 56, 996 (1952)
5. E.J. Slowinski, Jr., E.E. Gates, and E.E. Waring, J. Phys. Chem., 61, 808 (1957)
6. E.W. Hough, G.J. Heuer, and J.W. Walker, Trans. AIME, 216, 469 (1959)
7. W.L. Masterton, J. Bianchi, and E.J. Slowinski, Jr., J. Phys. Chem., 67, 615 (1963)
8. A.I. Rusanov, N.N. Kochurova, and V.N. Khabarov, Dokl. Akad. Nauk. SSSR, 202, 380 (1972)
9. C.S. Herrick and G.L. Gaines, Jr., J. Phys. Chem. 77, 2703 (1973)
10. R. Massoudi and A.D. King, Jr., J. Phys. Chem. 78, 2262 (1974)
11. C.L. Cutting, D.C. Jones, and R.H. Ottewill, J.Chem. Soc. 1955, 4067, 4076 (1955)
12. A. Hartkopf and B.L. Karger, Accounts Chem. Res., 6, 209 (1973)
13. T. Ikeda and T.Ozaki, Bull. Chem. Soc. Japan 23(2), 43 (1950)

References and Notes (cont.)

14. B.L. Karger, R.C. Castells, P.A. Sewell, and A. Hartkopf, J. Phys. Chem. 75, 3870 (1971)

15. R.Massoudi, Ph.D. Dissertation, University of Georgia, 1974; Diss. Abstr., 36 (5), 2246-b (1975)

16. The solubility of dimethyl ether in water is not well established. Beilstein (F.K. Beilstein, "Handbach der organischen Chemie," Springer-Verlag OHG, Berlin, 1, 281) lists a value of 37 for the Bunsen coefficient of this gas with water at 18°C; while the solubility of dimethyl ether in normal saline has been determined by Brody and coworkers to be 34.2 ml gas/ml solvent at 20.3°C. (A.W. Brody, K.P. Lyons, J.L. Kurowski, J.H. McGill, and M.J. Weaver, J. Appl. Physiology, 31, 125 (1971).

17. A. Weissberger, "Physical Methods of Organic Chemistry," 3rd ed, Part I, Interscience, New York, N.Y. 1959.

18. The densities of water were obtained from:
Handbook of Chemistry and Physics, R.C. Weast ed., 53 edition, CRC Press (1972), Pg F-11.

19. Isotherms for both propane and dimethyl ether at 0 and 13° were calculated using the nomograph found in:
G.W. Thomson, Ind. Eng. Chem. 35, 895 (1943)
Isotherms for dimethyl ether at the higher temperatures were calculated from the data found in:
W.S. Haworth and L.E. Sutton, Trans. Far. Soc. 2907 (1971)

20. Surface tensions for water at the various temperatures were taken from: J.L. Jasper, J.Phys. Chem. Ref. Data, 1 (4), (1972). In two instances involving propane at 37° and 50°; a short extrapolation of the capillary rise data to the zero pressure limit was used

References and Notes (cont.)

rather than a prior determined value against air.

21. J.H. van der Waals and J.C. Platteeuw, Advances in Chemical Physics 2, 1, (1959)

22. M.v. Stackelberg and W. Jahns, Z, Fur Electrochemie, 58, 162 (1954)

23. R.W. Gallant, Hydrocarbon Process. 47, 269 (1968)

24. R.R. Dreisbach ed., "Physical Properties of Chemical Compounds -II," Advances in Chemistry Series No.22, Amer. Chem. Soc., Washington, D.C. 1959

25. Handbook of Chemistry and Physics, R.C. Weast ed., 53 edition, C R C Press (1972), Pg. D-153

26. A detailed discussion relating the various thermodynamic properties measured in adsorption is found in: T.L. Hill, J.Chem. Phys. 17, 520 (1949)

27. C.Kemball and E.K. Rideal, Proc. Roy Soc., Ser A, 187, 53 (1943)

28. C.Kembal and E.K. Rideal, Proc. Roy Soc., Ser A, 187, 73(1946)

29. The average optical polarizabilities of dimethyl ether and propane are $\alpha = 5.16$ A^3 and $\alpha = 6.29$ A^3 respectively. H.A. Landolt and R. Bornstein, Zahlenwerte und Functionen. 6th ed. Springer, Berlin. 1951. Vol. 1, Pt. 2, p. 509

30. M. Blank and R.H. Ottewill, J.Phys. Chem. 68, 2206 (1964)

31. N.H. Fletcher, Phil. Mag. 18, 1287 (1968)

LATERAL TRANSPORT OF GLOBULAR MACROMOLECULES CONSTRAINED TO THE FLUID-FLUID INTERFACE

J. ADIN MANN, JR.
Case Western Reserve University

THOMAS R. McGREGOR
Dow Badische Company

ABSTRACT

The velocity autocorrelation function, $R_v(t)$, has been investigated for a dilute ensemble of spheres constrained to move in a fluid-fluid, two dimensional interface between bulk regions of much smaller molecules. Because of the divergence resulting from the long time behavior of $R_v(t)$ for disks, the behavior of constrained spheres was of special interest in computing estimates of lateral diffusion coefficients for the more general interface. Results are described. The methods of light scattering spectroscopy were used to produce experimental representations of certain integrals of the velocity autocorrelation function, but the long-time tail is a relatively small effect. The scattering process for polystyrene spheres moving in thin aqueous films was simulated using computer techniques in order to devise an optimal experimental design. The instrument built to study the dynamics of interfacial transport is described. First results concerning the autocorrelation functions of polystyrene spheres moving in ultra-thin films are described.

I. INTRODUCTION

The transport properties of interfaces are not well understood either at the macroscopic or microscopic level. Experimental methods are just being developed that measure molecular level events associated with the motion and collisions of particles constrained to move at an interface. The results are summarized in constitutive coefficients such as the surface diffusion coefficient $\bar{D}$ and the viscosity coefficient for shear, $\bar{\eta}$. These same coefficients should be consistent with continuum theory and experiments derived from such an analysis. The program is far from complete. Much of the background material supporting these contentions is found in a review article by Mann and Porzio(1).

Considerable effort has been invested in understanding the macroscopic methods for measuring surface viscosity(1). Even with the best continuum methods for measuring $\bar{\eta}$, as developed by Goodrich(2), there is a question as to whether the numbers are consistent with the only extant kinetic theory of surface viscosity(3) or even with any kinetic theory of surface viscosity. Unfortunately, the continuum instrument is not sensitive enough to measure the surface viscosity of the dilute gas monolayers for which statistical mechanical theories are most tractable. While the spin exchange effect of spin labeled surfactants is sufficiently sensitive to permit the determination of appropriate collision frequencies for dilute monolayers(4), special instrumentation must be built to make the determination quantitative.

The fact that an accurate model for the bulk diffusion coefficient can be constructed with a combination of classical fluid dynamics and a fluctuation-dissipation theorem suggests an experimental approach for the surface transport problem that can be done with only a modest investment of equipment and time. Further, it is possible to model the measurements to be described by the computer simulation techniques that has been named "molecular dynamics".

The purpose of this paper is the description of an apparatus for measuring the lateral diffusion coefficient of large test particles constrained to move in an ultra-thin film of small molecules.

II. THEORY

The theory will be outlined briefly in this section leaving much to be discussed elsewhere(5).

The starting point is simply the desire to understand from a molecular viewpoint the diffusion coefficient, $\bar{D}$, that appears in the mass flux equation

$$\vec{\bar{J}} = -\bar{D}\nabla\Gamma \tag{1}$$

where Γ is the number density of the insoluble surfactant spread on a surface and $\vec{\bar{J}}$ is the surface flux that results from a gradient in Γ. The surface is assumed isotropic so that $\bar{D}$ is a scalar.

A model for $\bar{D}$ is the classical Einstein result

$$\bar{D} = \lim_{t\to\infty} \frac{1}{2t} \langle (x(t) - x(0))^2 \rangle \tag{2}$$

where x is the coordinate of the test particle in the surface at time t and 0. The averaging is done over an appropriate ensemble. In terms of the velocity of the particle

$$\bar{D} = \lim_{t\to\infty} \int_0^t \left(1 - \frac{s}{t}\right) \langle v_q(s)\, v_q(0) \rangle\, ds \tag{3}$$

where v_q is a component of the velocity of the test particle at times s and 0. If $\langle v_q(s) v_q(0)\rangle$ goes to zero rapidly enough, then the form for $\bar{D}$ usually quoted is

$$\bar{D} = \int_0^\infty \langle v_q(s)\, v_q(0) \rangle\, ds \tag{4}$$

However, we wish to emphasize that such a result is valid only for a special class of correlation functions.

The classical theory of Brownian motion results in $R_v(t) = \langle v_q(t)\, v_q(0)\rangle$ decaying exponentially for all times much longer than the collision times. Equation (4) holds and the conventional formula obtains

$$\bar{D} = \frac{kT}{\zeta} \tag{5}$$

where ζ is the hydrodynamical friction coefficient for the test particle moving in the surface. All of this suggests that mass transport along a surface should be representable by a diffusion equation of the classical form derived from the field equation

$$\frac{d\Gamma}{dt} + \Gamma \nabla\cdot\vec{v} = \hat{n}\cdot[\vec{J}]_\Sigma$$

to be

$$\frac{\partial\Gamma}{\partial t} - \bar{D}\,\nabla\cdot\nabla\Gamma = 0 \tag{6}$$

where the substrate exchange term is zero, $\hat{n}\cdot[\vec{J}]_\Sigma = 0$, and convection has been neglected.

It was most surprising to read that this picture could be wrong fundamentally. There is strong evidence that R_V has a long-time tail such that neither Eq. (3) or Eq. (4) converge (5,6,8). If R_V does not provide a well defined $\bar{D}$, then it appears that the other constitutive coefficients are also ill defined. Stated most strongly, a conjecture is that a consistent 2D fluid dynamics is impossible.

Since the classical diffusion Eq. (6) has been applied with reasonable results in the analysis of several experiments in which an initial gradient in Γ was established and followed in time, the classical theory cannot be entirely wrong. However, the gradient experiments all have difficulties either with mass transfer between the substrate or surface convection(10). Also, the divergence effect is subtle and may require a precision beyond that possible with film balance experiments. These points have been discussed more fully elsewhere and will not be repeated here.

It is important to realize that the effect of the long-time tail in $R_V(t)$ is to make $\bar{D}$ dependent on t as well as the material state. Under conditions that are often observed in practice, $\bar{D}$ could be as much as a factor 3 larger than would be expected from three dimensional data even though it may appear that Eq. (6) is valid. It is important to develop experimental evidence about the long-time behavior of $R_V(t)$. That appears to be a very hard experiment to do in three dimension.

We(5) have suggested that the long-time tail should be observable in the motion of polystyrene spheres of diameter ca. 0.1 μm constrained to move in a free soap film of that thickness. The computation of Subramanian <u>et al</u>(9) is supportive of our conjecture. They report that the molecule dynamics of a large test particle 2 and 3 times the size of the surrounding bath particles. The velocity correlation function of the test particles showed the long-time tail effect starting very early in the time scale dictated by the characteristic relaxation time given by $\tau_B = M/\varsigma$ where M is the mass of the test particle. In fact, the tail was in by $2 \times \tau_B$. This effect would be easily observed by our light scattering experiments.

III. EXPERIMENT

The apparatus is described briefly as well as the materials and the data collection and processing procedures.

A. Apparatus

The output from a Spectra Physics 125 He-Ne Laser was directed accurately parallel along a triangular, precision optical rail bolted to a massive optical bench. The optical bench was supported by air cushions riding on a plywood (1" thick sheets) cinder block base. The weight was evenly distributed on the floor of an industrial acoustics company, double walled isolation chamber. This arrangement considerably attenuated (at least 50 db. at frequencies above a few Hertz) both building vibration as well as acoustical noise. The room was gounded to a copper water pipe and so served to shield the electronics from undesired pickup.

The laser beam was divided by a 50% beam splitter and directed vertically to normal incidence on the thin film. A lens of 25 cm focal length focused the beam to a ca. 100 μm diameter spot on the film through apertures that cleaned up beam clutter resulting from reflection and transmission through the various optical elements. An efficient spatial filter was not used on the input beam in this first version of the apparatus but will be a part of the next design. It is important in film experiments to have a very clean incident beam.

Since the incident pencil was accurately normal to the film, the reflected beam was divided by the beam splitter with the transmitted portion reflected into a spatial filter and then focused on a photo multiplier tube of common variety. The intensity of the reflected beam was monitored for the measurement of the film thickness. The equivalent water thickness was computed which assumes that the film is a uniform slab. Since the film thickness did not go below 100 nm for these experiments, correction for the surface layers was not important. Also, the number density of the particle was kept small enough so that correction in the refractive index used to compute film thicknesses was ignorable. In future experiments, the output from the photomultiplier tube will be conditioned, digitized and interfaced to the computer system used to collect and analyze correlation function data.

The base of the sample cell was machined from an aluminum cylinder 5 cm long and 5 cm in diameter and was center drilled to admit the incident beam. The film was formed in an aperture .5 cm in diameter machined in an aluminum disk 2 mm by 5 cm in diameter and sharpened from the top side to a fine, smooth edge against the flat bottom surface. The aperture disk was clamped to the top of the base. A glass slide was set into the base just below the aperture so that a 1 cm diameter hole in the slide could be moved across the aperture. A small amount of soap solution could be added to spread across the aperture on the glass slide. With a short movement of the slide, the hole would line up with the aperture leaving a film that was initially several micrometers thick. It drained slowly to an equilibrium film thickness that could be varied by adjusting the composition of the solution. A hollow needle was positioned into the film border so that solution could be added or withdrawn from the film as desired.

A hollowed out cone, 5 cm in diameter at the base and 5 cm high was machined to fit closely on top of the aperture ring. The top of the cone was drilled to pass the transmitted pencil into a Rayleigh horn. A slot was opened in the side of the cone so that the scattered radiation could be viewed from scattering angles of about 10° to 75° with respect to the transmitted pencil. The slot was covered by optical quality glass. The cone, aperture ring and base were kept in place tightly by machine screws. The entire cell was black anodized in order to control stray reflections. All openings to the interior of the cell could be closed by optical quality glass so that the film could be manipulated in a controlled atmosphere environment.

The cell was aligned with a Leitz micromanipulator fitted with an appropriate bracket. The film could be aligned accurately parallel to the gravity field. That close alignment obtained could be decided by observing the symmetry of the Newton's rings that develop in thick films.

The scattered radiation was detected through a pinhole-lens spatial filter that had an acceptance angle less than 1 degree around the set scattering angle. The spatial filter was set onto an accurately machined arm of a verticle moving Norelco x-ray goniometer. It was possible to position each component so that alignment to the illuminated region of the film was accurate to wihin a few micrometers.

The light beam emerging from the spatial filter aligned to view the scattered field was transmitted by a light pipe to an EMI photomultiplier tube (PMT) fixed to the optical bench. The PMT housing was designed for photon counting but by attaching a conditioning circuit directly to the BNC output of the PMT housing the analog mode could be used as well. At modest number densities there was enough scattered irradiation to allow the efficient use of the analog mode. This has the virtue of permitting an accurate block of the D.C. component of the photocurrent correlation function. However, extreme care must be used to insure that the role of the low frequency response occurs below 1 Hz (3 db. point was 0.2 Hz), otherwise the base line of the correlation function will go negative within the delay time span of the correlator. Correcting for this behavior would negate the advantage that blocking the D.C. component offers.

Either a SAICOR model 42 or model 43 was used to compute the current correlation functions usually in the clipped mode of operation for efficiency. The results were identical whether clipped or full correlation functions were computed.

The SAICOR instruments were modified so that 16 bits of memory for each channel could be read in an inverted twos complement code on the back connector of the correlator.

Therefore, the digital representation used in further data processing was resolved to about 1 part in 32000 since the current correlation functions are positive.

A unique system of control and data processing was designed and has applications far beyond the light scattering experiment. A Motorola M6800 microcomputer system was built up and programmed to handle six data channels (an arbitrary choice) in either a read or write mode. A separate channel was used to patch in a CRT terminal and yet another channel was tied into an acoustic coupler and over a telephone line to a highly configured Xerox, Sigma 7 computer. The transmission rate was 30 characters per second.

The unique feature of this system was that all of the manipulation of the microprocessor commands required for data acquisition and data processing was done through a very high level array oriented language called APL. For example, in this language non-linear least squares fitting procedures were programs consisting of six lines of code. Only certain primitive system functions had to be coded in the microprocessor language. As a result, software requiring several years to write using FORTRAN and assembler languages was done in weeks. Because of the conciseness and power of APL, new algorithms can be devised, implemented and debugged with surprising speed and while working with the data.

Once the data stream from the correlator had been sent through the microcomputer to the host and decoded, the cumulants of the correlation function were computed through second order. The first cumulant is the average diffusion coefficient.

The procedure was as follows. A film was formed and the thickness followed by recording the reflection irradiance. The scattered current correlation function was computed at various thicknesses for a short time compared to the change in thickness (usually two minutes or less). The cycle time for transmitting data to the host computer at 30 characters per second was about one minute. The total processing time was just slightly over a minute, depending on the time sharing load of the host. This was fast enough that the next collection could be initiated before the film had drained significantly. Therefore, average diffusion coefficients could be determined as a function of film thickness effectively.

B. Materials

The polystyrene spheres were prepared by standard methods by Krieger's(11) group. The dispersions were cleaned of extraneous surfactants and deionized over a mixed bed according to standard procedures. The particles used for our initial study were relatively large (.22 μm) with a narrow distribution around the average (± 10%). They had been well characterized by electron microscopy as well as three dimensional light scattering measurements.

The surfactant used was sodium dodecyl sulfate freshly extracted and recrystallized by standard methods. The films were drawn from solutions containing 0.5% by weight of the surfactant (above the CMC). The films were mobile and had lifetimes usually lasting hours.

The polystyrene particles were added to the surfactant solution to a volume fraction of 4×10^{-4}. If draining proceeds uniformly, then when the film is .22 μm thick, there will be approximately 65 μm^2/particle. Since the particles occupy an area of only .037 μm/particle, one should expect that the test particles are moving independently of each other. There were approximately 150 particles in the illuminated region. This is a large enough number so that the distortion to the correlation function as a result of the particles moving in and out of the region is ignorable.

V. RESULTS AND DISCUSSION

We found that the variation of $\langle D \rangle$ was small (ca 10%) as the film drained to the diameter of the particles. Crudely, it appeared that the long-time tail was effective only after $7 \times \tau_B$, so that the velocity correlation function had decayed nearly to zero very early during the period characteristic of the photo-current correlation function (ca $10^4 \times \tau_B$).

For qualitative comparisons, let N_B be the number of relaxation times, τ_B , before the long-time tail becomes evident in the velocity correlation function. Let N_1 be the number of relaxation times before the photo-current correlation function has decayed by half. Write

$$\bar{D}(N_1 \tau_B) = \int_0^{N_B \tau_B} \left(1 - \frac{s}{N_1 \tau_B}\right) R_v(s)\,ds + \int_{N_B \tau_B}^{N_1 \tau_B} \left(1 - \frac{s}{N_1 \tau_B}\right) R_v(s)\,ds$$

or with obvious definitions

$$\bar{D}(N_1 \tau_B) = \bar{D}_0(N_B \tau_B) + \bar{D}_1(N_B \tau_B, N_1 \tau_B)$$

and further define functions f_0, f_1 so that

$$\bar{D}(N_1 \tau_B) = D_c \times [\, f_0(N_B \tau_B) + f_1(N_B \tau_B, N_1 \tau_B)]$$

where D_c $(= kT/\varsigma, \varsigma = 6\pi\eta a)$ is the classical diffusion coefficient. An order of magnitude argument based on a long-time tail of the form 1/t turning on at $N_B \times \gamma_B$ suggests that for $N_B = 7$ $f_0 = .997$. The long-time distortion of $\bar{D}$ is in f_1 essentially.

The same crude model suggests the long-time behavior of f_1 to be 0.025 for $N_I = 1{,}000$, 0.07 for $N_I = 10^6$, 0.16 for $N_I = 10^{12}$. Clearly, the effect will be small and ignorable for many practical purposes when $N_B > 7$. Our initial experimental result suggests that this is the case.

We had expected convection to be a problem, in that for films several times thicker than the particles, tilting the film holder produced obvious motion. However, our alignment procedure apparently worked, since the effect of convection would be to increase significantly the apparent diffusion coefficient. It is possible that the relatively small variation of the correlation function, as the film thinned, was due to convection. Control experiments are planned to test this possibility.

The temperature control was crude for these experiments. Obviously, temperature gradients and the resulting surface tension gradients could set up convective patterns even though the alignment of the cell prevented gravity streaming. We intend to repeat these measurements with temperature gradients controlled to at least ± 1 millidegree/cm.

Clearly, particle sizes and media properties must be varied in order to establish the long-time tail behavior we have reported. These experiments are progressing and the results will be reported later in detail.

It would be especially revealing to study much smaller particles of uniform shape and size. Certain proteins may work well if their shape and sizes are undistorted by the surfactnats required to stabilize the films. A search for appropriate molecules has commenced.

Also, it is possible and would be desirable to set up a molecular dynamical computation analogous to that of Subramanian _et al_(9), but allow the bath particles the freedom to move out of the plane to the extent of the larger test particle. The test particle would still be confined to move only on a plane.

While there are questions concerning isolated particles yet to be examined by our methods, we are working toward experiments in which the number density of the particles is taken to the close packed limit. Since these systems are monolayers with respect to the particles, multiple scattering effects will be much smaller and easier to compute. Questions concerning long-range order and the particle-particle forces are to be studied.

ACKNOWLEDGEMENT

We thank Professor H.T. Davis for a number of helpful discussions on the molecular dynamics of Brownian motion in a hard sphere fluid. This project was supported by NIH through grant number GM-22031-01.

REFERENCES

1. Mann, J.A., and Porzio, K., "Capillarity: The Physical Nature of Fluid-Fluid Interfaces Including the Problem of Bio-membrane Structure", page 47, International Review of Science, Physical Chemistry Series Two, Vol. 7, Surface Chemistry and Colloids, M. Kerker, ed., Butterworths, London (1975).
2. Goodrich, F.C., Allen, L.H., J. Colloid Interface Sci. (1972) 40, 329, and Poskanzer, A., Goodrick, F.C., Birdi, K.S., Allen, L.H., "Abstract 136, Colloid Division," Spring Meeting, ACS, 1974.
3. Cooper, E.R., Mann, J.A., J. Phys. Chem. (1973) 77, 3024.
4. Mann, J.A. and McGregor, T.R., Advances in Chemistry Series, 144, "Monolayers", ed. by E.D. Goddard, ACS, 1975.
5. Mann, J.A., "Dynamics of Fluctuations in Interfaces: Lateral Motion of Surfactant Molecules," Kendall Award Symposium Honoring R.J. Good, Spring Meeting, ACS, 1976. Preprints available.
6. Alder, B.J. and Wainwright, T.E., Phys. Rev. A, 1 (1), 18 (1970).
7. Zwanzig, R. and Bixon, M., Phys. Rev. A. 2 (5), 2005 (1970).
8. Pomeau, Y. and Resibois, P., Phys. Reports 19 (2), 63 (1975).
9. Subramanian, G., Levitt, H.T., and Davis, H.T., J. Chem. Phys. 60 (2), 591 (1974).
10. Sakata, E.K. and Berg, J.C., I & EC Fundamentals, 8 (3), 570 (1969). See also Good, P.A. and Schecter, R.S., J. Coll. Inter. Sci. 40, 99 (1972).
11. Hiltner, P.A. and Krieger, I.M., J. Phys. Chem. 73, 2386 (1969).

DETERMINATION OF SURFACE AND INTERFACIAL TENSIONS FROM ROTATING MENISCI IN A VERTICAL TUBE

H. M. Princen and M. P. Aronson

Lever Brothers Company

The shape of the meniscus between a liquid and a lighter, immiscible fluid, contained in a vertical cylindrical tube, rotating about its axis, has been calculated by numerical integration of the appropriate form of the Laplace equation. Complete wetting of the tube wall by the lower liquid has been assumed. The interfacial tension can be determined from the change in the position of the bottom of the meniscus when the radius of the tube, the angular velocity, and the density difference between the fluid phases are known. Limitations on the use of rotating menisci in a vertical tube to measure interfacial tension are discussed. The results of preliminary experiments are presented. These results are in good agreement with the theoretical calculations.

I. INTRODUCTION

When a fluid drop, surrounded by a denser liquid, is subjected to solid-body rotation, the drop migrates to the axis of rotation and deforms. If the axis of rotation is perpendicular to the gravity field, the equilibrium shape of the drop is sensibly independent of gravity provided that the angular velocity and the viscosity of the outer liquid are not too low (1-3). The equilibrium shapes of spinning drops in the absence of gravity have been studied extensively by several workers (1, 2, 4-6). Careful experimentation has verified the dependence of these equilibrium shapes on parameters such as interfacial tension, density, drop volume and speed of rotation, which had been predicted theoretically (2, 3). Furthermore, since the shape of the drop is related to its interfacial tension, the equilibrium shape can be used to determine this tension. This method has proved invaluable for the measurement of low interfacial tensions in connection with tertiary oil recovery (7).

The configuration studied in this paper is that of an axially symmetric meniscus separating two fluid phases contained in a vertical cylindrical tube that is rotating about its symmetry axis. As the angular velocity of such a system is increased, the meniscus stretches out with the top of the meniscus moving up the tube while the bottom moves downward as depicted in Fig. 1. The position of the bottom of the meniscus relative to its position at zero rotation can be readily measured, and is related to its interfacial tension, among other variables. In this system, in contrast to that of drops rotating about a horizontal axis, the influence of gravity on the interfacial shape is generally not negligible, i.e, in this case both surface tension and gravity oppose the extension of the meniscus. This fact considerably complicates the mathematical analysis of the equilibrium shape of the system and possibly accounts for the limited number of investigations

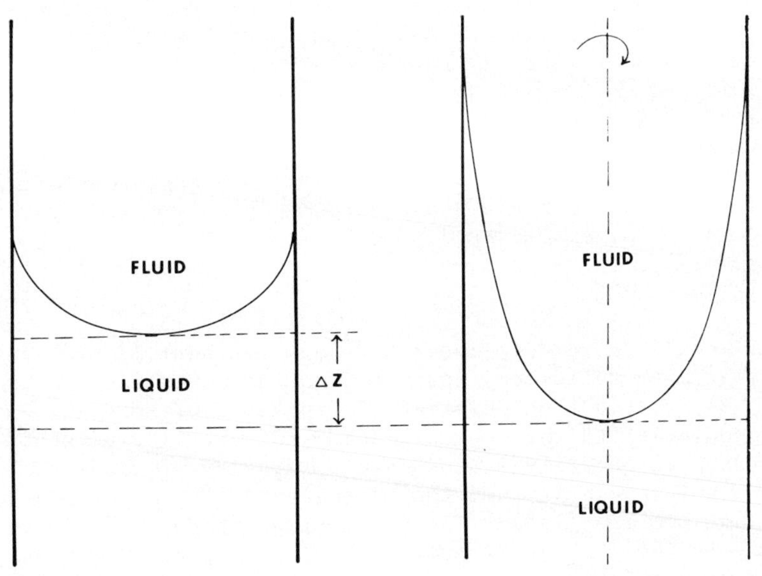

Fig. 1. Menisci in a vertical tube completely wetted by the liquid phase (θ=0). The left figure represents a stationary meniscus while the right figure represents a rotating meniscus.

of this configuration. The only studies we are aware of are those of Turkington and Osborne (8), Wasserman and Slattery (9) and Radoev (10), all of which deal with limiting solutions or cover a severely restricted range of conditions.

The purpose of the present study is to investigate this configuration in more depth and to generate solutions to the Laplace equation for the shape of the meniscus for a wide range of practical conditions. In addition, the potential use of rotating menisci for determination of surface and interfacial tensions is examined.

This paper presents a preliminary account of our work which is not yet concluded. A more detailed and complete presentation, particularly of the numerical and experimental results, will hopefully be published shortly.

II. FORMULATION AND METHODS OF SOLUTION

A. Profiles of Rotating Menisci

The profile of a rotating meniscus is determined from solutions of the Laplace equation of capillarity. The relevant coordinate system and notation for the configuration are shown in Fig. 2.

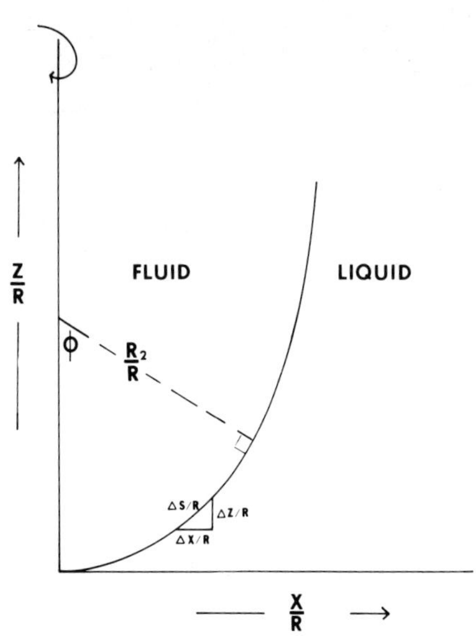

Fig. 2. Coordinate system of a rotating meniscus.

The Laplace equation appropriate for rotating axisymmetric interfaces in a vertical tube has been derived elsewhere (see Reference 11) and is given in Eq. (1).

$$\frac{1}{R_1} + \frac{1}{R_2} = \frac{2}{b} - \frac{(\rho_L - \rho_F)\omega^2}{2\gamma} X^2 + \frac{(\rho_L - \rho_F)g}{\gamma} Z \tag{1}$$

Here, $1/R_1$ and $1/R_2$ are the principal curvatures of the interface at point (X,Z) and are related to the profile through well known expressions from analytical geometry. By convention, the signs of these curvatures are positive if the centers of curvature lie in the upper fluid phase. The term 1/b is the curvature of the profile at the origin, while ρ_F and ρ_L are the densities of the upper fluid and lower liquid phases, respectively. γ is the interfacial tension, ω is the angular velocity and g is the gravitational constant. It should be noted that Eq. (1) neglects the influence of both gravity and angular velocity on interfacial tension, which is reasonable under usual experimental conditions (11).

When the radius of the tube, R, is selected as the unit of length, Eq. (1) can be written in a nondimensional form:

$$\frac{R}{R_1} + \frac{R}{R_2} = 2\left(\frac{R}{b}\right) - \frac{1}{2}(CR^2)\left(\frac{\omega^2 R}{g}\right)\left(\frac{X}{R}\right)^2 + (CR^2)\left(\frac{Z}{R}\right) \tag{2}$$

where C is a capillary constant defined as:

$$C = \frac{(\rho_L - \rho_F)g}{\gamma} \tag{3}$$

We have found it convenient to parametrize the profile in terms of the arc length S/R. This procedure avoids singularities and automatically accounts for points of inflection, i.e., changes in the sign of the curvature R/R_1. It can be shown that:

$$\frac{d(X/R)}{d(S/R)} = \cos\phi \tag{4}$$

and

$$\frac{d(Z/R)}{d(S/R)} = \sin\phi \tag{5}$$

There are several ways of relating the curvatures R/R_1 and R/R_2 to the profile of the interface, Z/R = F(X/R). Using the variables S/R, X/R, Z/R and ϕ, the radii of curvature, to be introduced into the Laplace equation, are given by:

and
$$\frac{R}{R_1} = \frac{d\phi}{d(S/R)} \tag{6}$$

$$\frac{R}{R_2} = \frac{\sin\phi}{X/R} \tag{7}$$

The system of Eqs.(2-7) can be solved as an initial value problem provided the parameters R/b, CR^2 and ω^2R/g are specified. The initial conditions are:

$$\frac{X}{R} = 0, \ \frac{Z}{R} = 0, \ \phi = 0 \ \text{at} \ \frac{S}{R} = 0 \tag{8}$$

It is now noted that for given values of the parameters CR^2 and ω^2R/g, there are an infinite number of profiles which are mathematically acceptable solutions to the Laplace Equation. Each of these profiles is uniquely defined by the value of R/b. However, for physical applications, it is clear that only profiles which actually intersect the vertical tube need be considered. This means that R/b is restricted to values for which the resulting profiles either just reach or go beyond X/R = 1. For this set of profiles, the contact angle, θ, at which the interface intersects the tube $(90^\circ - \phi)$ can still have any value. In the present study, we shall restrict ourselves to systems in which the tube is completely wetted by the liquid (θ=0), which uniquely determines the value of R/b (for given values of CR^2 and ω^2R/g).

B. Relation of Profile to Displacement of the Bottom of the Meniscus

In applications of rotating menisci to the measurement of interfacial tensions, it is the displacement of the bottom of the meniscus from its position at rest which is most conveniently measured experimentally. The relationship of this displacement to the meniscus profile is obtained through the following geometrical arguments.

Consider the profiles drawn in Fig. 3. The interface at rest is shown at the left of Fig. 3, while the rotating interface at equilibrium is shown at the right. If the position of the bottom of each meniscus, relative to a reference level that is independent of rotation, can be determined, then ΔZ can be obtained simply by subtraction. One such reference level for incompressible liquids is that level which the liquid/fluid interface would have if it were undeformed, i.e., planar; in other words, the level at which $V_1=V_2$ and $V_1'=V_2'$

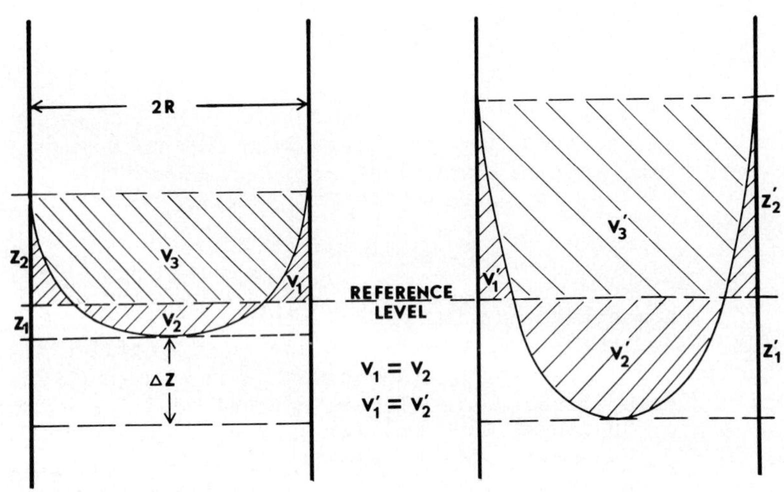

Fig. 3. Reference level for determining the displacement of the bottom of a rotating meniscus (right figure) relative to its position in a stationary tube (left figure).

(see Fig. 3). The positions of the bottom of the menisci relative to this reference, Z_1 and Z_1' are given by (see Fig. 3):

$$Z_1 = \frac{\pi R^2(Z_1+Z_2) - (V_2+V_3)}{\pi R^2} \qquad (9)$$

and

$$Z_1' = \frac{\pi R^2(Z_1'+Z_2') - (V_2'+V_3')}{\pi R^2} \qquad (10)$$

In the case of complete wetting, the volumes V_2+V_3 and $V_2'+V_3'$ are given by (see References 2 and 11 for more details):

$$\frac{V_2 + V_3}{R^3} = \frac{2\pi}{CR^2}\left[\frac{R}{b_o} + \frac{CR^2}{2}\left(\frac{Z_1 + Z_2}{R}\right) - 1\right] \qquad (11)$$

and

$$\frac{V_2' + V_3'}{R^3} = \frac{2\pi}{CR^2}\left[\frac{R}{b} - \frac{CR^2}{8}\left(\frac{\omega^2 R}{g}\right) + \frac{CR^2}{2}\left(\frac{Z_1' + Z_2'}{R}\right) - 1\right] \qquad (12)$$

where $1/b_o$ is the curvature at the bottom of the stationary meniscus. Equations (9-12) are now combined to yield the convenient expression for ΔZ given in Eq.(13).

$$\frac{\Delta Z}{R} = \frac{1}{4}\left(\frac{\omega^2 R}{g}\right) - \frac{2}{CR^2}\left(\frac{R}{b} - \frac{R}{b_o}\right) \tag{13}$$

where $\Delta Z = Z_1' - Z_1$. An equivalent form of Eq.(13) has been derived by Radoev (10) using a different procedure.

Equation (13) reveals that ΔZ is simply related to the profiles of the stationary and rotating interfaces through R/b_o and R/b.

C. Method of Solution

The set of Eqs.(2-8) can now be solved for given values of CR^2 and $\omega^2 R/g$. The solution which is desired for use with Eq.(13) is that profile for which:

$$\frac{X}{R} = 1 \text{ at } \phi = 90^{\circ} \tag{14}$$

The algorithm which we have used is as follows:

1. Values are chosen for CR^2 and $\omega^2 R/g$ and a trial value is assigned to R/b.
2. Equations (2-7) are solved as an initial value problem with Eq.(8) supplying the initial condition. A solution is sought on the interval $\phi = 0^{\circ}$ to $\phi = 90^{\circ}$.
3. If $X/R \neq 1$ at $\phi = 90^{\circ}$ or R/R_1 becomes negative (indicating an inflection point), R/b is altered and step 2 is repeated.
4. The algorithm is terminated either when, at $\phi = 90^{\circ}$, $X/R = 1$ within a desired tolerance, or when R/b is determined to the desired accuracy required for computing $\Delta Z/R$ by Eq.(13).

Fortunately, the iterative scheme outlined above is greatly facilitated by the dependence of the profile on R/b at fixed CR^2 and $\omega^2 R/g$. Let $(R/b)^*$ denote the value of R/b for which the profile satisfies Eq.(14). It turns out that profiles with $R/b < (R/b)^*$ either contain an inflection point or have $X/R > 1$ at $\phi = 90^{\circ}$. Conversely, profiles with $R/b > (R/b)^*$ have $X/R < 1$ at $\phi = 90^{\circ}$. This property allows us to alter the trial value of R/b in the correct direction depending either on the sign of R/R_1 or the value of X/R at $\phi = 90^{\circ}$.

Equations (2-7) were integrated numerically by computer

employing a 4th order Runge-Kutta routine with built-in step size adjustment. The absolute error in each integration step was chosen such that the resulting profile for a given value of R/b was invariant when this error was halved or further divided.

As an additional check on the accuracy of our numerical procedures, the profiles of a variety of rotating and non-rotating axisymmetric menisci were computed and compared to results published in the literature. These configurations include sessile drops (12-14), rotating drops in the absence of gravity (2, 15), and rotating menisci in a vertical tube (11). Our results were generally found to be in agreement with those in the literature within the cited numerical errors.

III. EXPERIMENTAL

A. Materials

Two systems were chosen for preliminary study. The first system studied was the meniscus formed between air and a 10^{-3}M aqueous solution of a nonionic surfactant, i.e., monodisperse octaoxyethylene dodecylether ($C_{12}EO_8$), obtained from Nikko Chemical Co. The surface tension of this solution was 33.1 dynes/cm as measured with a Wilhelmy plate. The CMC of $C_{12}EO_8$ is ca. 7 X 10^{-5} molar.

The second meniscus studied was that formed between water and benzene. The benzene was spectrographic grade obtained from J. T. Baker Comp. The water and benzene were mutually saturated before use. A literature value of the interfacial tension of 34.1 dynes/cm was used (16).

The water was double distilled, the last distillation being from alkaline potassium permanganate.

The CR^2 values (see Eq.(3)) calculated for these systems were 3.266 and 1.137, respectively.

B. Apparatus

The apparatus employed in the present study was originally designed to measure interfacial tensions from the shape of rotating drops in a horizontal tube and was kindly loaned to us by Professor S. G. Mason of McGill University, Montreal. The apparatus consisted of a precision bore, glass tube (~25 cm length and 0.602 cm internal radius) which was mounted on a rigid base through stainless steel end plugs fitted with ball bearings. The tube was driven by a variable speed stirring motor (G.T.-21, G.K. Heller Corp.). This assembly was mounted vertically on a wall and was carefully aligned. There was some play in the bearings which may have introduced a slight error by causing the tube to rotate slightly off-center.

The displacement of the bottom of the meniscus was measured with a cathetometer (Gaertner Scientific) which was carefully aligned. The accuracy of the cathetometer was ± 0.005 cm.

The speed of rotation was measured with a strobe light (Strobotac, General Radio Company) which was claimed to be accurate to ca. 1%.

C. Procedure

The glass tube was cleaned with chromic acid and vigorously rinsed with distilled water and then with the aqueous phase to be used in the experiment. The tube was filled about half with the aqueous phase. In the case of the benzene/water system a layer of benzene about 5 cm high was pipetted onto the aqueous phase. Care was taken to remove any air bubbles, trapped during filling.

The filled tube was attached to the drive and left to equilibrate for ca. 20 minutes while rotating at a speed of ca. 2,000 RPM. Rotation was stopped, and the position of the meniscus was measured after allowing sufficient time for drainage of liquid from the tube walls. At least 4 measurements of this rest position were recorded.

The tube was then rotated at about 2,500 RPM for approximately a minute, whereupon the speed of rotation was reduced somewhat and the experiment begun. The reason for starting the experiment at a relatively high velocity and working down in speed was to establish a receding meniscus and, thus, to have a better chance of satisfying the zero contact angle requirement described in the previous section. Unfortunately, this technique did not always prove successful (see below).

With the present apparatus, the speed of rotation was not sufficiently constant. To reduce the related error to a minimum, experiments were carried out by two workers - one to measure the position of the meniscus and the other to simultaneously measure the angular velocity.

For each run ca. 40 measurements of displacement as a function of angular velocity were recorded. These measurements were clustered about 7 speeds covering the range from $\sim$800-2,200 RPM.

During a run, rotation was stopped periodically and the static level was remeasured. This level was also recorded at the end of a run. Only at the lowest speed did the average of these levels differ by more than 0.2% of ΔZ.

III. RESULTS AND DISCUSSION

A. Theoretical

Examples of calculated profiles of rotating menisci are presented in Fig. 4. These profiles correspond to a $C^{\frac{1}{2}}R$ value of 1.1571. The values of $\omega(R/g)^{\frac{1}{2}}$ are indicated in the figure. All profiles meet the tube (X/R = 1) with zero contact angle. It appears from Fig. 4 that the profiles have a common intersection. However, further analysis has revealed that this is strictly not the case and is only apparent because of the limited range of rotational speeds used in Fig. 4.

The dependence of $\Delta Z/R$ on $C^{\frac{1}{2}}R$ at fixed $\omega(R/g)^{\frac{1}{2}}$ is shown in Fig. 5. Several points concerning this plot should be noted. For all values of $\omega(R/g)^{\frac{1}{2}}$, $\Delta Z/R$ approaches an asymptote as $C^{\frac{1}{2}}R$ increases. This limiting value is recognized as the first term in Eq. (13), i.e., $\omega^2R/4g$, and represents the displacement of the meniscus when capillary forces are negligible (11). Of fundamental importance in connection with the

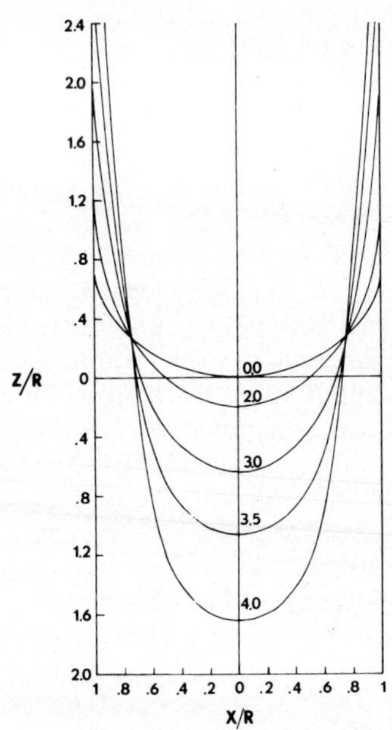

Fig. 4. Computed profiles of rotating menisci for $C^{\frac{1}{2}}R = 1.1571$. The values of $\omega(R/g)^{\frac{1}{2}}$ are indicated in the figure (0.0, 2.0, 3.0, 3.5 and 4.0).

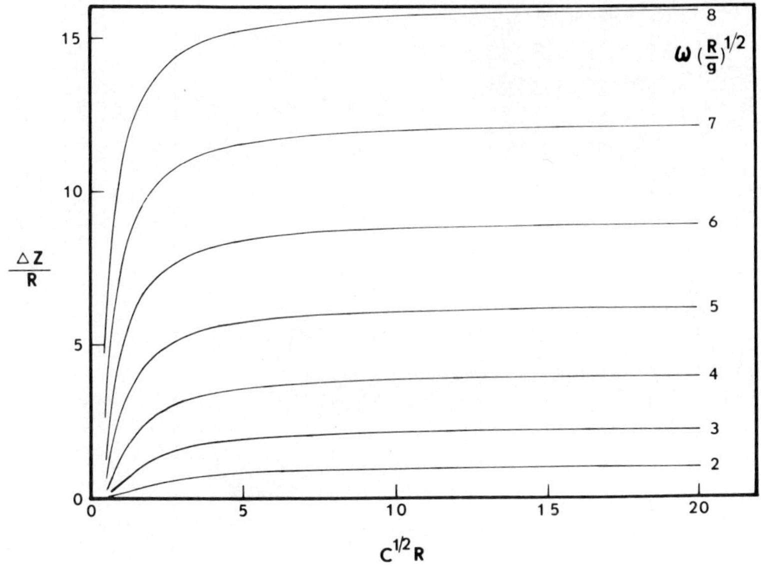

Fig. 5. Displacement of the bottom of the rotating meniscus, $\Delta Z/R$, as a function of $C^{\frac{1}{2}}R$ at constant $\omega(R/g)^{\frac{1}{2}}$.

use of rotating menisci for interfacial tension measurements is the fact that this asymptote is closely approached at fairly low $C^{\frac{1}{2}}R$ values. For example, at $C^{\frac{1}{2}}R>5$, $\Delta Z/R$ is within 10% of the limiting value.

The dependence of $\Delta Z/R$ on $\omega(R/g)^{\frac{1}{2}}$ at fixed $C^{\frac{1}{2}}R$ is shown in Fig. 6. The dashed curve represents the limiting value of $\Delta Z/R$ as $C^{\frac{1}{2}}R\rightarrow\infty$. Again, it is seen that the displacement becomes insensitive to $C^{\frac{1}{2}}R$ when this variable becomes greater than ca. 10.

In a future, more detailed report on this study, we intend to present all our numerical data in tabular form to facilitate their application to practical problems.

B. Experimental Results

Experimental results are compared with theoretical calculations in Fig. 7. Curve a is the calculated curve for the benzene/water system ($C^{\frac{1}{2}}R=1.137$); curve b for the 10^{-3} molar $C_{12}EO_8$/air system ($C^{\frac{1}{2}}R=3.266$). The solid points are the corresponding experimental results which, in most cases, are in agreement with theory within the experimental error.

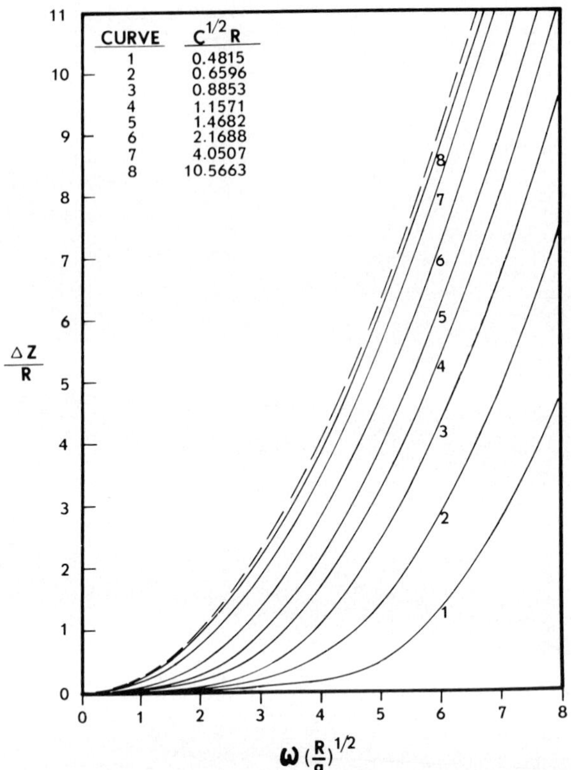

Fig. 6. Displacement of the bottom of the rotating meniscus, $\Delta Z/R$, as a function of $\omega(R/g)^{\frac{1}{2}}$ at constant $C^{\frac{1}{2}}R$.

Another, more practical way of looking at the agreement between theory and experiment is to compare the values of γ, derived from the experiments by means of our theoretical calculations, and the "true" values (34.1 dynes/cm for benzene/water and 33.1 dynes/cm for the nonionic solution/air systems). We find tensions of 35.0 ± 1.1 and 31.8 ± 0.7 dynes/cm, respectively. A preliminary analysis indicates that both the deviations in the mean and the standard deviations are consistent with the experimental errors in ω and ΔZ that can arise with our present set up. The method is capable of much higher accuracy provided the accuracy in the measurement of ω and ΔZ, and the stability of the motor drive, can be increased. There is no doubt that this can indeed be accomplished with some effort, although the method, by its very nature, does not appear attractive for low interfacial

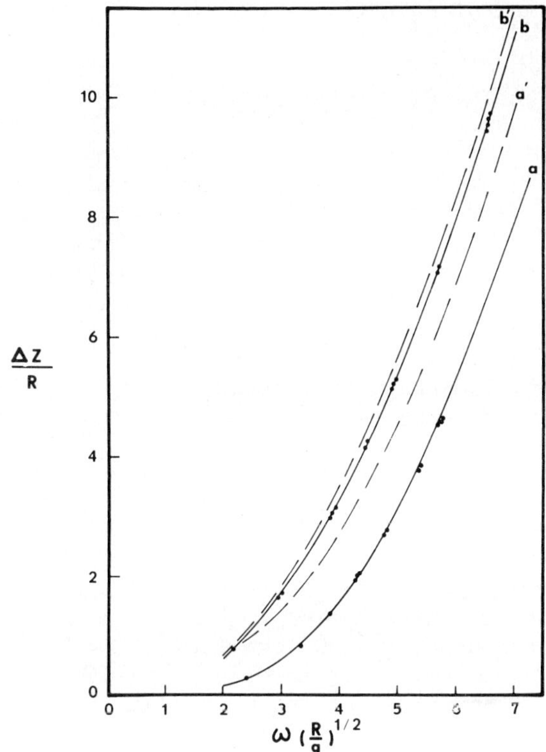

Fig. 7. Comparison between theory and experiment. Curve a is the theoretical curve for the benzene/water system ($C^{\frac{1}{2}}R$=1.137) while curve b is the theoretical curve for 10^{-3} molar $C_{12}EO_8$/air system ($C^{\frac{1}{2}}R$=3.266). The solid points are the experimental results. Curves a´ and b´ are theoretical curves for the respective systems using Radoev's approximation for R/b (10).

tensions (see below).

It should be pointed out that for the benzene/water system complete preferential wetting of the tube by water did not occur, although the receding contact angle was low. Nevertheless, this lack of complete wetting did not appear to strongly influence the results as judged by the agreement observed in Fig. 7 and in the computed value of the interfacial tension.

The dashed curves in Fig. 7, a´ and b´ , are theoretical curves in which an approximation, derived by Radoev (10), and applicable to high speeds, is used for R/b in Eq.(13). It is seen that this approximation of the true value of R/b, obtained from the actual solution of the Laplace equation, is not adequate.

C. Interfacial Tension From Rotating Menisci

The study reported here was undertaken, in part, to assess the use of rotating menisci to measure interfacial tensions. Although limited by the wettability requirement, the method is more convenient in certain respects than that using spinning drops in a horizontal tube. For example, fewer manipulations are generally involved and the volume of the fluid phase need not be accurately known (2, 3). Furthermore, no optical corrections are required (4).

Unfortunately, the shape of a meniscus rotating about a vertical axis depends only partially on surface tension forces. It is clear from Figs. 5 and 6 that the displacement $\Delta Z/R$ becomes less and less dependent on capillary forces as $C^{\frac{1}{2}}R$ increases. Above $C^{\frac{1}{2}}R=10$, the displacement is mainly governed by the balance of centrifugal and gravitational forces. As a practical matter, these effects limit the range of systems whose interfacial tensions can be accurately measured by rotating menisci. The technique would probably be unsuitable for systems of low interfacial tension or large density differences. For example, consider pairs of liquids having density differences in the range of 0.15-0.3 (e.g., octane/water or mineral oil/water). If the interfacial tension were ca. 0.1 dyne/cm, then the value of $C^{\frac{1}{2}}R$ in a tube of 0.5 cm radius would be in the range of 20 to 30. If the interfacial tensions were around 0.01 dyne/cm, then $C^{\frac{1}{2}}R$ would lie between 60 and 85! For such systems, the accuracy required in the measurement of the speed of rotation and the meniscus displacement would be excessive.

IV. REFERENCES

1. Vonnegut, B., Rev. Scient. Instr. 13, 6 (1942).
2. Princen, H. M., Zia, I. and Mason, S. G., J. Colloid Interface Sci. 23, 99 (1967).
3. Torza, S., Rev. Scient. Instr. 46, 778 (1975).
4. Silberberg, A., Ph.D. Thesis, Basel University (1952).
5. Rosenthal, D. K., J. Fluid Mech. 12, 358 (1962).
6. Ross, D. K., Austr. J. Phys. 21, 823 (1968).
7. Cayias, J. L., Schechter, R. S. and Wade, W. H., "Adsorption at Interfaces", ACS Symposium Series No. 8, 234 (1975).
8. Turkington, R. R. and Osborne, D. V., Proc. Phys. Soc. 82, 614 (1963).

9. Wasserman, M. L., and Slattery, J. C., Proc. Phys. Soc. 84, 795 (1964).
10. Radoev, B., Annuaire de l'Université de Sofia" Cl.d' Ochrida" Faculté de Chimie, 61, 361 (1966/1967) (In Bulgarian).
11. Princen, H. M., "Surface and Colloid Science", Vol. 2, Ch. 1, pp 62-80, E. Matijevic, Ed., Wiley-Interscience, New York (1969).
12. Bashforth, F. and Adams, J. C., "An Attempt to Test the Theories of Capillary Action", University Press, Cambridge, 1883.
13. Padday, J. F., Phil. Trans. Royal Soc. London, Ser. A 269, 265 (1971).
14. Lane, J. E., J. Colloid and Interface Sci. 42, 145 (1973).
15. Wade, W. H., Private Communication.
16. Donahue, D. J. and Bartell, F. E., J. Phys. Chem. 56, 480 (1952).

INTERFACIAL TENSION MEASUREMENT BY AN IMPROVED WILHELMY TECHNIQUE

F. Philips Pike and Chandrakant R. Thakkar
University of South Carolina

A new modification of the Wilhelmy technique has been developed for the measurement of interfacial tension, for the case where the lower liquid preferentially wets the Wilhelmy plate. From a balance of the forces existing on the static plate at equilibrium, it is shown that the interfacial force is a straight-line function of the depth of the upper fluid. Using a series of immersion depths, an extrapolation is possible to zero depth, from which the interfacial tension is derived.

In principle, an end correction must be applied, to correct for distortions of the meniscus where it turns around the plate ends. It is argued that the known end correction correlation for the gas-liquid system can be applied to the liquid-liquid case. With this assumption, this Wilhelmy technique becomes a primary measurement, and in addition the densities of the two fluids need not be measured except approximately. Unfortunately, there are many complications that arise from wettability problems of the plate, and the technique is an exacting one.

The new technique was used on the benzene-water system, generating the value of 34.69 dynes per cm at 29$\pm$ 1°c. From a careful survey of the literature values for the interfacial tension of the benzene-water system, and rejection of a portion of the values, the following equation was derived:

$$\gamma = 35.6544 - 0.037983t^{o}c - 0.00062274t^{2o}c.$$

The standard deviation was 0.4705 dyn/cm.

I. HISTORICAL PERSPECTIVE

Historically, physical measurements at interphase boundaries are among the most difficult of measurements to make, being very exacting in their requirements for freedom from contamination. They are also prone to wettability problems. In much of the early work, these problems were not sufficiently recognized and the resultant measurements are primarily of historical value. It is, therefore, difficult to assess the true values. For surface tension results, Harkins (1) argued that most of the data obtained before 1916 are in error, and suggested a separation of data by age. Morino (2) noticed that, if one put faith in some of the 'better' workers, the available data on benzene-air separated naturally into two groups, one of which he considered valid. Bonnet and Pike (3) found the same for nine common liquid pairs. When such arbitrary selections are made for surface tension values, for the more-studied liquids, comparisons between different authors provide a standard deviation of about 0.15 dyn/cm or better. When one examines interfacial tension measurements of the more common liquids, the standard deviation between various selected authors is typically no better than 1 dyn/cm. Even for the best-studied case, benzene-water, the standard deviation at room temperature is 0.47 dyn/cm. One can conclude that interfacial tension measurements are appreciably more prone to error than surface tension measurements, or perhaps some of the liquid-liquid techniques have not been sufficiently examined for flaws in procedure.

A major promise of the Wilhelmy technique, as applied to the liquid-liquid case, is that inherently it is a primary technique measuring the static, or equilibrium, value. There is a need for a meniscus end correction, similar or equal to that provided by Pike and Bonnet (4) for surface tension work. Fortunately, the end correction for liquid-liquid work appears to be small, hence even an approximate end correction might be quite satisfactory. But a review of the technique as a whole revealed serious wettability problems, and uncertainties in the precise details of the experimental aspects.

II. LITERATURE SURVEY ON PROBLEMS

Prior knowledge is assumed of the background information related to the vapor-liquid case, plus knowledge that in comparison the liquid-liquid case is extraordinarily prone to contamination problems. Beyond these aspects, the following problems about the Wilhelmy liquid-liquid case require study and reflection.

A. Physical Arrangements for the Plate

In marked contrast to the vapor-liquid case, in the liquid-liquid case it makes a very great difference whether the top phase or the bottom phase preferentially wets the plate.

The most common experimental technique suspends the Wilhelmy plate from a balance arm that can be precisely reset to an original position, without the use of a cathetometer. This makes it possible to separate out the interfacial tension force as the net difference in the forces above and below the top fluid. It is necessary to know the depth of the top fluid, to calculate a buoyancy term. This situation does not require a cathetometer, although one helps. Now, with this common situation, if the plate is preferentially wet by the lower phase, it turns out that the plate can and must be supported from above. This is the usual case, since the bottom phase is usually an aqueous phase, and usually wets the plate preferentially. Most papers illustrate this case. See Dognon (5), Ruyssen and Loos (6), Cheesman (7), Addison and Hutchinson (8), Jasper and Houseman (9), and Heertjes, de Smet and Witvoet (10).

Incidentally, it apparently is possible to actually use the above rearrangement with a heavy liquid such as chloroform or carbon tetrachloride floated on top of water. Such a system is, of course, not hydrodynamically stable. But, as Dognon (5) demonstrated, the procedure can be made to work if the top liquid spreads on water, and if it is thin enough.

If, on the other hand, the top phase preferentially wets the plate, circumstances require an awkward method of support of the plate that permits the plate to approach the interface from below. Ruyssen and Loos (6) have devised the proper support for this case, demonstrated its adequacy by actual practice, and explained the rationale very well. Heertjes, de Smet and Witvoet (10) also demonstrated that this mechanical support arrangement works.

The requirements imposed on the physical arrangements change abruptly if (a) a cathetometer is available and (b) it is possible to see clearly enough into the working zone. Now it is possible to support the Wilhelmy plate from above in either wetting case, for it is now not necessary to reset the plate edge exactly at the interfaces. Instead, the plate can be positioned to be sure that each meniscus is fully developed, in which case the edge of the plate may well be below an interface. With the cathetometer, the degree of immergence of the plate can be measured, and the required buoyancy terms can be calculated.

If, in case the bottom liquid is a non-polar liquid heavier than water (like carbon tetrachloride), and the top phase is polar (like water), the normal wetting tendency for glass, mica or metal plates is for the polar phase to preferentially wet the plate. Ordinarily, this situation leads to the use of the bottom plate support of Ruyssen and Loos (6). But by suitable surface coatings, such as carbon black or platinum black, the plate can be made hydrophobic. Now, the customary top plate support can be used.

B. Plate Wettability

From surface tension work, it has long been known that plate wettability can be greatly enhanced by roughening or polishing of the plate, and by cleanliness, presoaking, and choice of material or coating. While these techniques virtually always succeed in vapor-liquid cases, the problem is worse in liquid-liquid cases and adequate wettability may then be quite difficult or impossible. Incidentally, the usual plate is made of glass, and a cover glass at that, but platinum, gold, mica, and Teflon are generally satisfactory options.

1. Incompatibility of the Requirements.

A problem unique to liquid-liquid work arises simply because for a liquid-liquid interface to exist at all, the two fluids must be quite "unlike". Consider the case of benzene on top of water. If the water were to wet the plate preferentially, then the benzene would not, and vice versa. As the plate is lowered through the benzene, it acquires an adsorbed benzene layer. Now, when the plate touched the water interface and it was necessary that the water quickly wet the plate with a zero contact angle, the adsorbed benzene interfered greatly. While complete wetting would possibly occur in time as true equilibrium is approached, no one has reported this success. Instead, Dognon (5) reported that in this situation the plate was unevenly wetted and furnished unusable interfacial tension values.

a. Dognon's Solution. Dognon (5) has offered an ingenious and simple solution to this problem that is now virtually standard. Using the benzene-water system as the example, Dognon first wetted the (glass) plate with water, then removed the excess water by careful heating in air. He found that the wetted plate retained an incremental film of adhering water which, incidentally, was visibly apparent by the shiny, mirror-like appearance imparted to the plate where immersed in the benzene phase. Now, when the plate touched the water interface, a water meniscus was instantaneously and violently sucked up onto the walls. Even so Dognon did not claim that the contact angle became zero immediately. Instead, he called for time for equilibration of the system after the plate is finally in place.

b. Problems and Errors Involved. Unfortunately, Dognon gave no clues as to how to control the quantity of the adhering water film, nor what the errors amounted to. From our work, the initial water film amounts to about 1.36 mg/cm^2 on the plate, and contributes to the downward force on the balance arm. What happens to this film after the plate is in place is the big question. For the portion that ends up under the water meniscus, there is little question - it merely joins the bulk water phase and loses its former identity. For the part in contact with the benzene layer and benzene meniscus, it mainly depends on how dry the benzene was initially. If undersaturated, this film portion tends to dissolve and may disappear entirely, according to Ruyssen and Loos (6). As for the water film above the benzene meniscus, it also may or may not evaporate, depending upon the initial benzene dryness, and the time allowed for equilibration. What is needed is some technique to guarantee the constancy of the amount of water film on the plate at equilibrium, and some procedure to measure it. In this work, a behavior model is developed and used to generate a correction calculation. The correction raises the interfacial tension only about 1%.

c. Controlling the Initial Thickness of Water Film. The success of a given initial water film thickness depends in part on the initial dryness of the benzene. Besides this factor, if the initial film thickness is too small, the completeness of wetting may deteriorate. If too great, the correction involved may either become too large or too uncertain. Accordingly, the initial film should be controlled quantitatively. Dognon (5) merely spoke of drying the plate carefully (presumably in air) without heating. Thakkar (11) found it convenient to allow the suspended wetted plate to drain in a protected environment. Excess water tended to collect along the lower edge as a bead. From time to time this excess was removed by touching with a piece of filter paper, until no further drainage was visible. In addition, some experimentation was done with drying by use of infrared lamps. Guastalla (12) showed that, if the plate is dried too much, it wouldn't work perfectly even for the vapor-liquid case.

d. Controlling the Final Amount of Adhered Water. In addition to the initial amount of water adhering to the plate, two factors are involved - the initial dryness of the benzene and the time allowed for equilibration. Thakkar (11) generally employed a partially-saturated benzene, and allowed 45 minutes for equilibration, twice the time for readings usually to stabilize. Ruyssen and Loos (6) found that if the fluids were not substantially saturated before introduction of the plate, then the adhering water film tended to dissolve completely. As is seen by the results in Section VI, Thakkar obtained results that were quite self-consistent. Apparently, prior workers also felt that their results were satisfactorily consistent. But obviously, some better control is desirable.

e. Alternate Solution. A technique of some promise is suggested as an alternate to Dognon's solution. Using the benzene-water system as an example, first establish a water layer without any supernatant benzene. Now, condition the Wilhelmy plate with as little known surface moisture as possible, yet permitting the plate to be wetted properly upon contact with water. Next, position the plate at the water interface, complete with meniscus. In auxiliary apparatus, condition beforehand some purified benzene to a known water content. Using a pipet, layer a known volume of benzene carefully over the water with a minimum of disturbance. At this point, an upper, benzene meniscus appears. If the water content of the benzene layer is initially within a suitable range, the benzene meniscus should develop adequately. It would help if Addison's technique (see later) were used at this point to aid the development of both menisci. Finally adjust the plate edge to exactly the water interface and measure the sum of the surface and interfacial tension, plus the depth of the benzene layer. In comparison with Dognon's technique, the amount of final adhering water film should be much less, negligible or nearly so. Now all that is needed is the surface tension pull. Remove the plate, evaporate the benzene from it, condition to the same moisture content as before, then establish the benzene meniscus and measure the surface tension pull. Hopefully, this technique assures adequate wettability at the two menisci with an adsorbed water film so small as either to be negligible or easily estimated. Interestingly enough, Dognon (5) suggested the essence of this alternate procedure, although he apparently contemplated only adding a very thin layer of benzene (a matter of drops).

f. Conclusions. The inherent contradictions related to the wettability problems of the plate give rise to difficult problems. Fortunately, Dognon's ingenious technique holds great promise, in spite of the fact that a number of specific problems have surfaced. The present behavior model leads to a correction so small, about + 1%, that it doesn't have to be a precise model.

2. Addison's Technique.

In 1948, Addison and Hutchinson (13) reported a simple manipulative technique that greatly speeds up the adequate wetting of Wilhelmy plates in either the vapor-liquid or liquid-liquid case. They simply moved the plate beyond its resting position, causing the meniscus to wet more of the plate than needed, then retracted the plate. Now the meniscus would approach its equilibrium position by draining downward, aided by gravity, rather than the reverse. They do not specify any time in the extended position, but we have used 5 to 30 minutes. The technique is simple, easy and very useful.

3. Insoluble Films.

When a third component is present, and particularly when it is a large organic molecule, sometimes an insoluble film precipitates out on the plate. Alexand and Teorell (14) noted this behavior with either glass, mica or platinum plates, when proteins were added to an oil/water system. This precipitate led to severe wettability problems that they could find no way to alleviate. Later, Cheeseman (7) encountered the same troubles with other proteins. He found that by precoating the plate with soot, he could obtain reproducible wetting, and contact angles of zero. No doubt there are some 3 component systems for which a Wilhelmy plate cannot be wetted properly.

4. Heertjes' Case

Heertjes, de Smet and Witvoet (10) found that with platinum plates they could not measure the interfacial tension between water and insoluble alcohols. They attributed these wetting problems to the fact that the affinities between water and alcohols for platinum are almost the same. In other words, here is a case where the wetting preference for either phase is closely 1.00. Fortunately, these authors did find that if the platinum plate were coated with platinum black, incomplete wetting by one of the liquids could be prevented. Again, there are probably systems that cannot be adapted to the use of a Wilhelmy plate.

5. Adsorbed Ions

Tallmadge and Stella (15) encountered a damaging influence of the conventional chromate-sulfuric acid cleaning solution. They found that the weight of adhering water film on glass is considerably greater if chromate solution is used, than for other cleaning methods. Others, like Laug (16) have shown that traces of chromate strongly adhere to the glass and only desorb slowly. Sutherland (17) demonstrated that a large ion, hexadecyl sulfate, strongly and rapidly affected surface tension readings when using a mica Wilhelmy plate. Further, when both the large ion and traces of electrolytes were present, the results were influenced greatly by even the traces of electrolytes present in triple distilled water. Winstead (18) recommended the use of nitric acid for the cleaning of glassware, displacing acid chromate solutions. Davis and Rideal (19) advocated that, if chromate-acid solutions are used for cleaning, then a washing with syrupy phosphoric acid be used subsequently to remove the chromate, and a second washing with distilled water be used to remove the phosphate.

It is clear that the cleaning procedure for the plates must avoid unsuitable options. An empirical testing program is recommended.

C. Interfering Menisci

With either the Dognon or the present technique, there are two menisci present simultaneously on the Wilhelmy plate. As the thickness of the upper layer decreases, these two menisci mesh closer and closer together. There must come a point at which the two menisci interfere. Intuitively, this would occur at the minimum spacing that two unperturbed menisci would allow. This is 0.50 cm by calculation for benzene-water, and would mean that the minimum depth of benzene layer would be that same 0.50 cm. But our results show that consistent measurements were made at layer depths of 0.22 cm, while a depth of 0.13 clearly gave improper measurements. Obviously, then, some physical interference can occur without seriously changing the interfacial forces at the plate. But the unperturbed minimum spacing is a useful guide.

D. Time to Reach Equilibrium

It is inherent in a static test that sufficient time must be allowed to reach substantial equilibrium. For example, if the two layers were not sufficiently saturated with each other at the start, the thin water film on the plate soon dissolved causing the readings to drift away from equilibrium over the normal span of time. Ruyssen and Loos (6) made the above observation, and also stated that approximately 30 minutes sufficed normally for pure two-component systems with layer depths up to 0.75 cm. Greater depths require longer time. Also, for multi-component systems like paraffin-water, the time for equilibrium was many hours. The experience of Pike is that for presaturated benzene-water equilibrium was established within 20 minutes, but if a 3rd component were added (detergent) it took 30 - 40 hours. In this work, 45 minutes was allowed as a standard.

E. Plate Suspension

The method of suspension of the Wilhelmy plate is a sensitive point of technique. It is necessary to have the bottom edge exactly parallel to the interface before and during a measurement. A number of workers have used a semi-rigid wire attached to the plate. Such a wire can be easily bent to place the plate exactly horizontal before touching the interface. But often such bending leaves residual strains that tend to distort the shape but are not in themselves strong enough. Now when the plate touches the interface, the added tension force very often causes the plate to twist badly. Although the plate can generally be properly positioned by repeated trials, the process can be difficult. Alternately, the support can be made rigid enough, like a strong wire, or a glass rod. Another alternative is to suspend by means of a flexible thread for it does not retain stresses.

To tell if the plate is horizontal, before contacting an interface, a very sensitive technique can be employed. The plate edge is brought close to the interface, and that edge is observed in comparison with its mirror-image in the surface. The eye can detect the slightest deviation from parallelism, which means that the plate is exactly horizontal.

After contact, the plate has been subject to the relatively strong grasp of the meniscus. If the meniscus forces leave a permanent tilt to the plate, the tension readings are compromised. It is necessary to check carefully for horizontally during and after a measurement. Ruyssen and Loos (6) have explored some of the consequences of tilting.

F. End Correction for the Liquid-Liquid Case

The need for an end correction for the vapor-liquid case was reviewed by Pike and Bonnet (4), who developed by experiment a simple, predicting, dimensionless equation. The factors that require an end correction for the gas-liquid case apply equally to the liquid-liquid case. Other measurement techniques also call for corrections to the basic integrations of the Young-Laplace equation. In the case of the du Nouy ring technique and the drop weight/volume method, Alexander and Hayter (20) summarized the situation. For the drop/WV method, the stated correction factor is solely a function of a geometric factor. It is true that Alexander and Hayter (20), page 514, expresses the feeling that the original calibration for the gas-liquid case is likely to be several percent in error for the liquid-liquid case. But they do not justify this statement. And in practice, no distinction is made. In the Wilhelmy case, a 'several percent' error would be negligible, since the correction calculated for the benzene-water case is about + 0.4%. In the du Nouy ring technique, Alexander and Hayter (20) accept completely the idea that the correction is independent of the pairing of the fluids.

It seems reasonable to assume that the dimensionless end correction correlation of Pike and Bonnet (4) holds equally for the liquid-liquid case. Not only does the assumption appear appropriate, the corrections are in any case quite small and no great error seems likely.

III. THIS SPECIFIC TECHNIQUE

A. Summary

Two liquid layers of purified chemicals are first established in a controlled zone, with a vapor space above. These layers are presaturated to an appreciable but controlled degree dictated by experience. In simple terms, a dry carefully-cleaned Wilhelmy plate is touched to the upper surface of the top fluid, thus registering the surface tension force on a balance arm. The plate is now removed, dried, wetted thoroughly with the lower fluid, drained with the aid of filter paper and perhaps dried further in a controlled manner. The plate now possesses a definite adhering fluid film. It is immersed in the top fluid and touched to the liquid-liquid interface, registering the sum of the surface tension and the interfacial tension plus some buoyancy terms. Time is allowed for the attainment of equilibrium, then some. The difference between these two force measurements, properly corrected to ΔFc, measures the interfacial term plus a single buoyancy term. Care must be taken to assure a contact angle of zero for each meniscus each time. The details necessary to accomplish this form part of the necessary experience.

Measurements are made over a series of depths h of the top fluid, down to about half of the height difference between the two menisci involved, when they are developed alone and unperturbed. According to the model developed, the ΔFc term is a straight line function of h, down to values of h so low that the two menisci interfere in force terms. By analysis of the data, the nonvalid h measurements are located and discarded. Using the remaining data, the interfacial tension is calculated.

For the calculation procedure, the end correction is calculated from the known correlation. It is not necessary to measure accurately the fluid densities involved. In fact, it suffices to estimate them.

B. Points of Technique

The following pertains to the specific procedures used in this work.

1. Cleaning of the Plates

A standard procedure was adopted for the cleaning and storage of the glass plates. The plates were soaked in hot chromate-sulfuric acid solution for hours, until the cleaning solution formed a smooth, uniform film over all of the glass upon withdrawn. Generally it took about 3 hours to accomplish this, but often the plates were soaked overnight. Before intended use, the plates were washed slowly for 10 minutes with batches of syrupy phosphoric acid. Next the plates were flushed for about 2 minutes in running tap water, then rinsed twice in high quality distilled water. Finally, they were rinsed with redistilled acetone. If a dry plate were desired, it was dried by an infrared lamp. Plates to be used wet were stored under distilled water, properly covered. When a plate developed a non-wetting behavior, it was recleaned by the entire process.

2. Controlling the Adhering Water Film

Plates to be used wet were removed from storage under water, and suspended in air in a protected, dust-free area and allowed to drain. They were suspended in exactly the same manner as used for the surface property measurements. Excess water collected as a bead along the lower edge. This excess was removed from time to time by touching the water with a clean piece of absorbent paper that doesn't shed fibers. The touching was repeated until there was no visible bead after minutes of draining, say 5 minutes. Some plates were dried further by infrared lamps.

It would have been helpful to control the amount of residual water by weighing. The actual control was the appearance of the plate when immersed in benzene, the regularity of the menisci, and the history of the tension forces against time. More runs were discarded than retained. The average adhering film amounted to 30 mg.

3. Suspension of the Plates

The plates were suspended from a piece of cotton thread attached to a hook. The cotton thread had been boiled several hours in distilled water. The attachment to the glass plate was by one centered hole; two end holes would have been better. The plates were made level by the placement on the plates of small platinum wire bent pieces. The horizontality was checked from time to time by visual comparison of the bottom edge with its reflection from one of the working fluids.

4. Establishment of the Benzene Layer

The purified benzene was added by pipet, and the depth was measured by a cathetometer. The reproducibility of the depth measurements was obscured by poor visibility, but was estimated to be $\pm$ 0.2mm. For most of the runs, the initial benzene was close to, but not at, saturation.

5. Time to Reach Equilibrium

The time to substantially reach steady state was about 20 - 25 minutes. To be safe, the standard minimum time was set to be 45 minutes.

IV. EQUIPMENT AND SUPPLIES

A. Tensiometer

The apparatus employed was a commercial unit marketed by the Roller Smith Division, Federal Pacific Company, and known as the Rosano Tensiometer. Essentially it is an analytical torsion balance slightly modified to permit controlled suspension of the Wilhelmy plate. Its range was 1500 mg, readable to 0.1 mg.

B. Local Environment

A simple box-like wooden frame was built around the tensiometer, with walls of 10 mil polyethylene. Overall, the box was 24" high, 27" wide, and 18" deep. The front face was a moveable flap. The protection afforded against dust and air currents was most helpful.

C. Temperature Control

The normal control failed. During the runs it was 29 $\pm$ 1 ^{o}C.

D. Wilhelmy Plates

The Wilhelmy plates were Corning microscope slides made of hard glass specialized for this purpose. The plates were smooth, 4.992 cm long, 2.20 cm wide, and 0.015 cm thick, with factory-ground edges. These edges were much smoother than edges we made with a fine diamond saw.

E. Cathetometer

Vertical distances were measured by an Ealing Corp. Model No. 11-53520 cathetometer and measuring microscope. The usable range was 100 cm, readable to 0.01 mm.

F. Working Space

The working volume was essentially a jacketed Pyrex

vessel about 8.0 cm inside, and 250 ml volume. The jacketing unfortunately obscured the view of the working region. The size of the vessel was large enough to provide a negligible vessel correction according to Kawanishi, Seimiya and Sasaki (21).

G. Chemicals

The benzene employed was reagent-grade, purified further by three slow partial crystallizations. The distilled water was of very high purity. Ordinary distilled water was treated with acidic $KMnO_4$, and distilled in liter batches from all-Pyrex equipment, with front-and tail-end discard. It was stored in clean leached Pyrex flasks, closed with Pyrex stoppers. Acetone was used to flush and clean various equipment items. All acetone was redistilled before use, and carefully stored in glass.

V. MATHEMATICAL RELATIONSHIPS

There are four different physical relationships that are pertinent to the study, in regard to plate position with respect to the fluid layers, the existence of a meniscus, and the wetted condition of the hanging plate. Figure 1 illustrates these four positions or Cases, and serves to illustrate some of the variables.

The four Cases are designated by subscripts 0, 1, 2 and 3. By its use, the subscripts define these four circumstances, or Cases:

0 = the dry plate plus the supporting thread and hook, suspended in the vapor space above the two liquids.

1 = the plate wetted with an adhering, smooth water film, plus supporting thread and hook, suspended in the vapor space.

2 = the dry plate, plus supporting thread and hook, suspended with bottom edge exactly at the benzene-vapor surface, and wetted by an equilibrium benzene meniscus.

3 = The Wilhelmy plate functioning to measure interfacial tension. The plate, plus thread and hook, is supported with the bottom edge exactly at the benzene-water interface. It has two menisci, one above the other. In this model, of the original adhering water film (Case 1), only the part above the benzene meniscus remains. The part that initially contacted the benzene phase has mainly dissipated.

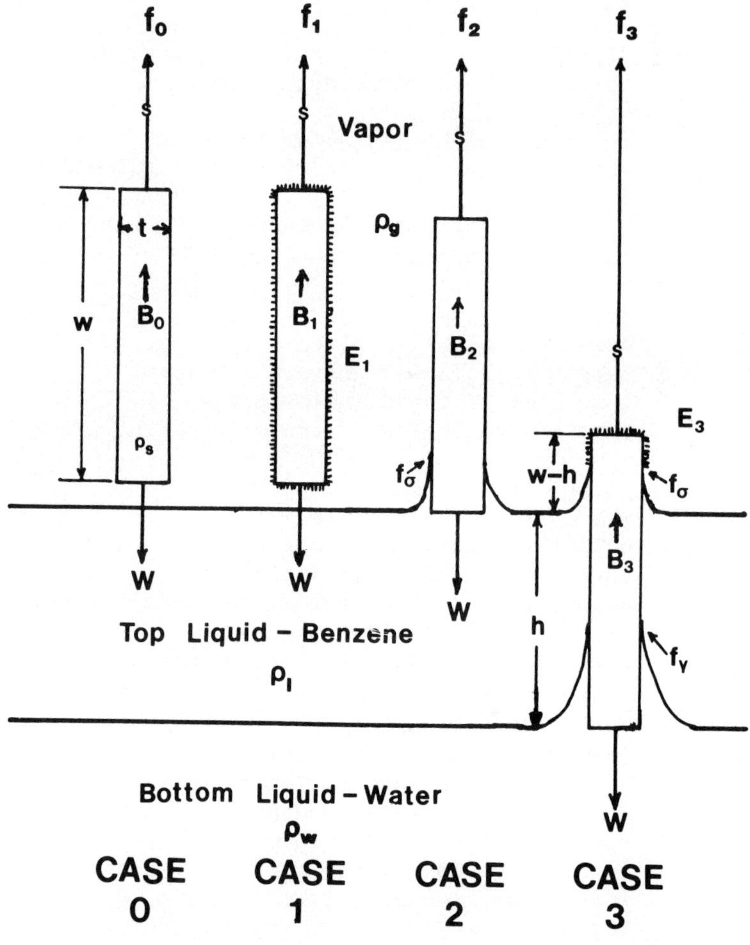

Fig. 1. Four cases illustrating four pertinent situations involving the Wilhelmy plate.

The symbols are defined as follows:

Symbol		Definition
f	=	upward force exerted by the balance arm when supporting a plate
F	=	difference in f between two cases
F_1	=	$F_1 = f_1 - f_0$ = Case 1 - Case 0
F_2	=	$F_2 = f_2 - f_0$ = Case 2 - Case 0
F_3	=	$F_3 = f_3 - f_1$ = Case 3 - Case 1
ΔF	=	$F_3 - F_2$ = difference in the F force terms between the two major working positions of the plate
ΔF_c	=	$\Delta F + \Delta E = \Delta F + f$ (water film) = corrected ΔF
B	=	upward buoyant force acting on plate, in Cases 0, 1, 2 and 3
W	=	inherent weight (mass) of the clean, dry plate
E_1	=	weight of complete adhering water film in Case 1.
E_3	=	weight of the partial water film on the plate in Case 1, according to Figure 1.
ΔE	=	$E_1 - E_3$
h	=	depth of benzene layer
w	=	height of Wilhelmy plate = 2.20 cm
ℓ	=	length of Wilhelmy plate = 4.992 cm
t	=	thickness of the Wilhelmy plate = 0.015 cm
A	=	area of the bottom or top edge of the plate $A = \ell t$
V	=	volume of the plate, $V = Aw$
p	=	wetted perimeter = $2(\ell + t)$
c	=	corner correction for surface tension, cm
c'	=	corner correction for interfacial tension, cm
P_d	=	downward hydrostatic pressure acting on the top of the plate, g/cm^2
P_u	=	upward hydrostatic pressure acting on the bottom of the plate, g/cm^2
ρ_g	=	density of the vapor
ρ_ℓ	=	density of the upper fluid
ρ_w	=	density of the bottom fluid
ρ_s	=	density of the solid plate
$\Delta\rho$	=	$\rho_\ell - \rho_g$
σ	=	surface tension, vapor-benzene, dyn/cm
γ	=	interfacial tension, liquid-liquid, dyn/cm
f_σ	=	downward force on the plate due to the surface tension pull (zero contact angle)
f_γ	=	downward force on the plate due to the interfacial tension pull (zero contact angle)

$$f_\sigma = \frac{1}{g}\left[p-4c\right]\sigma \;, \quad \cos\theta = 1$$

$$f_\gamma = \frac{1}{g}\left[p-4c'\right]\gamma \;, \quad \cos\theta = 1$$

$$g = \text{local gravitation constant} = 979.5 \text{ cm/sec}^2$$

A hydrostatic balance for Case 3 is as follows:

$$P_u = P_d + (w-h)\,\rho_g + h\rho_\ell = g/cm^2$$

The total downward force, neglecting the cross-sectional area of the adhering water film, is $P_d \times A$. The upward force = $P_u \times A$. Thus, the net buoyant force is $B_3 = (P_u - P_d)A$. Substituting for P_u,

$$B_3 = V\rho_g + h(\rho_\ell - \rho_g)A$$

Similarly, $B_o = V\rho_g = B_1 = B_2$

A force balance for Cases 3 and 1, in grams, give

$$f_3 = W + f_\sigma + f_\gamma + E_3 - B_3$$

and $f_1 = W + E_1 - B_1$

But $F_1 = f_3 - f_1$, thus

$$F_3 = f_\sigma + f_\gamma + (E_3 - E_1) - (B_3 - B_1)$$

Similarly, $f_2 = W + f_\sigma - B_2$

$$f_o = W - B_o$$

$$F_2 = f_1 - f_o = W + f_\sigma - B_2 - W + B_o$$

$$F_2 = f_\sigma - (B_2 - B_o)$$

Since $\Delta F = F_3 - F_2$

$$\Delta F = f\gamma - \Delta E - B_3 + B_1 - B_2 + B_o$$

This assumes that f_σ is the same for both Case 2 and Case 3, although there is a slight difference between the two, if Case 2 refers to dry benzene. Since here the benzene is nearly saturated, the difference is quite slight.

Now $$B_o = B_1 = B_2 = V\rho_g, \quad \text{and}$$

$$\Delta F + \Delta E = f\gamma - B_3 + V\rho_g$$

$$\Delta F_c = f\gamma - V\rho_g + h(\rho_\ell - \rho_g)A - V\rho_g$$

$$\Delta F_c = f\gamma + h\Delta\rho A = \Delta F + \Delta E$$

This is the final working equation, and is in the form

$$\Delta F_c = a - mh = \Delta F + \Delta E$$

where $$\Delta E = E_1 - E_3$$

The term E_1 is easily measured as $f_1 - f_o$. Experimentally, $E_1 = 30$ mg on the average. It would have been better to have measured E_1 for each run. In contrast, E_3 is difficult or impossible to measure.

In Case 3, h cm of the Wilhelmy plate is immersed in benzene out of the total plate height if w = 2.20 cm. It is safe to assume that the water film that had been on the plate, in the region now under the water meniscus, has lost its original meaning. It has merely merged with the bulk water. The region of the plate above the benzene phase is assumed to hold as much adhering water as it always did, although this could be questioned. If the original top layer of benzene were too dry, all of the water would evaporate from the plate surface. This occurred all too readily. As the water content of the initial benzene increased, it appeared that the plate surface in contact with the benzene dried faster than the top region. It was felt that one explanation of the many discarded runs is that not enough adhering water dissolved into the benzene. Therefore, for the successful runs, it is assumed that the residual water is all in the upper region. This assumption gave consistent results, which is reassuring. But obviously some more quantitative judgements are desirable.

At the moment the model is as follows. The height of the benzene meniscus by calculation and by measurement is 0.25 cm. The E_3 term is calculated by this equation:

$$E_3 = E_1 \left(\frac{2.20 - h - 0.25}{2.20}\right)$$

$$\Delta E = E_1 \left(\frac{h + 0.25}{2.20}\right) = 30 \left(\frac{h + 0.25}{2.20}\right)$$

As seen in Table 1, the ΔE term is significant at level of 2 to 10% of ΔF. But since we use, in effect, the value of ΔE extrapolated to $h = 0$, the final effect of having a ΔE term is to increase the interfacial tension about 1%.

VI. MEASUREMENTS AND RESULTS

TABLE 1
Measurements of Various Interfacial Factors at Various Depths of the Benzene Layer

	Measured		*Calculated*		
Depth h	F_3	F_1	ΔF	ΔE	ΔF_c
cm	*mg*	*mg*	*mg*	*mg*	*mg*
1.44	504.	294.	210.	23.5	233.5
1.01	543.	294.	249.	17.2	266.2
0.77	564.	294.	270.	13.9	283.9
0.69	575.	294.	281.	12.8	293.8
0.67	580.	294.	286.	12.6	298.6
0.64	582.	294.	288.	12.1	300.1
0.53	590.	294.	296.	10.7	306.7
0.25	622.	294.	328.	6.8	334.8
0.22	622.	294.	328.	6.4	334.4
0.13	614.	294.	320.	6.2	325.2

Temperature = 29 ± 1 ^{o}c

The correlating equation is $\Delta F_c = f\gamma - mh$. A plot of the data according to this equation is shown on Figure 2.

All the points seem consistent, excepting the one at $h = 0.13$ cm.

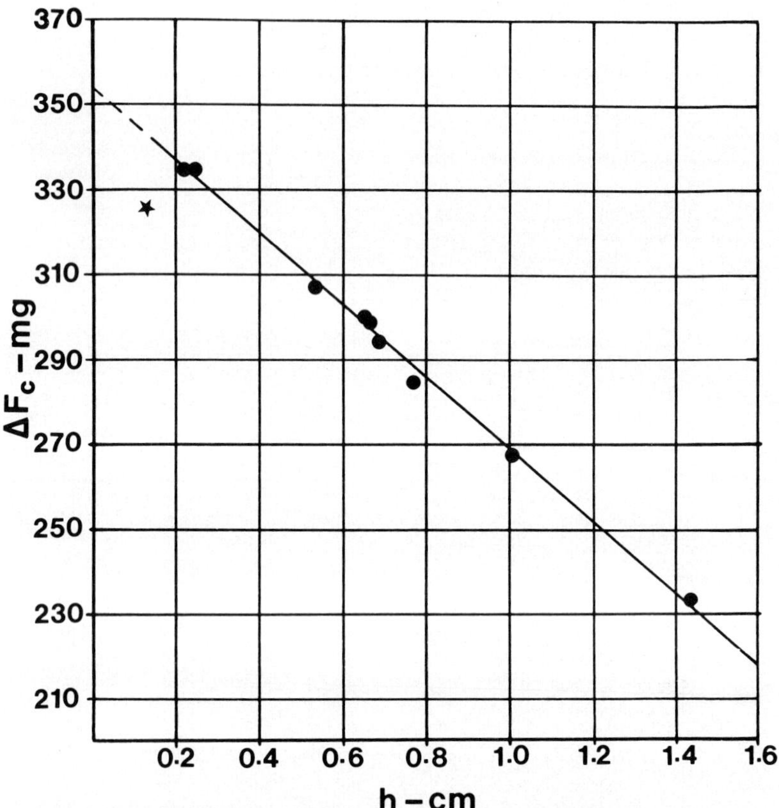

Fig 2. Illustration of ΔFc as a function of h, the depth of the benzene phase. The starred point represents meniscus interference.

Omitting this point, the figures were treated by the method of least squares to yield

$$\Delta Fc = 353.258 - 84.7755h = mg$$

The correlation coefficient was 0.9974, indicating a very good fit. The standard deviation in ΔFc was 2.445 mg, and

$$f_{\gamma} = 353.258 \text{ mg} = 0.35326g$$

Using the physical properties of the saturated benzene-water system at 30°C, and assuming that the interfacial tension was 34.02 dyn/cm, the dimensionless parameter of Pike and Bonnet (4)was calculated to be $T = 8.264 \times 10^{-4}$. From the given correlating equation,

$$4c' = 0.040 \text{ cm}$$

Turning to the end-corrected Wilhelmy equation, with $\cos\theta = 1.0000$, and $g = 979.5\ \text{cm/s}^2$.

$$\gamma = \frac{gf\gamma}{p-4c'}$$

$$\gamma = 34.69 \text{ dyn/cm, at } t = 29 \pm 1\ ^{\circ}\text{C}$$

The uncorrected value was 34.56, hence, the use of the end correction raised the indicated value by 0.14 dyn/cm, equivalent to 0.40%.

VII. THE BENZENE-WATER SYSTEM

An intensive literature survey was made on the interfacial tension of the benzene-water system. Unfortunately, the data scattered badly and many of the results are seriously in question. Guided primarily by the results of Harkins, Bartells, Addison, Hutchinson, Jennings and Padday, the data were separated into two classes, those to be taken seriously at this time (99 data points), the other (30 data points) judged to be primarily of historical value.

The accepted data were from references (6, 10, 11, 22 - 73) plus (74) as recalculated by (37), (75) assuming t = 20°C, (76) excepting 32.2 point, (77) as corrected by the author (78) excepting 32.9 value. Of the results of Jennings (79), the two values at 100°C and 176°C were excluded, although in retrospect they should have been included. The results of Cini, Ficalbi and Loglio (80) are most interesting to this compilation, for they obtained 41 values from 6.2°C to 48.9°C. However, their results were presented graphically and were not available as numbers until this study was concluded. In general, their values were within 0.2 dyn/cm of our correlation.

The literature values are plotted on Figure 3, with the solid points representing the accepted values, the other points the rejected values.

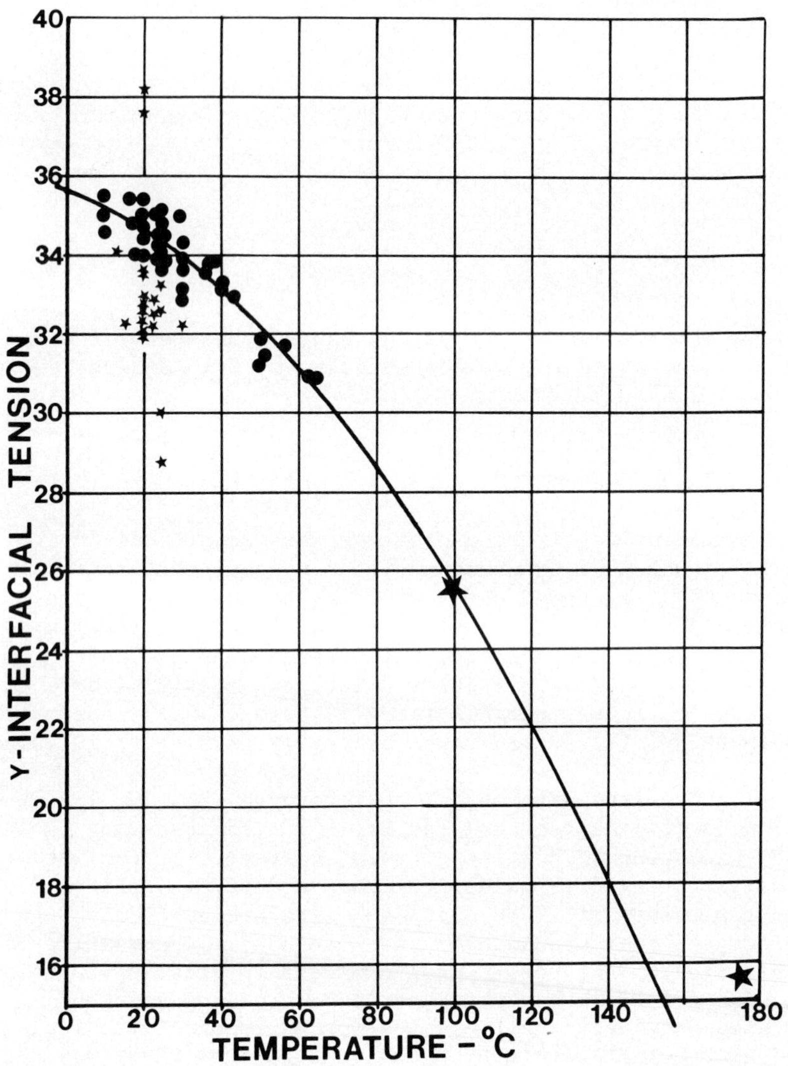

Fig. 3. Literature interfacial tension measurements as a function of temperature. The solid points are selected values. The starred points are rejected values. The large stars are points of Jennings (79).

The accepted data points were fitted by the method of least squares, using a statistical program compiled August, 1972, by A. J. Barr and J. H. Goodnight of North Carolina State University. The PROC REGR Subprogram of their SAS package was employed. The linear fit gave

$$\gamma = 36.3290 - 0.082246\ t^{o}c$$

with a standard deviation of 0.4793 at the mean value of 34.155 dyn/cm. The R squared value was 0.74431 and the F statistic was 276.54. The quadratic fit gave

$$\gamma = 35.6544 - 0.037983\ t^{o}c - 0.00062274\ t^{2\,o}c$$

with a standard deviation of 0.4705 at the mean. The R squared value was 0.75612 and the F statistic 145.72. The quadratic fit was statistically better. Incidentally, the predicting equations lose their meaning below the triple point of the benzene-water system, namely 5.55°c.

VIII. REFERENCES

1. Harkins, W. D. in "Physical Methods of Organic Chemistry", (A. Weissberger, Ed.), Vol. 1, Interscience Publ., New York, 1945.
2. Morino, Y., Sci. Pap. Inst. Phys. Chem. Res. (Japan) 23, 49 (1933).
3. Bonnet, J. C. and Pike, F. P., J. Chem. Eng. Data 17, 145 (1972).
4. Pike, F. P. and Bonnet, J. C., J. Colloid Interface Sci. 34, 597 (1970).
5. Dognon, A., Compt. Rend. 212, 854 (1941).
6. Ruyssen, R. and Loos, R., Meded. K. Vlaam. Acad. Wet. Lett. Schone Kunsten Belg. Kl.Wet.7, 5 (1945).
7. Cheesman, D. F., Arkiv Kemi Mineral. Geol. 22B, 1 (1946).
8. Addison, C. C. and Hutchinson, S. K., J. Chem. Soc. 930 (1948).
9. Jasper, J. J. and Houseman, B. L., J. Phys. Chem. 67, 1548 (1963).
10. Heertjes, P. M., de Smet, E. C. and Witvoet, W. C., Chem. Eng. Sci. 26, 1479 (1971).
11. Thakkar, C. R., "Interfacial Tension Measurement by an Improved Wilhelmy Technique", M. S. Engr. Thesis, Univ. South Carolina, 1973.
12. Guastalla, J., J. Chim. Physique 51, 583 (1954).

13. Addison, C. C. and Hutchinson, S. K., J. Chem. Soc, 930 (1948).
14. Alexander, A. E. and Teorell, T., Trans. Faraday Soc. 35, 727 (1939).
15. Tallmadge, J. A. and Stella, R., A. I. Ch. E. J., 14, 838 (1968).
16. Laug, E., Ind. Eng. Chem. And. Ed., 6, 111 (1934).
17. Sutherland, K. L., Aust. J. Chem. 12, 1 (1959).
18. Winstead, M., "Reagent Grade Water", South Bend Medical Foundation, South Bend, Indiana, 1967.
19. Davies, J. T. and Rideal, E. K., Biochem. Biophys. Acta 11, 165 (1953).
20. Alexander, A. E. and Hayter, J. B., "Physical Methods of Chemistry" (A. Weisberger and B. W. Rossiter, Eds.) Wiley-Interscience, 1971.
21. Kawanishi, T., Seimiya, T. and Sasaki, T., J. Colloid Interface Sci. 32, 622 (1970).
22. Hardy, W. B., Proc. R. Soc. London 88A, 303 (1913).
23. Lorant, O., Pfluger's Arch. Gesamte Physiol. Menschen Tiere 157, 211 (1914).
24. Harkins, W. D. and Humphrey, J. Am. Chem. Soc.38, 242 (1916).
25. Harkins, W. D., Clark, G. L. and Roberts, L. E., J. Am. Chem. Soc. 42, 700 (1920).
26. Harkins, W. D. and Cheng, Y. C., J. Am. Chem. Soc. 43, 35 (1921).
27. Reynolds, W. C., J. Chem. Soc. 119, 466 (1921).
28. Johansen, E. M., Ind. Eng. Chem. 16, 132 (1924).
29. Harkins, W. D. and McLauglin, J. Amer. Chem. Soc. 47, 1610 (1925).
30. Harkins, W. D. and Zollman, H., J. Am. Chem. Soc. 47,69 (1926).
31. Pound, J. R., J. Phys. Chem. 30, 791 (1926).
32. Bartell, F. E. and Miller, F. L., J. Am. Chem. Soc. 50, 1961 (1928).
33. Harkins, W. D. and Jordan, H. F., J. Am. Chem. Soc. 52, 1751 (1930).
34. Weber, L. J. and Traube, I., Biochem. Z. 219, 468 (1930).
35. Bartells, F. E. and Mack, G. L., J. Phys. Chem. 36, 65 (1932).
36. Mack, G. L. and Bartells, F. E., J. Am. Chem. Soc. 54, 936 (1932).
37. Bartell, F. E., Case, L. O. and Brown, H., J. Am. Chem. Soc. 55, 2769 (1933).
38. Speakman, J. C., J. Chem. Soc. 1449 (1933).
39. Carter, E. G. and Jones, D. C., Trans. Faraday Soc. 30, 1027 (1934).

40. Brown, F. M. and Elgin, J. C., Ind. Eng. Chem. Anal. Ed. 7, 399 (1935).
41. Stroganov, N. S., Protoplasma 24, 431 (1935).
42. Gibby, C. W. and Addison, C. C., J. Chem. Soc. 119 (1936).
43. Alexander, A. E. and Teorell, T., Trans. Faraday Soc. 35, 727 (1939).
44. Addison, C. C., Nature (London) 144, 783 (1939).
45. Dognon, A. and Abribat, M., Bull. Soc. Chem. Biol. 23, 62 (1941).
46. Zuidema, H. H. and Waters, G. W., Ind. Eng. Chem. Anal. Ed. 13, 312 (1941).
47. Bartells, F. E. and Davis, J. K., J. Phys. Chem. 45, 1321 (1941).
48. Transue, L. F., Washburn, E. R. and Kahler, F. H., J. Am. Chem. Soc. 64, 274 (1942).
49. Hutchinson, E., Trans. Faraday Soc. 39, 230 (1943).
50. Ward, A. F. H. and Tordai, L., J. Sci. Instrum. 21, 143 (1944).
51. Hutchinson, E., J. Colloid Sci. 3, 219 (1948).
52. Hutchinson, E., J. Colloid Sci. 3, 235 (1948).
53. Hutchinson, E., J. Colloid Sci. 3, 521, 531 (1948).
54. Douglas, H. W., J. Sci. Instrum. 27, 67 (1950).
55. Bartells, F. E. and C. W. Bjorklund, J. Phys. Chem. 56, 453 (1952).
56. Donahue, D. J. and Bartells, F. E., J. Phys. Chem. 56, 480 (1952).
57. Shewmaker, J. E., Vogler, C. E. and Washburn, E. R., J. Phys. Chem. 58, 945 (1954).
58. Shotton, E., J. Pharm. Pharmacol. 7, 990 (1955).
59. Pilpel, N., J. Colloid Sci. 11, 51 (1956).
60. Girifalco, L. A. and Good, R. J., J. Phys. Chem. 61, 904 (1957).
61. Harvey, R. R., J. Phys. Chem. 62, 322 (1958).
62. Shotton, E. and Kalyan, K., J. Pharm. Pharmacol. 12, 109 (1960).
63. Butler, E. B., J. Phys. Chem. 67, 1419 (1963).
64. Paul, G. W. and Mark deChazal, L. E., J. Chem. Eng. Data 12, 105 (1967).
65. Rehfield, S. J., J. Phys. Chem. 71, 738 (1967).
66. Chatterjee, A. K. and Chattoraj, D. K., J. Colloid Interface Sci. 26, 1 (1968).
67. Chatterjee, A. K. and N. F. Owens, Kolloid Z.Z.Polym. 233, 966 (1969).
68. MacRitchie, F. and Owens, N. F., J. Colloid Interface Sci. 29, 66 (1969).

69. MacRitchie, F., Trans. Faraday Soc. 65, 2503 (1969).
70. Johnson, M. C. R. and Saunders, Chem. Abstr. 74, 34967p (1971).
71. Heertjes, P.M., de Smet, E. C. and Belk, W. C. in "Proceedings of the International Solvent Extraction Conference, 1971" (J. G. Gregory, B. Evans, and P. C. Weston, Eds.), Vol. 1, 70. Society of Chemical Industry, London, 1971.
72. Luhning, R. W. and Sawistowski, H. in "Proceedings of the International Solvent Extraction Conference, 1971", (J. G. Gregory, B. Evans and P. C. Weston, Eds.), Vol. 2, 873. Society of Chemical Industry, London, 1971.
73. Padday, J. F. and Pitt, A. R., Proc. Roy. Soc. London 329A, 421 (1972).
74. Harkins, W. D., Brown, F. E. and Davies, E. C. H., J. Am. Chem. Soc. 39, 354 (1917).
75. Addison, C. C., Nature (London)144, 783 (1939).
76. Pliskin, I. and Treybal, R. E., J. Chem. Eng. Data. 11, 49 (1966).
77. Jasper, J. J., Nakoneczny j, M., Swingley, C. S. and Livingston, H. K., J. Phys. Chem. 74, 1535 (1970).
78. McCaffery, F. G. and Mungan, Necmettin, J. Can. Pet. Technol. 9, 185 (1970).
79. Jennings, H. Y., J. Colloid Interface Sci.24, 323 (1967).
80. Cini, R., Ficalbi, A., and Loglio, G., Inquinamento 12, 15 (1970).

IX. DISCUSSION

A. Professor F. M. Fowkes, Lehigh Univ.

Some years ago, Dr. Fowkes used an unpolished Teflon plate for oil/water interfacial tension measurements. The plate was first immersed in the aqueous phase, then the organic phase added. Now the plate was raised to cross the boundary, developing a meniscus of organic phase concave downward in the aqueous phase. Fowkes noted that the portion of the Teflon plate now immersed in the top phase possessed the same shiny, mirror-like appearance as described in this paper. He calculated that the contact angle for this physical combination was zero. By the use of Teflon, and not traversing the top phase first, he avoided some of the problems described. Fowkes also employed a surface cleaning technique to periodically remove trace impurities.

Response: This is another example of the existence of an adhering film of one phase on the plate, when immersed in a second liquid phase. It would appear that the calculation of interfacial tension in this case would also require some provision for the weight of the adhering liquid film.

B. Dr. Steven A. Shaya, Proctor and Gamble Co., Cincinnati, Ohio.

He is familiar with another technique to handle the wettability problem in the liquid-liquid case. Assuming that the lower phase is aqueous, the plate is first suspended or placed in this phase. Next the top layer is put in place. Now the plate is raised to either contact or enter the top phase. If the aqueous meniscus rises on the plate, the plate is raised into the top phase, bringing with it an adhering film of water. This technique assures a zero contact angle, and also avoids the water drainage problem of Dognon's (5) procedure.

Response: This technique certainly appears to be a useful option. However, it does create an adhering water film of a weight significant enough to require attention in the calculation procedure.

C. Professor A. W. Neumann, University of Toronto.

Dr. Newmann stated that the only way to assure that a measured interfacial tension value is correct, is to simultaneously measure the capillary rise on the plate. It should reach some specified height, else the run is invalid.

Response: It is felt that the statement made is somewhat illusory, for the height of rise and the weight of fluid suspended both are measures of the same interfacial tension. The only way one knows if the height of capillary rise is adequate is to compare the given run results with enough prior measurements to establish an acceptable range for the "maximum" height. Virtually all contamination lowers both the capillary rise and the weight suspended. The same experience that establishes an adequate concept of the 'correct' capillary rise also establishes an adequate concept of the 'correct' weight suspended, hence the 'correct' interfacial tension. There is a philosophical problem here that cannot be avoided.

D. Prof. Robert J. Good, SUNY, Buffalo.

In assessing literature results on benzene-water, it should be recognized that Harkins' results are simply too high, and a number of experienced experimentalists now recognize this. Good's conclusions derive from some unpublished work by the drop/weight method. His results were 0.8 to 1.0 dyn/cm lower.

The key point, he feels, is that Harkins' dimensionless correction correlation was based upon the liquid-vapor case, and simply doesn't apply closely to the liquid-liquid case. The events after the break of the liquid stem supporting the pendant drop are clearly dynamic, not static. Accordingly, the weight that falls must be influenced by inertial effects, hence by the viscosities of the two phases and the interfacial tension. But Harkins assumed only an influence of $\Delta\rho$, for a truly static situation.

Response: We accepted the corrections of Bartells (37) to some of Harkins results, but also accepted uncorrected the rest of Harkins results. Perhaps a second look is desirable.

A THEORY OF THE RATE OF RISE OF A LIQUID IN A CAPILLARY

S. Levine, P. Reed, E. J. Watson,
University of Manchester
and
G. Neale
University of Ottawa

The classic equation of Washburn and Rideal for the rate of penetration of a fluid into a capillary due to surface tension forces is based on the steady flow parabolic velocity profile of Poiseuille across a section of the capillary. However, such a model of fluid flow cannot apply at the two ends of the fluid column in the capillary, that is, in the regions of entry into the tube from the reservoir and in the vicinity of the advancing meniscus. Without a proper theory of the initial flow at entry, the impossible condition of an initial infinite acceleration is predicted. Szekely, Neumann and Chuang have attempted a more rigorous theory of capillary penetration in which the entry flow from a reservoir, and also dissipative effects due to circulation, are considered. They were able to remove the anomaly of the initial infinity in the fluid motion but their treatment of dissipative effects arising from circulation refers to high Reynolds number, whereas most wetting phenomena in capillaries occur at low Reynolds number. An improved theory of entry flow is presented in which the sink flow towards the capillary entrance from the reservoir is considered. Time-dependent solutions of the equations describing the rate of penetration are obtained for both vertical and horizontal flow, and the nature of the deviations from the Washburn equation are examined.

I. INTRODUCTION

The classic equation of Washburn (1) and Rideal (2) for the rate of penetration of a liquid into a cylindrical capillary due to surface tension forces is based on the steady-flow parabolic velocity profile of Poiseuille across a section of the capillary. It is well known that the above model of fluid flow cannot apply at the two ends of the fluid column in the capillary, that is, in the regions of entry from a reservoir and of the advancing meniscus. Without a proper theory of the initial flow at entry, the impossible condition of an infinite initial acceleration is predicted(3). Szekely, Neumann and Chuang (4) [S.N.C.] have attempted a more rigorous theory of capillary penetration in which the entry flow from a reservoir, and also dissipative effects due to circulation, are considered. They express the pressure effect of the flow from the reservoir converging on the entrance to the capillary as an increase in effective mass of fluid and therefore in effective length of fluid in the capillary tube. Consequently they are able to remove the anomaly of the initial infinity in the fluid motion. They postulate that dissipative effects arising from circulation occur through the formation of a vena contracta (5) at the entrance to the capillary, but their treatment refers to high Reynolds number whereas most wetting phenomena in capillaries are described in terms of low Reynolds number. The problem of entry flow into channels and pipes has been investigated by a number of authors (6,7,8) but most of this work has dealt with steady state flow at high Reynolds number.

To illustrate the difficulties in determining the entry flow conditions we consider two hypothetical limiting cases (i) and (ii), with a vertical cylindrical tube, the bottom end of which is brought at time t=0 into contact with the horizontal surface of a reservoir of incompressible liquid. Let a be the radius of the cylinder, z the distance of rise in the capillary at time t, ρ the liquid density, μ its viscosity and γ the adhesion tension. If the rising meniscus is a liquid/gas interface then $\gamma = \gamma_{sg} - \gamma_{s\ell}$ where γ_{sg} and $\gamma_{s\ell}$ are the interfacial tensions of the solid/gas and solid/liquid interfaces respectively. In case (i) we imagine that the liquid in the reservoir has zero velocity and that it instantaneously receives its fully developed velocity on entry into the capillary. Then the rate of change of total momentum associated with the capillary flow is

$$\rho\pi a^2 \lim_{\Delta t \to o} \left[\frac{(z + \Delta z)\, \frac{d}{dt}(z + \Delta z) - z\frac{dz}{dt}}{\Delta t} \right] = \rho\pi a^2 \frac{d}{dt}\left(z\frac{dz}{dt}\right) . \qquad (1.1)$$

Assuming the Poiseuille viscous dragging force and ignoring contact line hysteresis effects, the equation of motion for the advancing fluid front is

$$\rho\pi a^2 \frac{d}{dt}\left(z\frac{dz}{dt}\right) = 2\pi a\gamma - 8\pi\mu z\frac{dz}{dt} - \rho\pi a^2 gz \tag{1.2}$$

where the last term is due to gravity. In case (ii) we suppose that the element of liquid of volume $\pi a^2 \Delta z$ added to the capillary from the reservoir in time Δt has the fully developed velocity dz/dt, with the rest of the reservoir at zero velocity. Then (1.1) is replaced by

$$\rho\pi a^2 \left[\lim_{\Delta z \to o} (z + \Delta z)\right] \times \left[\lim_{\Delta t \to o} \frac{\frac{d}{dt}(z + \Delta z) - \frac{dz}{dt}}{\Delta t}\right] = \rho\pi a^2 z\frac{d^2 z}{dt^2} \tag{1.3}$$

which becomes the left-hand side of the equation of motion (1.2). This is the form used by Rideal (2).

The effect of curvature in the meniscus is ignored in (1.2) and one should interpret z as the equivalent length of the liquid column, defined by the requirement that

$$\pi a^2 (dz/dt) = \text{volume flux of liquid along capillary tube.} \tag{1.4}$$

This flux is independent of z but a function of time for an incompressible liquid. Also no account is taken in (1.2) of the departure from Poiseuille flow at the meniscus where both tangential and normal stresses should be continuous. In the limit of zero velocity the last-named stress condition becomes Laplace's equation relating pressure difference to meniscus curvature. If Young's equation holds then γ is identified with $\gamma_{lg} \cos \theta_c$ where γ_{lg} is the liquid/gas interfacial tension and θ_c is the contact angle. In practice this will not apply due to roughness and heterogeneity of the solid surface. These properties will tend to alter the effective adhesion tension γ. But in addition there will be energy dissipation due to hysteresis in the contact angles at the three-phase line between the meniscus and the solid cylinder. Provided any contact angle dependence on velocity can be ignored, such dissipative effects can be included in the force term $2\pi a\gamma$ in (1.2) by the appropriate alteration in the value of γ.

II. ENERGY BALANCE METHOD

S.N.C. (4) do not start with the equation of force balance (1.2) but rather with the equation of energy balance which reads (5):

Rate of change of (kinetic energy + potential energy) = net rate of input of (kinetic energy + potential energy + pressure energy) - rate of work done on the surroundings - rate of work dissipated irreversibly. (2.1)

However, there are errors or misprints in the results obtained by S.N.C. on the basis of the relation (2.1). To apply (2.1) we consider a fixed volume of the cylindrical tube extending from the position of entry $z = 0$ to a distance z_o situated in the gas phase just <u>above</u> the meniscus. Let us absorb into the potential energy the term due to surface tension. Then the left-hand side of (2.1) is

$$\frac{d}{dt}\left[\frac{1}{2}\rho\pi a^2 z\left(\frac{dz}{dt}\right)^2 + \frac{1}{2}\rho\pi a^2 z^2 g - 2\pi a\gamma z\right] \tag{2.2}$$

where γ includes energy losses associated with contact angle hysteresis. For a horizontal capillary the term involving g would be absent. Let w = mass flow rate = density x volume flux. At the entry $z = 0$, $w = \rho\pi a^2$ (dz/dt) whereas at $z = z_o$ in the gas phase we can assume $w = 0$. Thus

$$\text{net input of kinetic energy} = \frac{1}{2}\rho\pi a^2\left(\frac{dz}{dt}\right)^3 . \tag{2.3}$$

The net input of potential energy is zero because at entry $z = 0$, and $w = 0$ at exit $z = z_o$. The net input of pressure energy is

$$\left[\frac{p}{\rho} w\right]_{z=0} = (p)_{z=0}\, \pi a^2 \frac{dz}{dt} . \tag{2.4}$$

Subtracting from (2.4) the rate of work done against the external (atmospheric) pressure p_o, we have the contribution to the right-hand side of (2.1), namely

$$\pi a^2 (p_{z=0} - p_o)\frac{dz}{dt} . \tag{2.5}$$

S.N.C equate (2.5) to the rate of change of kinetic energy in the reservoir due to the flow converging on the entry and find that (2.5) equals

$$\rho\pi a^2 h \frac{d^2z}{dt^2}\frac{dz}{dt} \tag{2.6}$$

where $h = 7a/6$, which is equivalent to increasing the column length z to an effective length $z + h$. We shall however improve upon their calculation of (2.5) in section III. The rate of work done against the surroundings includes work against the frictional forces at the tube wall, which is

$$-8\pi\mu z \left(\frac{dz}{dt}\right)^2 \tag{2.7}$$

assuming Poiseuille flow.

The work of Seyer and Catania (8) on dissipative energy losses due to circulation at a sudden contraction in a circular cylinder is used to determine the rate of work dissipated irreversibly at the entry to the capillary tube. Denoting the two sections of the tube by 1 and 2 with the direction of flow from 1 to 2, let V_1 and V_2 be the component of the liquid velocity parallel to the axis of the tube, p_1 and p_2 the corresponding pressures and $\hat{E}_v$ the rate of energy dissipation per unit mass flow rate at the tube contraction. If $\langle\cdot\cdot\rangle$ denotes average across a tube section, then in steady state flow the energy balance relation is (Ref. 5, eqn. (7.3.14)

$$0 = \hat{E}_v + \frac{1}{2}\left[\frac{\langle V_2^3\rangle}{\langle V_2\rangle} - \frac{\langle V_1^3\rangle}{\langle V_1\rangle}\right] + \frac{1}{\rho}(p_2 - p_1) \, . \tag{2.8}$$

Continuity of mass flow requires that the contraction factor β in cross-sectional area be given by

$$\beta = \frac{a_2^{\,2}}{a_1^{\,2}} = \frac{\langle V_1\rangle}{\langle V_2\rangle} \tag{2.9}$$

where a_1 and a_2 are the radii of the two circular sections. Defining a dimensionless pressure loss across the contraction by

$$\phi = \frac{2(p_1 - p_2)}{\rho\langle V_2\rangle^2} \tag{2.10}$$

(2.8) can be written as

$$\frac{\hat{E}_v}{\langle V_2\rangle^2} = \frac{1}{2}(\phi - \xi_2 + \beta^2\xi_1) \tag{2.11}$$

where

$$\xi_i = \langle V_i^3\rangle/\langle V_i\rangle^3;\ i = 1,2. \tag{2.12}$$

According to Seyer and Catania (8), experiments suggest the form

$$\phi = K + \frac{K'}{Re} \tag{2.13}$$

where Re is the Reynolds number and where K and K' are functions of β alone for Newtonian flow (*typically* K *will be about 2 and* K' *between 10 and 500, although there is some uncertainty as to their actual values).* In wetting of fine capillaries Re << 2000 and we can expect the flow to be laminar. For Poiseuille flow pattern $\xi_1 = \xi_2 = 2$ and we can write (2.11) as

$$\frac{\hat{E}_v}{\langle V_2\rangle^2} = \frac{1}{2}\left(K'' + \frac{K'}{Re}\right) \tag{2.14}$$

where $K'' = K - 2(1-\beta^2)$. In the case where the tube of radius a_1 becomes an infinite reservoir $\beta = 0$, and then $K'' = K - 2 \approx 0$. Putting $a_2 = a$, $\langle V_2\rangle = dz/dt$ and

$$Re = \frac{2\rho a}{\mu}\frac{dz}{dt} \tag{2.15}$$

the rate of energy dissipation at entry in the capillary tube from the reservoir is

$$\hat{E}_v \text{ x mass flow rate} = \hat{E}_v\left[\rho\pi a^2 \frac{dz}{dt}\right]$$
$$= \frac{1}{2}\rho\pi a^2\left[K''\left(\frac{dz}{dt}\right)^3 + \frac{K'\mu}{2\rho a}\left(\frac{dz}{dt}\right)^2\right]. \tag{2.16}$$

Using (2.2)-(2.7) and (2.16), the energy balance equation (2.1) leads to the equation of motion

$$\rho a^2(z+h)\frac{d^2z}{dt^2} = 2a\gamma - 8\mu z\frac{dz}{dt} - \rho a^2 gz$$
$$- \frac{1}{2}\rho a^2\left[K''\left(\frac{dz}{dt}\right)^2 + \frac{K'\mu}{2\rho a}\left(\frac{dz}{dt}\right)\right]. \tag{2.17}$$

Equation (2.17) differs from that of S.N.C. in two respects. First they have no dissipation term in K', which is proportional to the velocity dz/dt. Second, in place of (2.11) they assume

$$\frac{\hat{E}_v}{\langle V_2 \rangle_2^2} = \frac{1}{2} e_v \tag{2.18}$$

where e_v is the friction loss factor which they choose as 0.45, corresponding to turbulent flow at high Reynolds number. We need to identify e_v with our K" but S.N.C. obtain $2 + e_v$ in their equation of motion as a result of errors in their treatment of the energy balance relation (2.1). At the low Reynolds number characteristic of wetting processes, the omitted term involving K' in the dissipation energy rate (2.16) is the *dominant* one. Assuming that the form (2.14) is valid, and that (2.1) can be equated to (2.8), it is the task of fluid mechanical theory to determine the constants h, K', and K" and an approximate method is described in the next section.

III. FLUID DYNAMICAL THEORY OF RATE OF WETTING

We introduce cylindrical polar co-ordinates (r,ψ,x) and corresponding fluid velocity components $(v,0,u)$. The origin **O** is on the axis of the tube at entry position; x is distance along the vertical axis into the tube, r is measured from the axis normal to it in a horizontal plane and ψ is the the angle of rotation about the axis. By symmetry the flow conditions are independent of angle ψ (see Figure). The velocity components v,u and the pressure p are functions of x,r and time t. For an incompressible fluid the equation of continuity is

$$\frac{\partial}{\partial x}(ru) + \frac{\partial}{\partial r}(rv) = 0 \tag{3.1}$$

and the Navier-Stokes equation for component u reads

$$\frac{\partial u}{\partial t} + u\frac{\partial u}{\partial x} + v\frac{\partial u}{\partial r} = -\frac{1}{\rho}\frac{\partial p}{\partial x} + \nu\left[\frac{\partial^2 u}{\partial r^2} + \frac{1}{r}\frac{\partial u}{\partial r} + \frac{\partial^2 u}{\partial x^2}\right] - g \tag{3.2}$$

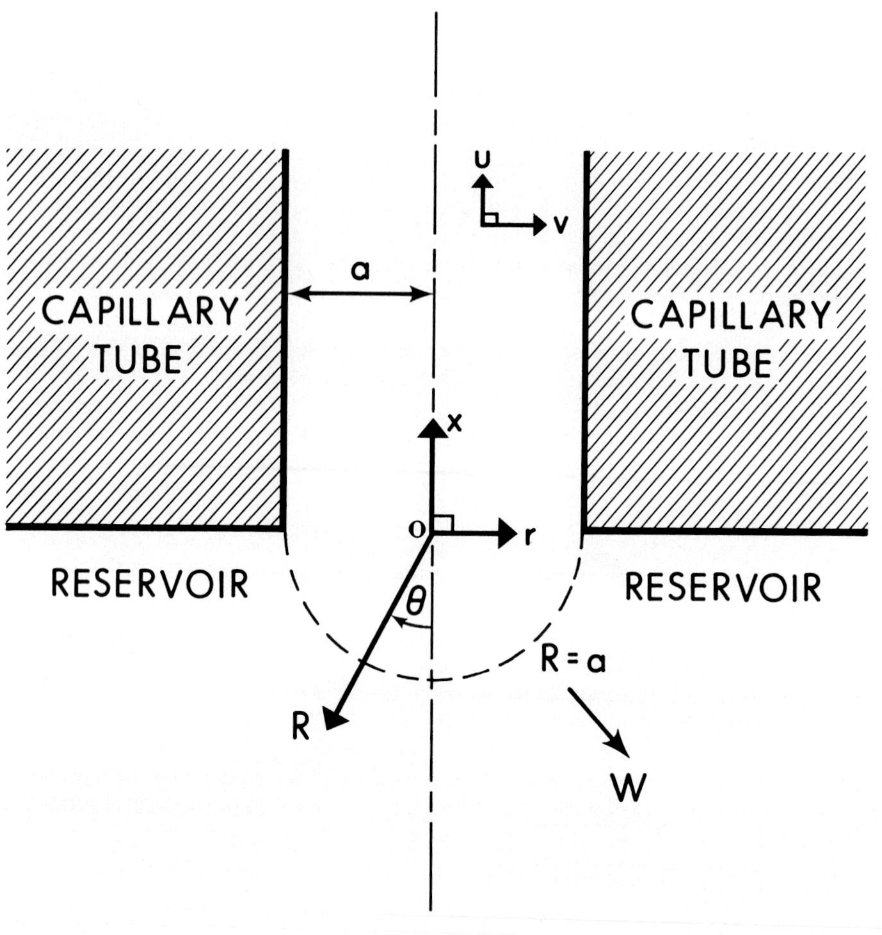

THE CO-ORDINATE SYSTEM FOR ENTRY FLOW INTO A CAPILLARY

where $\nu = \mu/\rho$, the kinematic viscosity. On multiplying (3.2) by r and using (3.1) it may be verified that

$$\frac{\partial}{\partial t}(ru) + \frac{\partial}{\partial x}(ru^2) + \frac{\partial}{\partial r}(ruv) = -\frac{r}{\rho}\frac{\partial p}{\partial x}$$

$$+ \nu\left[\frac{\partial}{\partial r}\left(r\frac{\partial u}{\partial r}\right) + r\frac{\partial^2 u}{\partial x^2}\right] - gr \quad . \tag{3.3}$$

Integrating with respect to r across a horizontal section of the tube from r = 0 to r = a and making use of the boundary (no-slip) condition

$$u = 0 \quad \text{at} \quad r = a \tag{3.4}$$

we obtain

$$\frac{\partial}{\partial t}\int_0^a rudr + \frac{\partial}{\partial x}\int_0^a ru^2 dr = -\frac{1}{\rho}\int_0^a r\frac{\partial p}{\partial x}dr$$

$$+ \nu a\left.\frac{\partial u}{\partial r}\right|_{r=a} + \nu\frac{\partial^2}{\partial x^2}\int_0^a rudr - \frac{1}{2}ga^2 \quad . \tag{3.5}$$

The volume flux along the tube is given by (1.4), which reads here

$$\pi a^2\frac{dz}{dt} = 2\pi\int_0^a rudr \tag{3.6}$$

where the equivalent length of the column of liquid z is a function of t only. It follows that the last integral in (3.5) vanishes and we have

$$\frac{1}{2}a^2\frac{d^2z}{dt^2} + \frac{\partial}{\partial x}\int_0^a ru^2 dr = -\frac{1}{\rho}\int_0^a r\frac{\partial p}{\partial x}dr + \nu a\left.\frac{\partial u}{\partial r}\right|_{r=a} - \frac{1}{2}ga^2 \quad . \tag{3.7}$$

On integrating (3.7) with respect to x from x = 0 to x = z(t), we obtain

$$\frac{1}{2}a^2 z \frac{d^2z}{dt^2} + \left[\int_o^a ru^2 dr\right]_{x=0}^{x=z(t)} = -\frac{1}{\rho}\int_o^a r\left[p(z,r,t)-p(0,r,t)\right]dr$$

$$+ \nu a \int_o^{z(t)} \left(\frac{\partial u}{\partial r}\right)_{r=a} dx - \frac{1}{2}ga^2 z(t) \tag{3.8}$$

which is an exact relation.

If Poiseuille flow is assumed

$$u = 2\left(1 - \frac{r^2}{a^2}\right)\frac{dz}{dt} \tag{3.9}$$

which satisfies the continuity condition (3.6). Since u is independent of x, the second term on the left hand side of (3.8) is zero. Also

$$\left.\frac{\partial u}{\partial r}\right|_{r=a} = -\frac{4}{a}\frac{dz}{dt} \quad . \tag{3.10}$$

In Poiseuille flow the radial velocity component $v = 0$ and therefore from the Navier-Stokes equation for the v-component, $\partial p/\partial r = 0$, i.e. p is independent of r and we can write in the relation (3.8), $p(z,r,t) = p(z,t)$. If the meniscus is a hemi-spherical surface of radius a with contact angle θ_c and $p(z,t)$ is identified with the liquid pressure at the meniscus, then Laplace's equation yields

$$p(z,t) = p_o - \frac{2}{a}\gamma_{lg}\cos\theta_c = p_o - \frac{2\gamma}{a} \tag{3.11}$$

using Young's equation. Where the meniscus departs from a spherical shape (at larger a) and/or Young's equation does not apply, the second expression for $p(z,t)$ in (3.11) is interpreted as defining the quantity γ. If now the pressure at entry $p(0,t)$ is approximated by the atmospheric pressure p_o and use is made of (3.9)-(3.11), then the equation of motion (3.8) reduces to

$$\rho\pi a^2 z \frac{d^2z}{dt^2} = 2\pi a\gamma - 8\pi\mu z\frac{dz}{dt} - \rho\pi a^2 gz \quad . \tag{3.12}$$

This is the limiting case (ii) described in the Introduction.

To obtain a better approximation we abandon the approximation that $p(0,t) = p_o$ but retain Poiseuille flow and Laplace's equation (3.11). Equation (3.8) now yields

$$\frac{1}{2}a^2 z \frac{d^2 z}{dt^2} = -\frac{1}{\rho}\left[\frac{1}{2}a^2\left(p_o - \frac{2\gamma}{a}\right) - \int_o^a rp(0,t)\,dr\right] - 4\nu z\frac{dz}{dt} - \frac{1}{2}ga^2 z \tag{3.13}$$

and there remains to evaluate the integral in (3.13). We introduce spherical polar co-ordinates R,θ,Φ in the reservoir with origin also at **O**, such that the polar axis is a continuation of the capillary tube axis (see Figure). It is supposed that the fluid velocity vector $\underline{V}$ for $R>a$ (i.e. beyond the hemispherical region of the reservoir adjacent to the bottom of the capillary) has a component W in the R direction only to describe a sink flow towards entry into the capillary. The boundary conditions are that $W \to 0$ and the pressure $p^* \to p_o$† as $R\to\infty$ and also that the volume flux inwards across $R = a$ is given by the continuity condition

$$-2\pi a^2 \int_o^{\pi/2} \sin\theta \; W_{R=a} \; d\theta = \pi a^2 \frac{dz}{dt} \quad . \tag{3.14}$$

A simple solution of the continuity equation

$$\frac{\partial}{\partial R}\left(R^2 W\right) = 0 \tag{3.15}$$

which satisfies (3.14) is

$$W = -\frac{1}{2}\frac{a^2}{R^2}\frac{dz}{dt} \quad . \tag{3.16}$$

The Navier-Stokes equation for component W, depending on R and t only, reads

$$\frac{\partial W}{\partial t} = -\frac{1}{\rho}\frac{\partial p^*}{\partial R} + \nu\left[\frac{1}{R^2}\frac{\partial}{\partial R}\left(R^2\frac{\partial W}{\partial R}\right) - \frac{2W}{R^2}\right] \tag{3.17}$$

neglecting the non-linear inertia terms (creeping flow).

---

† p^* is a modified pressure defined by $p^* = p - R\cos\theta\rho g$, and accounts directly for hydrostatic (gravity) effects. If the fluid is stationary, $p = p_o + R\cos\theta\rho g$ whence $p^* = p_o$.

Substitution of (3.16) into (3.17) and integration from $R = R$ to $R = \infty$ yields

$$p^* = p^*(R,t) = p_o - \frac{1}{2}\frac{\rho a^2}{R}\frac{d^2z}{dt^2} \tag{3.18}$$

which satisfies the necessary condition of incompressible flow $\nabla^2 p^* = 0$.

The velocity field in the hemispherical region of the reservoir $R \leq a$ is unknown but we apply the momentum balance equation to circumvent this difficulty as follows. For the system in the hemisphere:

Rate of change of total momentum in system = Flux of momentum entering - flux of momentum leaving + sum of forces acting on system. (3.19)

The stress tensor at the hemisphere $R = a$ has components

$$\sigma_{RR} = -p^*_{R=a} + 2\mu\left(\frac{\partial W}{\partial R}\right)_{R=a}, \quad \sigma_{R\theta} = 0 \quad . \tag{3.20}$$

Hence, the force in the x-direction exerted per unit area across $R = a$ by the fluid in the reservoir outside $R = a$ is

$$-\sigma_{RR}\cos\theta = \left(p_o - \frac{1}{2}\rho a\frac{d^2z}{dt^2} - \frac{2\mu}{a}\frac{dz}{dt}\right)\cos\theta \tag{3.21}$$

using (3.16) and (3.18). Integration over the hemisphere yields the force term

$$X_1 = 2\pi a^2\int_o^{\pi/2}(-\sigma_{RR})\sin\theta\cos\theta d\theta = \pi a^2\left(p_o - \frac{1}{2}\rho a\frac{d^2z}{dt^2} - \frac{2\mu}{a}\frac{dz}{dt}\right) . \tag{3.22}$$

The corresponding force term in the x-direction exerted by the fluid in the capillary tube across the circular base $x = 0$ is

$$X_2 = -2\pi\int_o^a rp(o,t)dr \tag{3.23}$$

The flux of momentum in the x-direction entering the hemisphere across R = a is

$$2\pi\rho a^2 \int_0^{\pi/2} \sin\theta \; W^2_{R=a} \cos\theta d\theta = \frac{1}{4}\pi\rho a^2 \left(\frac{dz}{dt}\right)^2 \tag{3.24}$$

using (3.16). Assuming Poiseuille flow in the capillary tube the flux of momentum leaving at x = 0 is

$$2\pi\rho \int_0^a ru^2 dr = \frac{4}{3}\pi\rho a^2 \left(\frac{dz}{dt}\right)^2 \tag{3.25}$$

There remains to estimate the left-hand side of equation (3.19). The acceleration inside the hemisphere being unknown, we only find a 'reasonable mean' acceleration. At x = 0, the acceleration in the x-direction is taken to be that given by (3.9), i.e.

$$\frac{du}{dt} = 2\left(1 - \frac{r^2}{a^2}\right)\frac{d^2z}{dt^2} \tag{3.26}$$

Hence, the flux of acceleration in the x-direction across x = 0 is

$$2\pi \int_0^a ru \frac{du}{dt} dr = \frac{4}{3}\pi a^2 \frac{dz}{dt}\frac{d^2z}{dt^2} \tag{3.27}$$

and since the flux of volume equals $\pi a^2 (dz/dt)$ the mean acceleration is taken as $(4/3)(d^2z/dt^2)$. The acceleration at R = a radially outwards is taken to be

$$\left(\frac{DW}{Dt}\right)_{R=a} = \left(\frac{\partial W}{\partial t} + W\frac{\partial W}{\partial R}\right)_{R=a} = -\frac{1}{2}\frac{d^2z}{dt^2} - \frac{1}{2a}\left(\frac{dz}{dt}\right)^2 \tag{3.28}$$

where we differentiate 'following the motion', as in the Navier-Stokes equations. Then the flux of acceleration in the x-direction inwards across R = a is

$$2\pi a^2 \int_0^{\pi/2} \left(\frac{DW}{Dt}\right)_{R=a} \sin\theta \; W_{R=a} \cos\theta d\theta = \frac{1}{4}\pi a^2 \frac{dz}{dt}\left(\frac{d^2z}{dt^2} + \frac{1}{a}\left(\frac{dz}{dt}\right)^2\right) \tag{3.29}$$

using (3.16) and (3.28). Dividing by the volume flux $\pi a^2(dz/dt)$ the mean acceleration at R = a equals $(ad^2z/dt^2 + dz/dt)/4a$. Taking the mean acceleration in the hemisphere as the mean of these two mean accelerations over R = a and x = 0, the left-hand side of (3.19) equals

$$\text{Mass of system x mean acceleration} = \frac{2}{3}\rho\pi a^3\left(\frac{19}{24}\frac{d^2z}{dt^2} + \frac{1}{8a}\left(\frac{dz}{dt}\right)^2\right). \tag{3.30}$$

The momentum balance equation (3.19) therefore reads

$$X_1 + X_2 = \frac{2}{3}\rho\pi a^3\left(\frac{19}{24}\frac{d^2z}{dt^2} + \frac{1}{8a}\left(\frac{dz}{dt}\right)^2\right) - \frac{1}{4}\rho\pi a^2\left(\frac{dz}{dt}\right)^2 + \frac{4}{3}\rho\pi a^2\left(\frac{dz}{dt}\right)^2 \tag{3.31}$$

from which we derive

$$\int_0^a r\, p(0,t)\, dr = \frac{a^2 p_o}{2} - \mu a\frac{dz}{dt} - \frac{37}{72}\rho a^3\frac{d^2z}{dt^2} - \frac{7}{12}\rho a^2\left(\frac{dz}{dt}\right)^2 . \tag{3.32}$$

Thus, the equation of motion (3.13) becomes

$$\rho a^2\left(z + \frac{37}{36}a\right)\frac{d^2z}{dt^2} = 2a\gamma - 8\mu z\frac{dz}{dt} - \rho a^2 gz - \frac{1}{2}\rho a^2\left[\frac{7}{3}\left(\frac{dz}{dt}\right)^2 + \frac{4\mu}{\rho a}\left(\frac{dz}{dt}\right)\right]. \tag{3.33}$$

It is encouraging to note that (3.33) is *identical in form* to the macroscopic energy balance equation (2.17), with h = (37/36)a, K' = 8 and K" = 7/3. Observe that these values of K' and K" are consistent with the experimental values discussed in Section II.

Equation (3.33) may be expressed in non-dimensional form as

$$\left(Z + \tau\right)\frac{d^2Z}{dT^2} + \left(Z + \frac{9\tau}{37}\right)\frac{dZ}{dT} + \frac{7}{6}\left(\frac{dZ}{dT}\right)^2 + \lambda Z = 1 \tag{3.34}$$

where $Z = (\alpha/\sqrt{\beta})z$, $T = \alpha t$, $\alpha = 8\mu/\rho a^2$, $\beta = 2\gamma/\rho a$, $\tau = 37a\alpha/(36\sqrt{\beta})$ and $\lambda = g/(\alpha\sqrt{\beta})$. By way of example, $\alpha = 3200$, $\beta = 28800$, $\tau = 0.097$, $\lambda = 0.0018$ for H_2O at 20^oC in a capillary of radius 0.005 cm. The non-linear differential equation (3.34) needs to be solved numerically, subject to the boundary conditions that at $T = 0$, $Z = 0$ and $dZ/dT = 0$. However, we can obtain a simple analytical solution valid for large Z by using a method of successive approximations, which is outlined briefly below.

As discussed in the paragraph following (2.18), most wetting processes in capillaries (and certainly those at large Z) occur at such low Reynolds number that the $(dZ/dT)^2$ term in (3.34) may be neglected in relation to the dZ/dT term. Thus, upon introducing the transformation $U = dT/dZ$ (from which $d^2Z/dT^2 = -[(dU/dZ)/U^3]$) into (3.34) we obtain, for low Reynolds number,

$$-(Z + \tau)\frac{dU}{dZ} + \left(Z + \frac{9\tau}{37}\right)U^2 = (1 - \lambda Z)\,U^3 \quad . \qquad (3.35)$$

[N.B. the high Reynolds number approximation of (3.34), in which the $(9\tau/37)$ term alone is absent, has been solved in similar fashion elsewhere (9)].

A first approximation of (3.35) for large times is obtained by neglecting the first term in this equation, thus

$$U = U_o = (Z + 9\tau/37)/(1 - \lambda Z) \qquad (3.36)$$

which is a generalization of the Washburn equation (incorporating low Reynolds number dissipation losses via the parameter τ). A solution of (3.35) which is suitable for $Z \lesssim 1/\lambda >> 1$ is

$$U = U_o + U_1 + U_2 + U_3 + \cdots \qquad (3.37)$$

where U_1 and U_2 are given by

$$-(Z + \tau)\frac{dU_o}{dZ} + (Z + 9\tau/37)\,2U_oU_1 = (1 - \lambda Z)\,3U_o^2U_1 \qquad (3.38)$$

$$-(Z + \tau)\frac{dU_1}{dZ} + (Z + 9\tau/37)(U_1^2 + 2U_oU_2)$$

$$= (1 - \lambda Z)(3U_oU_1^2 + 3U_o^2U_2) \qquad (3.39)$$

to linear terms in U_1 and U_2 respectively. From (3.38), (3.39) and (3.36) we obtain

$$U_1 = -\frac{(Z + \tau)\quad(1 + c\lambda\tau)}{(Z + c\tau)^2\quad(1 - \lambda Z)} \tag{3.40}$$

$$U_2 = -\frac{(Z + \tau)\quad(1 + c\lambda\tau)}{(Z + c\tau)^5\quad(1 - \lambda Z)}\,F \tag{3.41}$$

where $F = -2\lambda Z^2 + ([2c-3]\lambda\tau+3)Z + (c\lambda\tau^2 + [4-c]\tau)$ and where $c = 9/37$. Recall that $\lambda = 0$ for horizontal capillaries. It may be verified that the series (3.37) converges rapidly for $Z \lesssim 1/\lambda \gg 1$. Thus, the following tables display the values of U_o, U_1 and U_2 for different values of Z for the particular case of water at 20°C in a (vertical or horizontal) capillary of radius a = 0.005 cm. In this case the first three terms of the infinite series (3.37) are sufficient for Z > 4.

Vertical Capillary
(τ = 0.097, λ = 0.0018)

Z	U_o	U_1	U_2
2	2.031	-0.514	-0.393
3	3.040	-0.341	-0.115
4	4.053	-0.255	-0.048
6	6.090	-0.170	-0.014
8	8.141	-0.128	-0.006
10	10.208	-0.102	-0.003

Horizontal Capillary
(τ = 0.097, λ = 0.0)

Z	U_o	U_1	U_2
2	2.024	-0.512	-0.348
3	3.024	-0.339	-0.106
4	4.024	-0.253	-0.045
6	6.024	-0.168	-0.014
8	8.024	-0.126	-0.006
10	10.024	-0.100	-0.003

It should be noted that for a given liquid at a given temperature, the parameters τ and λ are functions only of the capillary radius a. Thus, for water at 20°C, $\tau = 0.0069/\sqrt{a}$

and $\lambda = 1022a^{5/2}$. Hence, for values of a of practical interest ($a \lesssim 0.005$ cm.), the effect of the parameter λ in (3.36)-(3.41) will be negligible (this is borne out by the tabulated predictions) and only the effects of τ need be considered. However, the preceding numerical results are unsatisfactory for small Z. Under these conditions (i.e. at small times after entry of the liquid into the capillary) a numerical solution of the complete basic equation (3.34) is required, and this is currently underway.

Further refinements to the analysis presented here are currently being considered. In particular, the departure from Poiseuille flow in the capillary near the entrance and in the vicinity of the moving meniscus appears to be amenable to mathematical treatment. Also receiving attention is the departure from the conventional equilibrium Laplace equation (3.11) associated with unsteady state flow conditions.

IV. REFERENCES

1. Washburn, E.W., Phys. Rev., 17, 273 (1921).
2. Rideal, E.K., Phil. Mag., 44, 1152 (1922).
3. Levine, S. and Neale, G., Faraday Trans. II, 71, 12 (1975).
4. Szekely, J., Neumann, A.W. and Chuang, Y.K., J. Colloid Interface Sci., 35, 273 (1971).
5. Bird, R.B., Stewart, W.E. and Lightfoot, E.N., "Transport Phenomena", Wiley, New York, 1960.
6. van Dyke, M., J. Fluid Mech., 44, 813 (1970).
7. Wilson, S.D.R., J. Fluid Mech., 38, 793 (1969); 96, 787 (1970).
8. Seyer, F.A. and Catania, P.J., Can. J. Chem. Eng., 50, 31 (1972).
9. Levine, S. and Neale, G., Proc. International Conference on Colloid and Surface Science, Vol. 1, Edited by E. Wolfram, (International Union of Pure and Applied Chemistry), p. 761 (1975).

ACKNOWLEDGMENT: We are indebted to Tioxide International Limited for a postdoctoral research assistantship to P. R.

EVALUATION OF THE NORMAL-STRESS CONDITION FOR CURVED LIQUID-GAS INTERFACES

by Charles B. Weinberger and John A. Tallmadge
Department of Chemical Engineering, Drexel University

1. ABSTRACT

Application of theoretical expressions for interfacial forces to a specific geometry often poses uncertainties questions when applied to a curved surface, especially where flow effects are important. The case considered here is the normal-stress, boundary condition between (a) an incompressible liquid having important viscous effects on the surface as well as in the bulk and (b) a gas having negligible viscous effects. The geometry chosen is that of a thin, liquid film such as found in entrainment coating of solids by withdrawal from liquid baths. The normal stress terms which are present are evaluated in terms of different sets of parameters. The needs and uses for the formulations are discussed.

2. INTRODUCTION

Theoretical equations frequently pose uncertainties or questions when applied to a specific geometry. The authors found such a case in one application, namely the use of an interfacial boundary condition to predict the location of a meniscus deformed by flow (1, 2), and would like to share their result with other workers in interfacial phenomena. It is believed that the results presented below are most useful for curved interfaces where flow effects are important.

The equation of interest here is that showing the relationship between the normal stresses at a liquid-gas interface. The normal boundary condition at a curved interface between an incompressible Newtonian liquid and a gas with negligible viscosity is given by Batchelor (3). One equivalent form is (1, 2):

$$P_G - P_L = \sigma c = 2\mu \frac{\partial w}{\partial s} \qquad (1)$$

Here P is pressure, σ is surface tension, c is curvature, μ is bulk viscosity of the liquid, w is surface velocity along the interface, and s is distance along the interface. The term $\partial w / \partial s$ describes a stretching motion along the interface. Equation 1 is restricted to the interface.

For those interested in other problems in interfacial phenomena, we note by comparison with Scriven (4) that Equation 1 applies for cases of negligible effects due to surface tension gradients, surface viscosity, surface elasticity, and interfacial mass transfer.

The problem encountered here is the evaluation of the term, $\partial w/\partial s$.

3. THE STRETCH TERM, $\partial w/\partial s$

Consider the evaluation of the stretch term in the free coating of flat sheets - a process which involves the continuous, vertically-upward withdrawal of a flat belt from a rectangular bath of a wetting liquid. The emergence of the solid support from the liquid bath produces an enlarged meniscus near the bath surface and, at some distance above the bath, results in some finite coating thickness. A diagram of a typical experimental apparatus to obtain free coating data is shown in Soroka and Tallmadge (5).

The specific study in which evaluation of the stretch term posed a problem was the prediction of the size of the deformed meniscus; the size was used in turn for the prediction of flow fields in the films and rectangular baths. For two-dimensional flow, Lee and Tallmadge (1) evaluated the flow term using the numerical approximation $\Delta w/\Delta s$. In this paper, we are interested in obtaining evaluations in analytical form.

The laboratory coordinate system selected here was chosen to be consistent with that of earlier workers (6). We take y as the horizontal distance from the belt, x as vertically upward, and the origin at the belt surface (y=o) and the extrapolated bath surface (x=o). The meniscus is described by (a) the thickness profile, h(x), taken in the y-direction, and (b) the angle, θ, between the upward tangent to the interface (ds) and the vertical. Thickness h is very large near the bath surface, decreases through the meniscus, and reaches an asymptotically constant value at large height. In this notation, the curvature is given simply by

$$c = \frac{(d^2h/dx^2)}{[1 + (dh/dx)^2]^{3/2}} \tag{2}$$

The notational system chosen here includes s as upward along the interface, with dx and-dy as components of ds, and with u and v as the x and y components of the upward surface velocity w.

The first question faced here is "What is the stretch term expressed in terms of the x component of velocity (u), the thickness (h), and their x derivatives?" The analytical answer to this question should facilitate numerical computation of the two-dimensional flow field, since these parameters are the ones which are usually calculated and stored as functions of position. The analytical answer would be useful, not only for two-dimensional studies (1), but also for calculations using one-dimensional models (7), wherein y-components of velocity are often neglected. Such an answer would also clarify when the stretch term $\partial w/\partial s$ may be rewritten as du/dx, as well as the magnitude of any neglected quantity.

4. THE X DIRECTION EXPRESSION

In accordance with Equation 1, evaluation of the stretch term is restricted to the interface. Thus w (s) and the partial derivative may be rewritten as the total derivative:

$$\frac{\partial w}{\partial s} = \frac{dw}{ds} \tag{3}$$

Along the interface, the following relationships hold:

$$dx = ds \cos\theta \tag{4}$$

$$u = w \cos\theta \tag{5}$$

Here $\theta = \theta(s)$. Differentiating Equation 5,

$$du = \cos\theta \, dw - w \sin\theta \, d\theta \tag{6}$$

and combining Equation 6 with 4 leads to:

$$\frac{dw}{ds} = \frac{du}{dx} + w \tan\theta \frac{d\theta}{ds} \tag{7}$$

Equation 7 shows an additive, correction term to du/dx. Equation 7 may be rewritten in terms of the first derivative of h ($dh/dx \equiv m$) and other parameters by noting that:

$$-\tan\theta = dy/dx = dh/dx \equiv m \tag{8}$$

$$\cos\theta = \frac{1}{(1+m^2)^{1/2}} \tag{9}$$

$$-\frac{d\theta}{ds} = \frac{dm/dx}{(1+m^2)^{3/2}} \tag{10}$$

$$w = u\ (1+m^2)^{1/2} \tag{11}$$

Thus by substitution of these parameters, another form of Equation 7 is

$$\frac{dw}{ds} = \frac{du}{dx} + \frac{u(m)(dm/dx)}{(1+m^2)}$$

$$= \frac{du}{dx} + \frac{u(dh/dx)(d^2h/dx^2)}{1+(dh/dx)^2} \tag{12}$$

5. LIMITING FORMS OF EQUATION 12

For calculation purposes, we are particularly concerned with the relative size of each term term near the top and the bottom of the meniscus. In the thin film near the top of the meniscus, where thickness h approaches a finite constant and m and dm/dx tend to vanish, Equation 12 tends toward

$$\frac{dw}{ds} = \frac{du}{dx} \text{ for large } x \tag{13}$$

For very large x, the surface velocity w tends toward a constant, so that dw/ds vanishes.

Equation 13 can be derived by using an order of magnitude analysis, where u(x) is of the order (Q/hb) and flux Q and width b are constants. For this case, du/dx is of the order (-u/h)(dh/dx), so that dw/ds at large x is of the order:

$$\frac{dw}{ds} \sim \frac{du}{dx}\left(1 - h\ \frac{d^2h}{dx^2}\right)$$

On the other hand, near the bath surface and the bottom of the meniscus, the derivative du/dx becomes very small so that Equation 12 tends toward

$$\frac{dw}{ds} = \frac{u(dh/dx)(d^2h/dx^2)}{1+(dh/dx)^2} \text{ for small } x \tag{14}$$

In this region, u tends to vanish and the h derivatives tend to grow with decreased s. Thus the term in Equation 14 is indeterminate. To explore the behavior of dw/ds at small x, it appears useful to examine the asymptotic behavior for large y, as discussed below.

6. THE Y DIRECTION EXPRESSION

To obtain the expression for the stretch term in the y direction, we proceed in a fashion analogous to that used for the x direction. The corresponding expressions to Equations 4, 5, and 7 are

$$-dy = ds \sin\theta \tag{15}$$

$$-v = w \sin\theta \tag{16}$$

$$\frac{dw}{ds} = \frac{dv}{dy} - \frac{w}{\tan\theta}\frac{d\theta}{ds} \tag{17}$$

For large y, θ approaches π/2, so that for finite w,

$$\lim_{y\to\infty} \frac{w}{\tan\theta}\frac{d\theta}{ds} = 0 \tag{18}$$

Thus Equation 18 reduces to

$$\frac{dw}{ds} = \frac{dv}{dy} \quad \text{for large y or small x} \tag{19}$$

Equation 19 shows the limiting behavior of the stretch term near the bath surface. We note parenthetically that the size of the surface velocity w near the bath surface will depend on boundary conditions. For example, with infinite baths, w will vanish; but for one finite bath case with a source term at the surface (1), w will approach a constant. In either case, dw/ds will vanish near the bath surface.

If desired, Equation 17 may be written in terms of h derivatives, using the following expressions

$$-\sin\theta = \frac{m}{(1+m^2)^{1/2}} \tag{20}$$

$$w = v\,(1+m^2)^{1/2}/m \tag{21}$$

Thus Equation 17 becomes

$$\frac{dw}{ds} = \frac{dv}{dy} - \frac{v(dm/dx)}{m^2(1+m^2)}$$

$$\frac{dw}{ds} = \frac{dv}{dy} - \frac{v(d^2h/dx^2)}{(dh/dx)^2\left[1+(dh/dx)^2\right]} \quad (22)$$

7. DISCUSSION OF X AND Y EXPRESSIONS

The stretch term in the equation expressing the normal-stress boundary condition, Equation 1, has been evaluated for rectangular coordinates, as shown in Equation 12 and 22. It is clear that other terms beside du/dx and dv/dy are needed, but that these terms are small in certain limiting cases. It appears that the results given above may be applied directly to both two-dimensional and one-dimensional models of free coating. They may also be applied by extension to other geometries where menisci are deformed by flow.

8. WAVY FLOW

In an investigation of wavy flow in falling films (where flux Q is considered as given), Penev et al (8) report the following normal-stress boundary condition as their Equation 4:

$$P_G - P_L = \sigma c + 2\mu\left(\frac{\partial u}{\partial x}\right)\frac{\left[1+(\partial h/\partial x)^2\right]}{\left[1-(\partial h/\partial x)^2\right]} \quad (23)$$

Equation 23 reduces, for steady flows of thin films where $\partial h/\partial x$ is small compared to unity, to

$$P_G - P_L = \sigma c + 2\mu \frac{du}{dx} \quad (24)$$

Since Equation 24 is restricted to the interface, where u = u(x) for steady flow, the total derivative is identical to the partial derivative. Equation 24 is identical to the combined form of Equations 1 and 13, so there is agreement here for the special case of thin films. Equation 23 was derived from Equation 60-14 of Landau and Lifshitz (9).

If applied to free coating, Equation 23 appears to exhibit a singularity at the point where $\partial h/\partial x$ is unity. However, the behavior of $\partial u/\partial x$ at that point must also be examined, since it may tend to zero more rapidly than the denominator term. The point may also be related to the surface stagnation point in free coating, where w=o (10), even though the locations of each of these two singularities seem to differ in the general case. The location of minimum pressure (11) along the interface is also of interest here, at least for comparison purposes.

9. FREE COATING FLUX

In a theoretical prediction of the constant thickness h_o (and thus flux Q), which occurs in free coating, Spiers et al (7) used the following normal-stress boundary condition (their Equation 14) for their one-dimensional approximation:

$$P_G - P_L = \sigma c + 2\mu \frac{\partial u}{\partial x}(x,h) \qquad (25)$$

Since Equation 25 is restricted to the interface, the dependence of the stretch term upon both x and h can be reduced to a dependence upon either x or h. Thus Equation 25 is identical to Equation 24. Since Equation 25 contains no correction factors for the stretch term, it should be limited to the thin film region for free coating.

Spiers et al referred to the work of Williamson (12) who, in studying flow fields of separation of tapes from an adhesive fluid, reported three interface equations (2-19, 2-23, 2-24) which may be summarized as

$$P_G - P_L = \sigma c + 2\mu \left(\frac{\partial u}{\partial x}\right)_{y=h} \qquad (26)$$

The subscript on the partial term indicates the interface restriction in an explicit manner.

11. DISCUSSION

The normal-stress boundary condition for one-dimensional models has been given in terms of $\partial u/\partial x$ by several authors, as noted in Sections 8 and 9 above. In this work, by explicitly identifying u as the x - component of the surface velocity w, this component depends only upon the x - coordinate of the position on the free surface. This allows the partial derivative to be written in terms of ordinary, or total, derivatives. Presentation of the free surface boundary condition in such a form appears to clarify computational aspects of the stretch term for both thick and thin films.

ACKNOWLEDGMENT

Discussions with William Krantz have been very helpful in recognizing and defining the problems in evaluation of the stretch term.

NOTATION

c	curvature, Eqn. 2
h	meniscus thickness
m	dh/dx
P_G	pressure, gas phase
P_L	pressure, liquid phase
s	distance along interface, upward
u	x component of w, along interface, Eqn. 5
v	y component of w, along interface, Eqn. 16
w	surface velocity, tangent to interface
x	vertical coordinate, upward from bath surface
y	horizontal coordinate, distance from belt

Greek Letters

μ	viscosity
σ	interfacial tension, liquid-gas
θ	angle between surface and vertical, Section 3

REFERENCES

1. Lee, C. Y., and Tallmadge, J. A., AIChE J., 20, 1079 (1974).

2. Lee, C. Y., "Meniscus Flow Fields and Profiles in Free Coating", Ph.D. Dissertation, Drexel Univ., Philadelphia, Pa. (1974).

3. Batchelor, G. K., "An Introduction to Fluid Dynamics," Chapter 3, Cambridge University Press (1962).

4. Scriven, L. E., Chem. Engr. Sci., 12, 98 (1960).

5. Soroka, A. J., and Tallmadge, J. A. AIChE J., 17, 505 (1971).

6. Lee, C. Y. and Tallmadge, J. A., Ind. Engr. Chem. Fundam. 13, 356 (1974).

7. Spiers, R. P., Subbaraman, C. V., and Wilkinson, W. L. Chem. Engr. Sci., 29, 389 (1974).

8. Penev, V., Krylov, V.S., Boyadjiev, C. H., and Vorotilin, V. P., Int. J. Heat Mass Transfer, 15, 1395 (1972).

9. Landau, L. D. and Lifshitz, E. M., "Fluid Mechanics", Chapter 7, Addison-Wesley, Reading, Mass (1959).

10. Lee, C. Y. and Tallmadge, J. A., AIChE J. 19, 865 (1973).

11. Lee, C. Y. and Tallmadge, J. A., AIChE J. 20, 1034 (1974).

12. Williamson, A. S., J. Fluid Mech. 52, 639 (1972).

SURFACE TENSION AND GAS PERMEABILITY DATA FOR SOLUBLE LIGNINS AT AN AIR-LIQUID INTERFACE

Karl F. Keirstead
Reed Paper Limited

The adsorption of both unrefined and processed soluble lignins at an air-liquid interface has been studied by measuring the changes in surface tension with time. The maximum bubble-pressure method was used for time intervals ranging from 1 to 100 seconds. The sessile bubble method was used for longer periods of times. The latter method also permits a calculation of bubble volume. The permeability of the bubble surface lamella to the enclosed air has been studied by determining the bubble volume at various intervals.

SOLUBLE LIGNINS

With the exception of some special preparations, the soluble lignins mentioned in this paper are commercial preparations designated as Lignosol products, commonly used as dispersing and emulsifying agents by reason of their strong adsorption at interfaces.

Lignosulphonates (LS) are derived from acid sulphite pulping of soft woods while kraft lignins are derived from the kraft pulping of either soft or hard woods. Further chemical or physical treatment gives rise to a variety of products of widely different surface active properties.

The approximate composition of calcium base sulphite liquor is shown in Table 1.

TABLE 1
Approximate composition of calcium base spent sulphite liquor (1).

Lignosulphonates	55	Sugar acids and residues	12
Hexose sugars	14	Resins and extractives	3
Pentose sugars	6	Ash	10

In the purified lignosulphonate, XD-65, 35% of the low molecular weight material has been removed by a commercial resin diffusion process which reduces the sugar content from 20 to 9.5%.

Kraft lignin contains lignin, hemicelluloses and acids derived from carbohydrates and is purified by precipitation upon acidification and separation by filtration. Kraft lignins contain a variety of functional groups such as carbonyls, carboxyls, aliphatic and phenolic hydroxyls.

Lignosulphonates differ from kraft lignins in molecular weight and polydispersity. Lignosulphonates exhibit an immense range in molecular weight and have a great polydispersity. McCarthy and co-workers have reported their molecular weights as ranging from 400-150,000 (2), while those of kraft lignin are lower, 1,050-1,600, and have a moderate polydispersity (3).

The lignosulphonate molecule in solution has a gel-like structure consisting of linear chains cross-linked to give a netted approximately spherical structure. The ionizing sulphate groups are attached mostly to the surface of the microgel.

Gardon and Mason (4) showed that lignosulphonates behave as expanding polyelectrolytes. Goring and co-workers found a relation between adsorption on titanium dioxide and the molecular weight of a lignin. Maximum adsorption was produced by LS at weight average molecular weights between 10,000 and 40,000 (5).

The properties of lignin monolayers at the air-water interface have been studied by Luner and Kemft (6). Force-area and potential-area isotherms were obtained by several lignins spread on water as monomolecular films. Surface potential and surface moments suggested that lignin monomers were standing on their phenolic and benzylic acid groups (kraft lignin). It appears obvious that a change in the conformation of the lignin molecule must take place to account for the difference between the spherical structure in solution and the elongated hydrophile-lyophile structure at the air-water interface which results in a reduction of surface tension.

GAS DISPERSION IN A GEL

The original basis for the present work was a study of the formation and stability of lignosulphonate produced bubbles in bubble sensitized gelled blasting agents, essentially concentrated solutions of ammonium and sodium nitrate thickened with cross-linked guar gum. The function of the bubbles is to increase the sensitivity of the composition to detonation by shock. A matrix density of 1.05 derived from an initial density of 1.4 contains 25% entrained air. At the end of the mixing, zero time, the bubble diameters may range from 25 to 100 microns. Diffusion of gas from high pressure regions (the smaller bubbles) to low pressure regions (the larger bubbles) is responsible for a marked change in bubble size distribution, which will eventually reach equilibrium.

The tendency of the larger bubbles to rise and collect at the top as a foam is balanced by the yield value of the cross-linked guar thickener.

Diffusion of gas from bubble to bubble and through the continuum may eventually lead to a collection of foam at the surface and a concomitant increase in the gel density. Local microscopic variations in the yield value of the gel may lead to random microscopic movements of a single bubble along with some corresponding local micromovement of the continuum.

The surface active properties of the surfactant affect the rate of aeration, the bubble size distribution and the gas permeability of the bubble membrane. Adsorption of a lignin polymer on the surface of the bubble membrane could be expected to affect the rate of gas diffusion. Bubble coalescence also gives rise to an increase in bubble diameter and may be a function of the bubble surface plasticity.

The complexity of change of bubble size in a gel matrix led to the study of surface tension and gas permeability of single bubbles in lignin aqueous solutions.

MOLALITY AND NORMALITY OF LIGNOSULPHONATES

The concentration of surface active solutions is usually expressed in terms of normality or molality. Because of the variations in molecular weight and acid groups of soluble lignins, the concentrations in this report are given as percent weight/volume. The corresponding molality and normality may be visualized by remembering that a 1% w/v having an average molecular weight of 10,000 is 0.001 molar. On the basis of 3 acid groups per 1,000 molecular weight, the corresponding normality is 0.000033.

SURFACE TENSION BY THE MAXIMUM BUBBLE-PRESSURE METHOD

Typically the surface tension of a surfactant diminishes with time. The drop is rapid at first according to the substance, the concentration and the pH. At the end of 30 minutes or so, the rate of decrease becomes very slow and the curve expressing the phenomenon becomes asymptotic to the axis of the abscissa.

The dynamic surface tension of a purified lignosulphonate, XD-65, is shown in Figure 1.

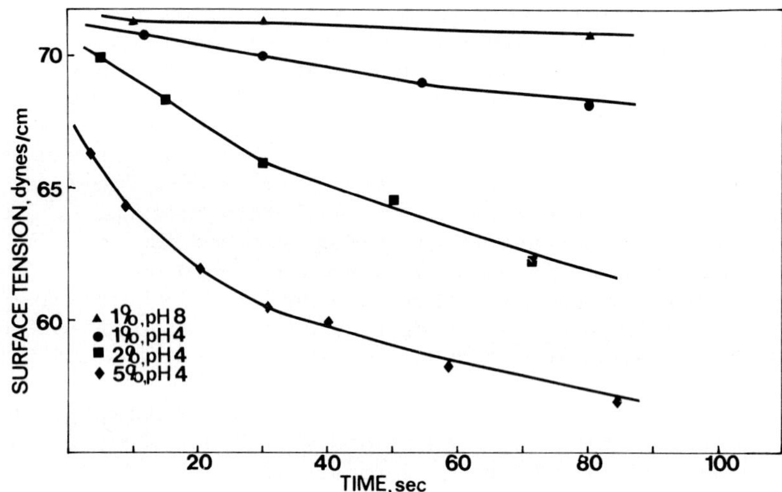

Fig. 1. Effect of ageing time on surface tension of a purified lignosulphonate (XD-65).

Typically the surface tension is less at pH 4 than at pH 8. Figure 4 shows the effect in greater detail. As the concentration is increased the 80 second surface tension drops markedly. Another feature is the difference between the rates of change over the first 30 seconds.

The surface tension values were determined by the maximum bubble-pressure as developed by S. Sugden and others (7). In principle, a series of single bubbles is formed at the tip of a small capillary at gradually reduced pressure and a gradually longer surface age. A second larger capillary is used to correct for differences in the level of the sample solution.

Chemical treatment typically produces a shift in the surface tension measured at a given pH and concentration. For example, a mild alkaline treatment lowers the surface tension while a severe alkaline treatment increases it. These treatments may also typically increase the molecular weight. However two samples subjected to the same type of chemical treatment may have the same molecular weight but different surface active properties. It is suggested that surface tension data can be used as an additional parameter to characterize a lignin.

FOAM FRACTIONATION

Foam fractionation has been used to remove a low concentration of kraft liquor from kraft mill effluent (8). The foam fractionation of a soluble lignin, either lignosulphonate or kraft lignin, produces a fairly wide range of surface active components. The first fraction generally has the lowest surface tension and the greatest foaming tendency, both in terms of initial foam height and foam life. The bottom residue has the highest surface tension and is relatively non-foaming. Re-fractionation of the first and bottom fraction produces an even greater spread in surface tension values.

The results of the foam fraction of an un-refined sodium lignosulphonate are shown in Figure 2, as determined by the maximum bubble-pressure method. The curve F-O shows the surface tension values of the product before fractionation. The 10% concentration of the samples shown in Figure 2 accentuates the differences between the fractions.

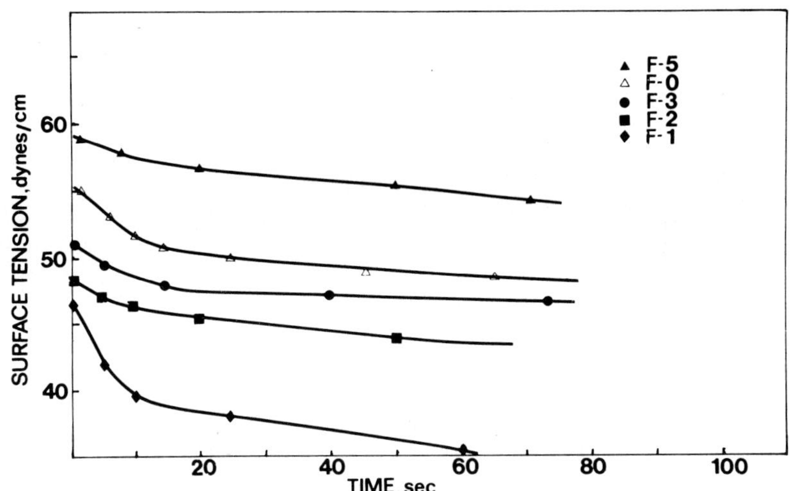

Fig. 2. Effect of surface age on the surface tension (maximum bubble pressure method) of fractions of sodium lignosulphonate (10%, pH 6.2).

Lignin, at the air-water interface, may be envisioned as consisting of ellipsoid or semi-linear molecules that differ in their polar asymmetry. Diffusion or preferential bonding with adjacent molecules leads to a selective orientation of the more asymmetric molecules resulting in a time dependent lowering of the surface tension.

SESSILE BUBBLE METHOD

For the observation of surface tension changes over a period of hours and even days, the sessile bubble method has many advantages. Time lapse measurements can be made on the same undisturbed bubble until diffusion of gas through the bubble lamella reduces the size of the bubble to a dimension that is impractical to measure. The sessile drop method has been used extensively to determine the surface tension of molten solids and in the form of a large drop in an immiscible liquid to determine interfacial tension. The sessile bubble and drop are subject to the same geometry and the same method of calculation of surface tension from the measured parameters of the contour.

N.K. Adams and H.L. Shute used both the maximum bubble-pressure method and the sessile bubble method to study the effect of surface age on the surface tension of paraffin chain salts (9). Further details were obtained by G.C. Nutting and F.A. Long using the same method (10). The latter authors made some observations of the differences between values obtained by the sessile bubble and the ring method. More recently the sessile bubble method was used by T. Sakai to measure the surface tension of molten linear polyethylene (11).

The authors mentioned above used direct observation with a microscope to measure the essential parameters of the drop (bubble) whereas we have followed the photographic method and the method of calculation described in The Supplement to the Tables of Bashford and Adams by D. White, Department of Energy, Mines and Resources (12).

The bubble was held on the end of a vertical rod, of either 5mm and 8mm in diameter. One aim of the photography was to obtain a print of small grain size that would permit optimum measurement with a travelling microscope permitting both vertical and horizontal movement and reading to 0.001 cm. Readings were made from Polaroid prints or negatives and occasionally from slide transparencies. The larger bubbles were photographed with either a constructed microscope or with a 35mm camera furnished with a macro lens and extension tubes. The magnification factor was determined from the known diameter of the rod holder which always appeared in the photograph.

Figure 3 shows the reduction in surface tension for solutions of Lignosol D-10 for times of 4-5 hours, at pH 4,6 and 8 (2.5% w/v).

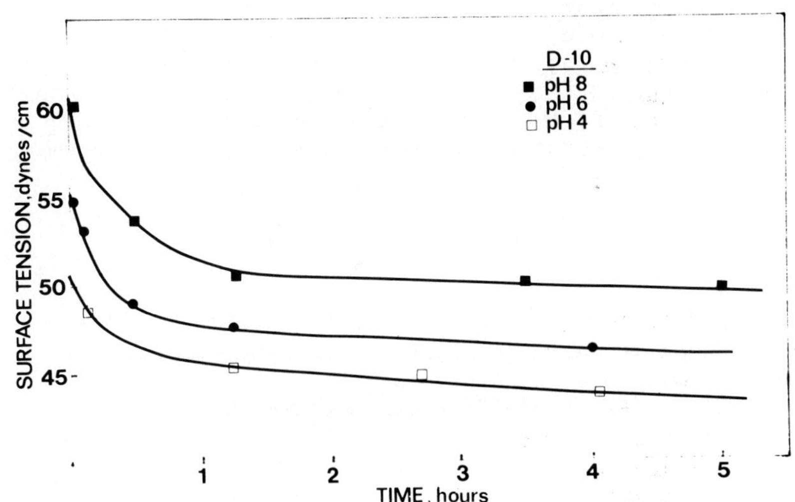

Fig. 3. Effect of pH on the surface tension of a purified and modified lignosulphonate (sessile bubble).

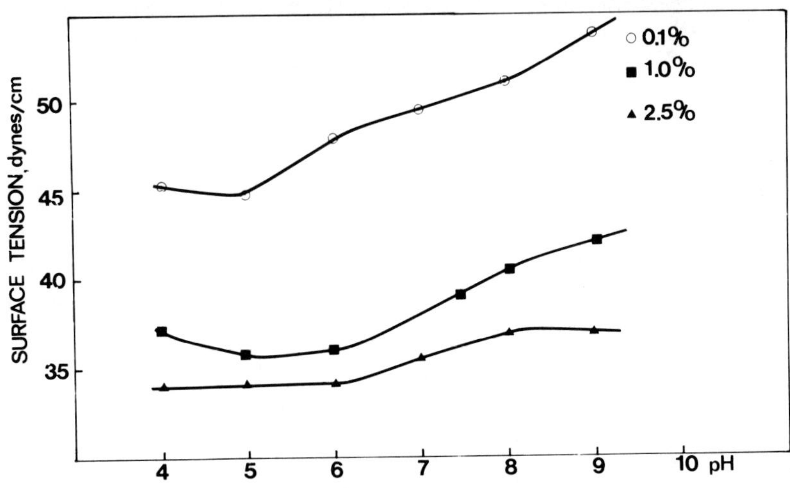

Fig. 4. Effect of pH on the surface of a purified and modified sodium lignosulphonate (SFX-65).

The effect of pH on the surface tension of SFX-65 over the pH range 4 - 9 is shown in Figure 4. Here the determinations were made with the ring method; an electronically controlled tensiometer. The small inflexion at pH 6 may be related to a significant change in light adsorption that takes place in that region. The change in light adsorption may indicate some change in the internal structure of the molecule. The change in surface tension as a function of pH shown in Figure 4 is small as compared with sodium laurate (10). At 0.001 N, the surface tension of sodium laurate changes from about 34 to 60 dynes/cm as the pH is changed from pH 7 to pH 10.

THE EFFECT OF FOAM FRACTIONATION ON MOLECULAR WEIGHT

The relation between the molecular weight and the surface tension is shown in Figure 5. An unpurified lignosulphonate (XD) and a purified lignosulphonate were foam fractionated. Surface tensions were measured by the sessile bubble method at a surface age of 30 minutes. Molecular weight was measured by the diffusion method (13). The molecular weight and the surface tension of the product before fractionation is indicated by the vertical arrow.

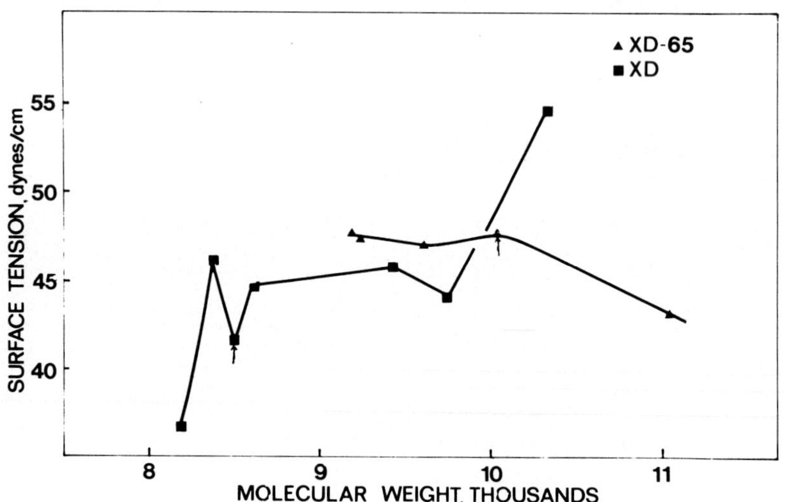

Fig. 5. Surface tension versus the molecular weight of sodium lignosulphonate (purified and unpurified).

It may be observed that purification has changed the average molecular weight of XD from about 8,400 to about 10,000 for XD-65. It is evident that both that molecular weight and the surface tension are related in these samples but in an opposite manner. The surface tension of XD has increased with the molecular weight while the surface tension of XD-65 has generally decreased with molecular weight. For the purpose of this relationship, the order of fractionation is not indicated. We have not been able to show a difference in surface tension for high and low molecular weight samples prepared by conventional means (resin diffusion and ultra-filtration). It must be remembered that XD contains many surface active components which would influence the preparation of foam fractionated samples.

SURFACTANT MICELLATION

Surface active agents are unique in that colloidal aggregates form spontaneously by association of molecules or ions, and that these aggregates are in reversible thermodynamic equilibrium with the surrounding environment. It is also generally accepted that the aggregates or micelles begin to form only when a definite concentration range is reached and that this range is relatively narrow. The possibility of aggregation in lignins has been discussed by Goring (14). Another view is expressed by Schwartz et al (15). For a typical surfactant of molecular weight 300, the micelle has a particle weight of about 10,000 to 15,000. This is the order of magnitude of the molecular weight of many polymeric lignins. In simplest terms the polymeric surfactants already exist as covalently bound micelles. Considering, however, that soluble lignins comprise a wide diversity of substances, it is possible that a lignin of lower molecular weight and of relatively low solubility may possibly exhibit micellar formation.

The plot of surface tension vs log concentration has been used to determine the critical micelle concentration of a surfactant (16). Figure 6 shows a plot of surface tension vs log concentration as determined by the tensiometer method. When triplicate readings were averaged there is is some scattering of points but no clear indication of a definite break in the plot. Similar results were obtained for a kraft lignin.

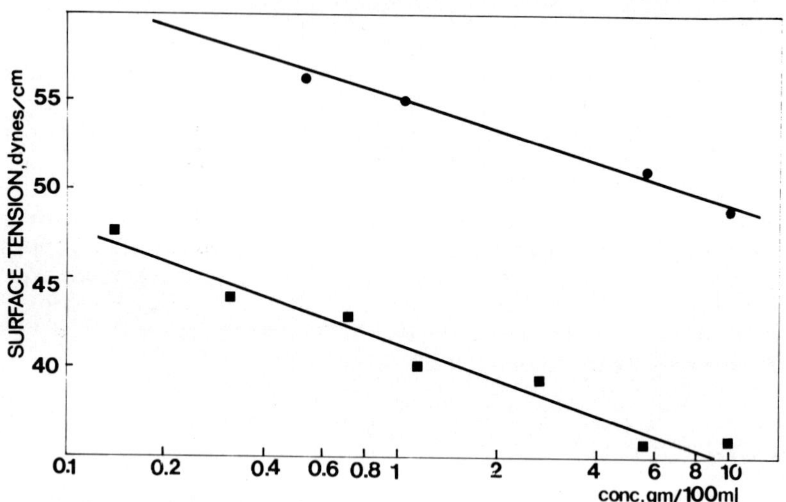

Fig. 6. Surface tension versus the logarithm of the concentration of two fractions of a calcium lignosulphonate.

PERMEABILITY OF ENCLOSED BUBBLES TO GASES

The permeability of soap films resting at a gas/liquid interface has been studied by Mason and co-workers (17). Strong indications were found that diffusion actually takes place through pores between surfactant molecules. Permeation requires that the molecule of gas pass through a hole in the monolayer. The hole may originate from the kinetic energy of the striking molecule, from fluctuations in the molecular density in the monolayer or both. Clearly the permeability of a bubble is affected by the size, conformation and orientation of the polymer adsorbed at the air/liquid interface.

Figure 7 shows the volumes of three bubbles calculated from the same prints as for the surface tension values shown in Figure 3. Rates of volume change over the first hour for pH 4, 8, and 6 respectively; 18,50 and 60 cc x 10^4.

Typical permeability data for a variety of LS are shown in Table 2.

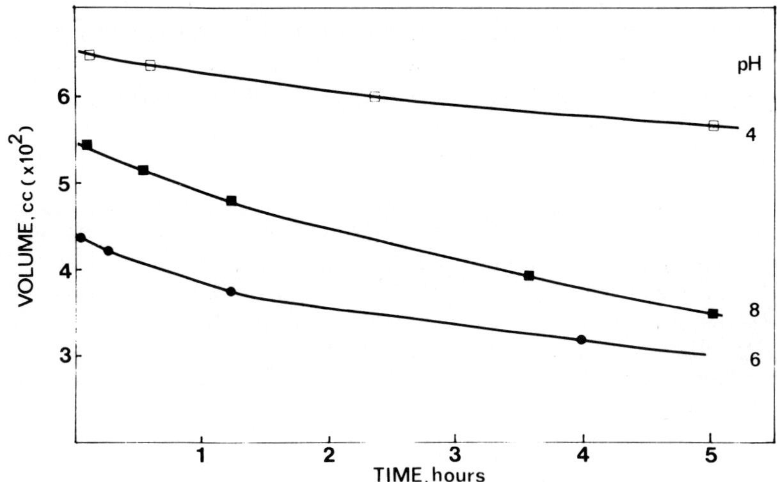

Fig. 7. Gas permeability versus pH of a purified and processed lignosulphonate (D-10), 2.5% w/v.

TABLE 2
Permeability Data for Various Lignosulphonates

Lignosulphonate Salt	*Permeability,-dv/dt cc/hr x 10^4*	*Conc. % w/v*	*pH*
NH_4, (TSD)	30.0	5.0	5.0
Na, Purified, (XD-65)	18.0 (S-1)	5.0	6.7
NH_4, Purified and Processed, (DR-15)	42.0	2.5	4.0
NH_4, Alk. Hydrolysis, (TSF)	11.0	2.5	4.0
Na, Purified and Modified, (SFX-65)	40.0	5.0	4.0

SUMMARY

The surface activity of soluble lignins at an air/liquid interface is markedly time dependent and is a function of lignin modification, pH and concentration. Surface tension values at a given pH and concentration can be used along with average molecular weight to characterize a lignin.

Foam fractionation produces a diversity of surface active components. Time dependency of the surface tension and the diversity of foam fractionated components lead to the concept of polydispersity in respect to polar asymmetry within a given lignin sample.

The author thanks Reed Paper Ltd. for permission to publish. The author is grateful to Dr. Denis W.G. White and A. Mar for helpful advice and discussion, and to Dr. J.V. Benko who performed the molecular weight determinations. The author is indebted to Department of Mines, Laval University for the use of their Fisher Autotensiomat. The work was supported (in part) by the Defence Research Board of Canada. (Grant W5).

BIBLIOGRAPHY

1. Wenzl, H.F.J., Paper Trade J. 149 (22) 47 (1965).
2. Felicetta, V.F., Ahola, A., and McCarthy, J.L., J. Am. Chem. Soc. 77, 3470 (1955).
3. Rezanowich, A., Yean W.Q. and Goring, D.A.I. Svensk Papperstidn., 66, 144 (1963).
4. Garden J.L., and Mason, S.G., Can. J. Chem. 33, 1477 (1955).
5. Rezanowich, A., Jaworzyn, J.F., and Goring, D.A.I., Pulp & Paper Mag. Can. 62, T-172 (1961).
6. Luner P., and Kempf, U., Tappi, 53, 2069 (1970).
7. Sugden, S., J. Chem. Soc. (1922), 858; (1924), 27.
8. Brasch, D.J., Appita 28 (1), 29 (1974).
9. Adam, N.K., and Shute, H.L., Trans. Faraday, Soc. 34; 758 (1938).
10. Nutting, G.C., and Long, F.A., J. Am. Chem. Soc. 63, 84 (1941).
11. Sakai, T., Polymer 6, 659 (1965).
12. White, D.W.G., "Supplement to the Tables of Bashforth and Adams", Department of Energy, Mines and Resources, Ottawa, 1967.
13. Benko, J.V., Tappi, 44, 766 (1961); 47, 508 (1964).
14. Goring, D.A.I., in "Lignins", (K.V. Sarkanen, and C.H. Ludwig, Eds.), p. 751. Wiley-Interscience, New York, 1971.
15. Schwartz, A.M., and Perry, J.W., "Surface Active Agents", Vol. 1, p. 313. Interscience, New York, 1949.
16. Zimmels, Y., and Lin, I.J., Colloid & Polymer Sci., 252, 594, (1974).
17. Princen, H.M., Overbeek, J. Th. G., and Mason, S.G., J. Colloid Interface Sci. 24, 125 (1967).

THE SURFACE TENSION OF BINARY MIXTURES OF NITROETHANE WITH THE HYDROCARBONS n-HEXANE, CYCLOHEXANE, METHYLCYCLOHEXANE, BENZENE, AND TOLUENE AT 30°C.[1]

Richard S. Myers
Delta State College

G. Patricia Angel and H. Lawrence Clever[2]
Emory University

ABSTRACT

The surface tension of mixtures of n-hexane, cyclohexane, methylcyclohexane, benzene, and toluene with nitroethane was measured by the maximum bubble pressure method at 30.0°C. Individual and composite ideal isotherms for the liquid-vapor (air) interface were calculated. The adsorption isotherms are compared with isotherms calculated earlier for ethanol and 2-methylbutanol-2 (t-pentyl alcohol) mixtures with all of the same hydrocarbons except n-hexane. The non-ideal nature of the bulk liquid hydrocarbon-polar molecule mixtures is presented and the surface tension and adsorption isotherms as a function of bulk liquid activity are discussed qualitatively.

I. INTRODUCTION

The present work is one in a series of papers reporting on the properties of binary mixtures of small polar molecules with hydrocarbons. Reported here are the surface tension and density of several nitroethane-hydrocarbon mixtures at 30°C. The results from the nitroethane mixtures are compared with the results of an earlier study (1) on the surface tension and density of mixtures of ethanol and of 2-methylbutanol-2 (t-pentyl alcohol) with all of the same hydrocarbons except n-hexane.

---

1. Supported in part by a Grant from the National Science Foundation.
2. Author to whom correspondence should be addressed.

II. EXPERIMENTAL

A. Materials

Nitroethane, Matheson, Coleman, and Bell Co., dried over $CaSO_4$, distilled under reduced pressure through a 60 cm. helice packed column, bp 57°C at 97 torr. n-Hexane, Phillips Petroleum Co., 99 mol %; Cyclohexane, Fisher Scientific Co., Reagent Grade; and Methylcyclohexane, Phillips Petroleum Co., 99 mol % were each shaken with concentrated H_2SO_4, washed with water until neutral to litmus, dried over $CaSO_4$ and distilled at atmospheric pressure from over CaH_2. Benzene, Phillips Petroleum Co., pure grade, 99 mol % and Toluene, J.T. Baker Chemical Co., Analyzed Reagent grade were each shaken with concentrated H_2SO_4, washed with water until neutral to litmus, dried over $CaSO_4$, and distilled from over CaH_2 through a 60 cm packed column at atmospheric pressure.

B. Sample Preparation and Density Measurement

The mixtures were prepared in specially designed mixing bottles (2) which allowed the bulk liquid composition to be corrected for material in the vapor phase by assuming Raoult's law. Densities were determined in a calibrated 16 cm^3 density bottle with a constant-bore capillary neck of 1.00 ± 0.01 mm which was graduated at 1 mm intervals for a length of 2 cm. The density bottle was calibrated with freshly distilled water. Material was transferred directly from the mixing bottle to the density bottle by a gas tight syringe. The temperature of the density measurement was controlled to 30.00 ± 0.02°C.

C. The Surface Tension Measurement

The surface tension was measured by the maximum bubble pressure technique on an apparatus of the type described by Quayle (3). The surface tension bubbler was calibrated by use of highly purified samples of benzene, n-heptane, and n-octane. The bubbler air was presaturated with sample vapor before being drawn through the sample to prevent surface temperature and composition changes due to evaporation. The sample was transferred to the presaturation device and to the surface tension bubbler directly from the mixing bottle by means of a gas tight syringe. The temperature was controlled to 30.00 ± 0.02°C. for the surface tension measurement.

III. RESULTS

A. Surface Tension and Excess Surface Tension

The composition, density, surface tension, and excess surface tension are given in Table 1.

TABLE 1
Hydrocarbon - Nitroethane Mixtures: Composition, Density, Surface Tension, and Excess Surface Tension at 30°C.

Nitroethane, Mole Fraction X	*Density, g cm^{3}*	*Surface[1] Tension dyne cm^{-1}*	*Excess Surface Tension dyne cm^{-1}*
(1 - X) n-Hexane + X Nitroethane			
0.1264	0.6805	18.07	-1.16
0.2726	0.7138	18.48	-2.80
0.3609	0.7385	18.42	-4.10
Liquid - Liquid Immiscible			
0.6500	0.8495	18.87	-7.71
0.7764	0.9030	20.42	-7.94
0.8713	0.9555	23.58	-6.11
(1 - X) Cyclohexane + Nitroethane			
0.1364	0.7925	24.11	-0.72
0.2381	0.8090	24.38	-1.24
0.3745	0.8380	24.59	-2.08
0.5058	0.8696	24.88	-2.80
0.6179	0.8995	25.15	-3.40
0.7565	0.9444	26.22	-3.40
0.8720	0.9840	28.30	-2.21
(1 - X) Methylcyclohexane + X Nitroethane			
0.1328	0.7800	23.22	-0.70
0.1378	0.7810	23.39	-0.57
0.2372	0.7985	23.30	-1.53
0.3808	0.8274	23.57	-2.52
0.4991	0.8550	23.64	-3.48
0.6229	0.8890	24.09	-4.11
0.7531	0.9306	24.90	-4.44

TABLE 1 (Continued)

(1 - X) Methylcyclohexane + X Nitroethane			
0.8678	0.9755	27.24	-3.10
(1 - X) Benzene + X Nitroethane			
0.1243	0.8855	27.73	-0.25
0.2555	0.9040	28.00	-0.51
0.3693	0.9225	28.38	-0.58
0.5034	0.9440	29.08	-0.42
0.6352	0.9675	29.32	-0.71
0.7461	0.9870	30.00	-0.47
0.8735	1.0120	30.24	-0.75
(1 - X) Toluene + X Nitroethane			
0.1269	0.8740	27.91	-0.03
0.2541	0.8920	28.38	-0.08
0.3706	0.9100	28.80	-0.13
0.5004	0.9315	29.18	-0.28
0.6248	0.9540	29.48	-0.49
0.7474	0.9783	30.24	-0.23
0.8785	1.0075	30.62	-0.38
0.8798	1.0080	30.90	-0.11

1. Component pure liquid surface tensions are in Table 4.

The surface tension as a function of mole fraction is shown in Fig. 1.

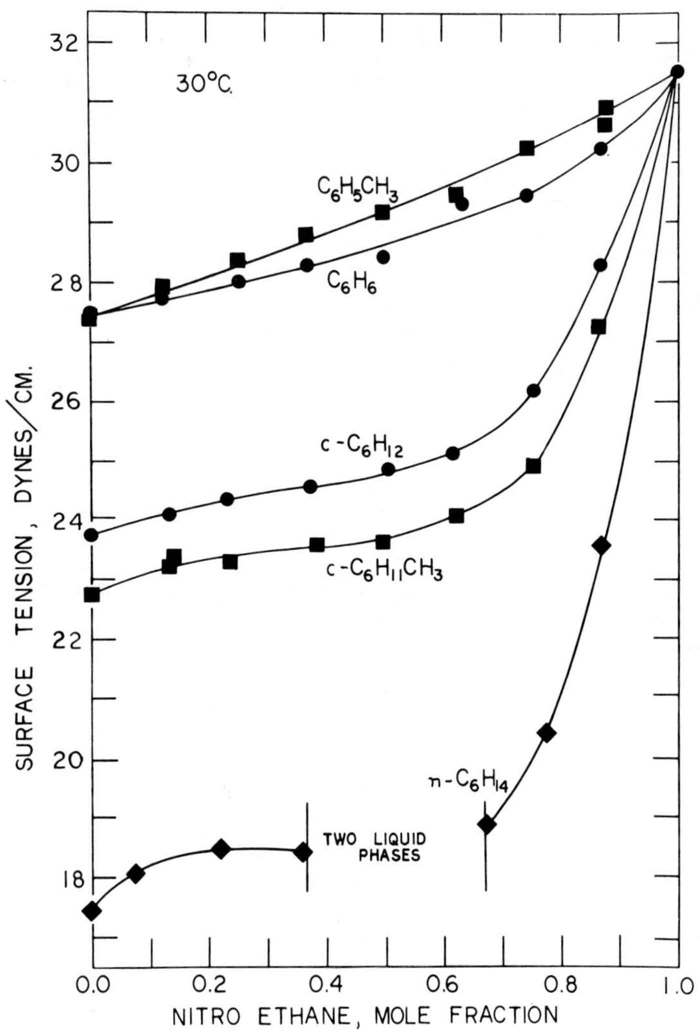

Fig. 1. Surface tension as a function of mole fraction at 30°C. for the hydrocarbon - nitroethane mixtures.

The excess surface tension is defined as

$$\sigma^E = \sigma - (X_1\sigma_1^{\,o} + X_2\sigma_2^{\,o}) \qquad (1)$$

where σ is the solution surface tension, $\sigma_1^{\,o}$ and $\sigma_2^{\,o}$ are the hydrocarbon and nitroethane pure liquid surface tensions, respectively and X_1 and X_2 are the hydrocarbon and nitroethane mole fractions, respectively. The excess surface tension values were fitted to the excess property equation

$$Y^E = X_2(1-X_2) \sum_{n=o}^{m} A_n(1-2X_2)^n \qquad (2)$$

by the method of least squares. In the equation Y^E is the excess property, X_2 is the nitroethane mole fraction, and A_n and n are parameters varied to give the best fit. The equation parameters for the excess surface tension are given in Table 2 and the excess surface tension as a function of mole fraction is shown in Fig. 2.

TABLE 2
Excess Surface Tension of the Nitroethane Solutions: Parameters for the Excess Properties Equation.

Nitroethane with:	A_0	A_1	A_2	A_3	A_4	*Std. Dev.*
n-Hexane	-23.945	28.200	-18.925	3.155	5.835	0.07
Cyclohexane	-10.862	10.911	- 9.244	-2.023	10.021	0.08
Methylcyclohexane	-13.585	14.501	-13.854	0.792	16.412	0.08
Benzene	- 2.063	- 0.698	- 3.956	5.917	-	0.11
Toluene	- 1.184	2.538	2.737	-2.806	- 5.446	0.10

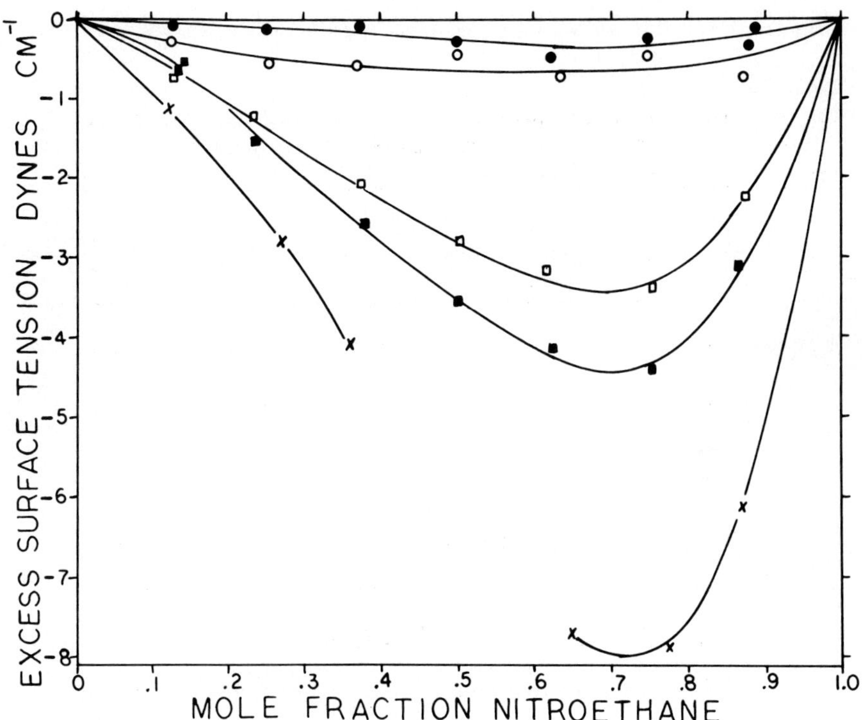

Fig. 2. Excess surface tension as a function of mole fraction at 30°C. for the hydrocarbon - nitroethane mixtures.

B. Density and Excess Volume

The composition and density data of Table 1 were used to calculate excess volume of mixing for each of the solutions.

$$V^E = V - V_{ideal} \quad (3)$$

where $$V_{ideal} = X_1V_1^o + X_2V_2^o \quad (4)$$

In equations 3 and 4 V^E is the excess volume, V the solution molar volume, V_{ideal} the solution ideal molar volume, V_1^o and V_2^o the molar volumes of the pure hydrocarbon and nitroethane components, respectively and X_1 and X_2 the hydrocarbon and nitroethane mole fractions, respectively. The parameters of the least squares fit of equation 2, the excess properties equation, to the excess volumes is given in Table 3.

TABLE 3.
Excess Volume of the Hydrocarbon - Nitroethane Solutions: Parameters for the Excess Properties Equation.

Nitroethane with:	A_0	A_1	A_2	A_3	*Std. Dev.*
Cyclohexane	3.279	0.330	-1.231	-	0.08
Methylcyclohexane	2.104	-0.061	0.571	1.166	0.01
Benzene	-0.118	0.099	-0.169	-	0.03
Toluene	-0.580	-0.234	0.136	0.578	0.01

C. Individual and Composite Isotherms

The Gibbs adsorption equations for surface excesses

$$\Gamma_2^{\ 1} = -(X_2/RT)(\delta\sigma/\delta X_2) \tag{5}$$

and

$$\Gamma_2^{\ N} = X_1\ \Gamma_2^{\ 1} \tag{6}$$

were used to obtain the composite isotherm $\Gamma_2^{\ N}$ defined as the excess of solute in the surface of unit area over a region of the bulk liquid containing the same number of moles of all species. The individual isotherms Γ_1 and Γ_2 represent a surface concentration per unit area. To calculate the individual isotherms one must assume a model of the orientation of the molecules at the interface, area occupied per molecule and whether there is monolayer or multilayer adsorption. We have assumed spherical molecules and monolayer adsorption and calculated Γ_1 and Γ_2 from

$$\Gamma_2 = \left(\Gamma_2^{\ 1} + \frac{X_2}{X_1 a_1}\right) \Bigg/ \left(1 + \frac{X_2 a_2}{X_1 a_1}\right) \tag{7}$$

and

$$\Gamma_1 = (X_1/X_2)(\Gamma_2 - \Gamma_2^{\ 1}) \tag{8}$$

where a_1 and a_2 are the surface areas per molecule of hydrocarbon and nitroethane, respectively. The spherical molecule cross sectional areas calculated from liquid molar volumes are 36.4, 32.1, 35.8, 28.2, and 31.7 square Ångstroms for n-hexane, cyclohexane, methylcyclohexane, benzene, and toluene, respectively. Nitroethane occupies 24.3 $Å^2$. The individual surface concentrations are extended to the pure liquid surface concentrations.

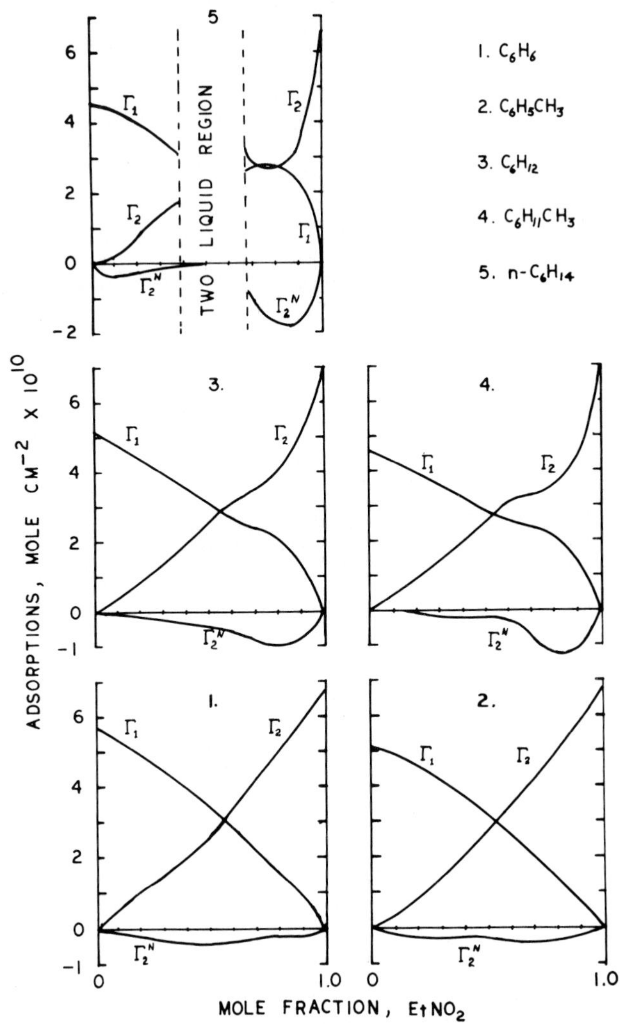

Fig. 3. Ideal individual and composite isotherms as a function of mole fraction at 30°C. for the hydrocarbon - nitroethane mixtures.

The individual and composite isotherms as a function of mole fraction are shown for each of the five systems in Fig. 3. The composite isotherms are compared directly on a larger scale in Fig. 4. The parallel order of increasing maximum magnitude of excess surface tension (Fig. 2) and increasing maximum magnitude of composite adsorption (Fig. 4) is expected.

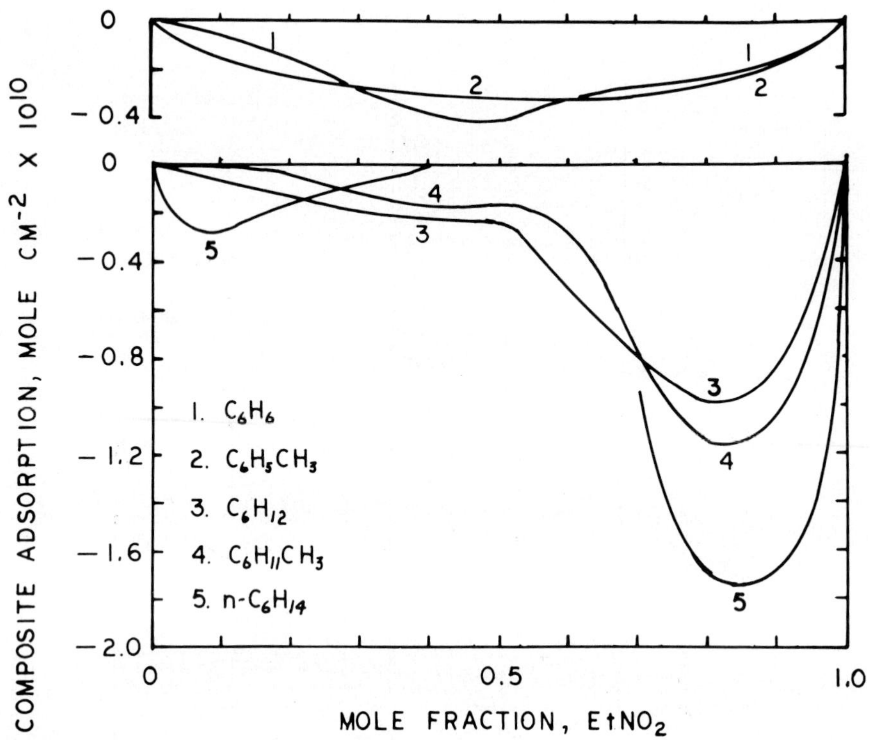

Fig. 4. A comparison of the ideal composite isotherms as a function of mole fraction at 30°C. for the five hydrocarbon - nitroethane mixtures. The surfaces are deficient in nitroethane.

IV. DISCUSSION

A. Comparison of Composite Isotherms of Hydrocarbon Mixtures with Nitroethane, Ethanol, and t-Pentyl Alcohol

Recently we reported the result of surface tension measurements of ethanol and t-pentyl alcohol mixtures with some of the same hydrocarbons (1). The composite isotherms Γ_2^N for nitroethane, ethanol, and t-pentyl alcohol mixtures with the hydrocarbons cyclohexane, methylcyclohexane, benzene, and toluene are compared in Fig. 5. Table 4 compares the pure liquid surface tensions and the difference in pure liquid surface tensions for the components of the 12 mixtures. In the alcohol mixtures the alcohol is the component of lower surface tension, and in the nitroethane mixtures the hydrocarbon is the component of lower surface tension. The composite adsorption isotherms of Fig. 5 show order and magnitude that is in approximate proportion to the systems difference in the pure component surface tensions.

TABLE 4
Comparison of the Pure Liquid Surface Tensions[a] *and of the Difference in Pure Liquid Surface Tensions, 30°C.*

		$\Delta\sigma = \sigma^o_{hydrocarbon} - \sigma^o_{polar\ comp.}$		
Hydrocarbon		*Nitroethane (31.50)*	*Ethanol (21.48)*	*t-Pentyl Alcohol (21.89)*
n-Hexane	(17.45)	-14.05	-	-
Cyclohexane	(23.74)	- 7.76	2.26	1.85
Methylcyclohexane	(22.76)	- 8.74	1.28	0.87
Benzene	(27.48)	- 4.02	6.00	5.59
Toluene	(27.41)	- 4.09	5.93	5.52

a. *The pure liquid surface tensions (dyne cm^{-1}) are in ( ).*

The nitroethane composite isotherm is negative. The values represent the surface deficiency of nitroethane, the component of higher surface tension. Both of the alcohols show a surface excess in concentration except the cycloalkane - ethanol solutions which are deficient in alcohol between about 0.85 and 1.0 mole fraction ethanol. There is a possibility that the minimum in the cycloalkane - ethanol surface tension and the resulting negative composite isotherm for solutions concentrated in ethanol is due to trace water contamination. A systematic study of the effect of water in these systems is planned.

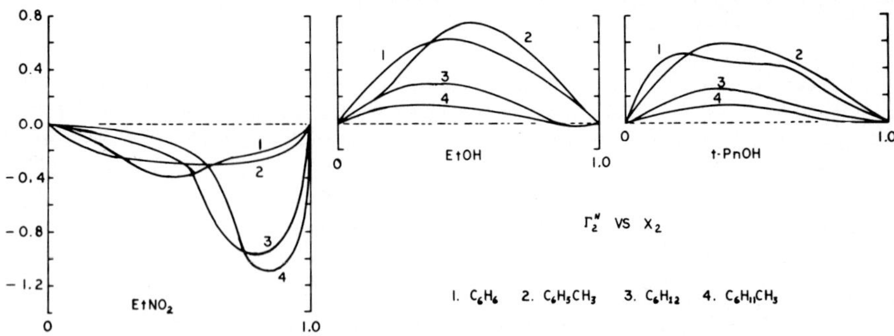

Fig. 5. A comparison of the ideal composite isotherms as a function of mole fraction at 30°C. for mixtures of cyclohexane, methylcyclohexane, benzene, and toluene with nitroethane (left) ethanol (center), and 2-methylbutanol-2 (t-pentyl alcohol) (right).

B. Bulk Liquid Nonideal Character

The composite adsorption isotherms were calculated assuming ideal solution behavior. Obviously these mixtures of hydrocarbon and polar molecules are far from ideal. The excess volume, excess Gibbs energy, activity coefficients, and excess enthalpy of mixing are known for many of the twelve solutions compared here. Literature sources for the properties of the bulk liquid mixtures are given below.

1. *Density and Excess Volume*

The references are hydrocarbon - nitroethane, this work, and (5), and hydrocarbon - alcohol (4).

2. *Activity Coefficients and Excess Gibbs Energy*

The references are benzene - nitroethane (5), toluene - nitroethane (6), benzene - ethanol (7) toluene and methylcyclohexane - ethanol (8), cyclohexane - ethanol (9), and benzene - t-pentyl alcohol (4).

3. Enthalpy of Mixing

The references are benzene - nitroethane (10), toluene - nitroethane (6), cyclohexane and methylcyclohexane - nitroethane and benzene and toluene - t-pentyl alcohol (11), toluene and methylcyclohexane - ethanol (8), and cyclohexane - ethanol (12).

The data from the references are summarized in Figs. 6 and 7. Figure 6 compares the excess volume function V^E/X_1X_2 as a function of mole fraction for the 12 mixtures. Fig. 7 compares the excess thermodynamic functions G^E/RTX_1X_2, H^E/RTX_1X_2, and TS^E/RTX_1X_2.

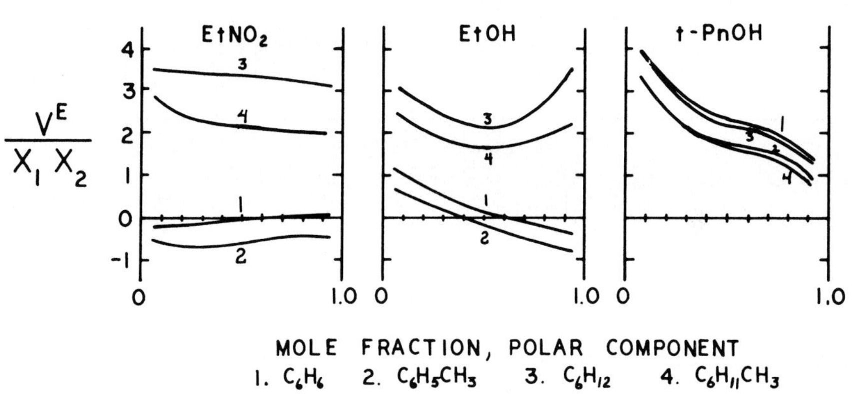

Fig. 6. Nonideal character of the bulk liquids. A comparison of the excess volume of mixing function V^E/X_1X_2 as a function of mole fraction at 30°C. for mixtures of cyclohexane, methylcyclohexane, benzene, and toluene with nitroethane (left), ethanol (center), and t-pentyl alcohol (right).

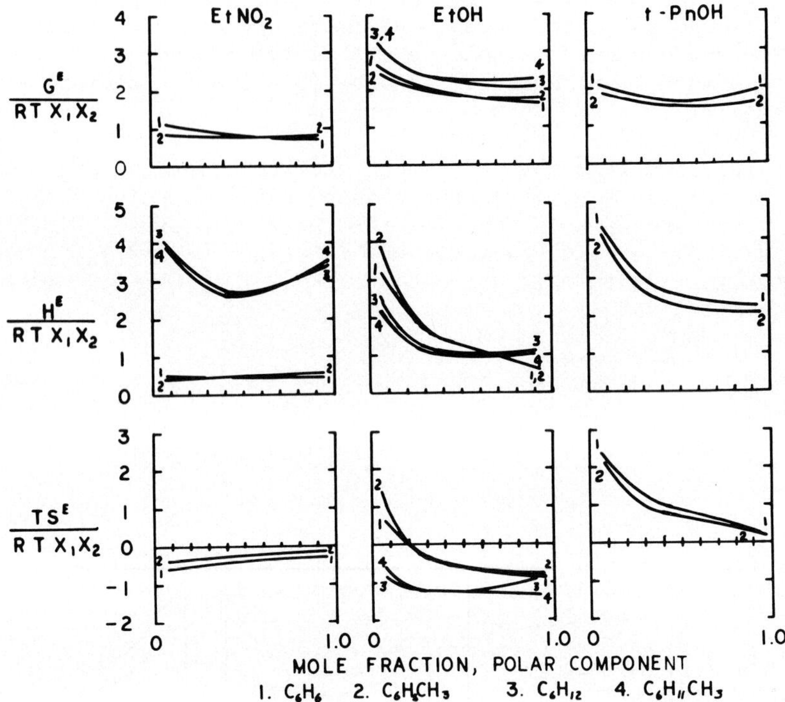

Fig. 7. Nonideal character of the bulk liquids. A comparison of the excess thermodynamic properties of mixing functions G^E/RTX_1X_2, H^E/RTX_1X_2, *and* TS^E/RTX_1X_2 *for mixtures of cyclohexane, methylcyclohexane, benzene, and toluene with nitroethane (left), ethanol (center), and t-pentyl alcohol (right).*

Although the hydrocarbon mixtures with ethanol and nitroethane show similar order and magnitude of excess volume there are also some striking differences in behavior. There is a definite minimum in the cyclohexane - ethanol curves and the aromatic hydrocarbons with nitroethane and with ethanol have opposite sign slopes. The t-pentyl alcohol results show significant differences from both nitroethane and ethanol solutions in that benzene and cyclohexane solutions and toluene and methylcyclohexane solutions show similar behavior, while in the other system the aromatic hydrocarbon solutions and the cycloalkane solutions are the similar pairs.

There are enthalpy of mixing data for ten of the twelve solutions and excess Gibbs energy of mixing data for eight of the twelve solutions. All enthalpy of mixing and excess Gibbs energy of mixing presently available for these solutions are endothermic, and all of the solutions show positive deviations from Raoults law. The behavior of G^E/RTX_1X_2 when plotted against mole fraction can tell us something of the nature of the nonideal nature of the solution. When G^E/RTX_1X_2 is zero for all values of mole fraction the solution is ideal, when G^E/RTX_1X_2 shows horizontal linear behavior which is independent of temperature the solution is regular, non-horizontal linear or curved plots of G^E/RTX_1X_2 mean increasingly complex systems. Obviously these systems are quite complex.

C. Bulk Liquid Activity and Surface Properties

A frequently used relation (13,14) between the bulk liquid and surface activity of a component is

$$\ln a_i^s = \ln a_i + \frac{A_i}{RT} (\sigma - \sigma_1{}^o) \tag{9}$$

where a_i^s is the surface activity, a_i the bulk activity, A_i the area per molecule in the surface, and σ_i^o the pure liquid surface tension of component i and σ the solution surface tension. All of the systems compared here show positive excess Gibbs energy of mixing, $a_i^s > a_i$ for the solution component of lower surface tension, and $a_i^s < a_i$ for the solution component of higher surface tension.

The ethanol - hydrocarbon mixtures show some unusual behavior. Toluene - ethanol excess surface tensions appear positive when plotted as a function of mole fraction, however, when plotted against bulk activity the values are negative as they are for the other systems. The cyclohexane and methylcyclohexane - ethanol surface tensions show a minimum when plotted against mole fraction. The minimum also appears when the surface tension is plotted against bulk liquid activity.

Our attempts to calculate composite adsorption isotherms as a function of activity rather than mole fraction have had only limited success. Preliminary calculations of composite isotherms on an activity basis have been carried out for ethanol solutions of benzene, toluene, and cyclohexane, and for t-pentyl alcohol solutions of benzene and toluene. The results indicate the activity composite isotherms are about double the magnitude of the mole fraction isotherms with the maximum shifted toward the component of lower surface tension for benzene and toluene solutions of both alcohols. The activity effect is even more pronounced for the cyclohexane - ethanol solution. The adsorption increases almost eight fold and the maximum is shifted from 0.4 to 0.65 mole fraction ethanol. We hope to carry out activity composite isotherm calculations for all of the systems for which activity data is available in the near future.

V. REFERENCES

1. Myers, R. S., and Clever, H. L., *J. Chem. Thermodynamics* 6, 949 (1974).
2. Battino, R., *J. Phys. Chem.* 70, 3408 (1968).
3. Quayle, O. R., *Chem. Rev.* 53, 439 (1953).
4. Myers, R. S., and Clever, H. L., *J. Chem. Thermodynamics* 2, 53 (1970).
5. Derrick, M. E., and Clever, H. L., *J. Colloid and Interface Sci.* 39, 593 (1972).
6. Orye, R. V., and Prausnitz, J. M., *Trans. Faraday Soc.* 61, 1338 (1976).
7. Brown, I., Fock, W., and Smith, F., *J. Chem. Thermodynamics* 1, 273 (1969).
8. Hwa, S. C. P., and Ziegler, W. T., *J. Phys. Chem.* 70, 2572 (1968).
9. Washburn, E. R., and Handorf, B. H., *J. Amer. Chem. Soc.* 57, 441 (1935).
10. Clever, H. L., and Hsu, K. Y., *J. Chem. Thermodynamics* in press.
11. Hsu, K. Y., and Clever, H. L., *J. Chem. Eng. Data* 20, 268 (1975).
12. Goates, J. R., Snow, R. L., and James, M. R., *J. Phys. Chem.* 65, 335 (1961).
13. Butler, J. A. V., *Proc. Roy. Soc.* (London) 135A, 348 (1932).
14. Eriksson, J. C., *Ark. Kemi* 26, 49 (1966).

THE INFLUENCE OF SUBSTRATUM SURFACE PROPERTIES ON THE ATTACHMENT OF A MARINE BACTERIUM

M. Fletcher
Marine Science Laboratory
Menai Bridge, Gwynedd, Wales

G.I. Loeb
Naval Research Laboratory
Washington, D.C.USA

ABSTRACT

Attachment to solid surfaces is an important part of the life cycle of many types of organisms, including many marine forms such as barnacles, bryozoa and bacteria, among others. The parameters influencing attachment, however, have been most extensively studied with cultured cells derived from animal tissues. The results of those studies indicate that attachment and enhanced growth of tissue cells occurs on high energy, negatively charged surfaces. We show here, however, for marine bacteria, that the opposite preference may be demonstrated, and cite data which raises the possibility that other marine forms may also not follow the tissue cell model. We postulate the involvement of non-polar interactions and increased importance of charge repulsion in these cases.

I. INTRODUCTION

Attachment to a solid surface enhances the growth and development of many aquatic organisms and cells in tissue culture, and indeed is an essential part of the life cycle in certain cases. Many marine invertebrates (the barnacle, for example) have free-swimming larval stages in which dispersion throughout large regions and exploration of possible attachment sites occurs prior to attachment and meta-morphosis to the sedentary adult form (1). Aquatic micro-organisms have been observed to proliferate upon immersed surfaces (2), as do oral bacteria on teeth (3), and recently cultured tissue cell lines have been found to exhibit anchorage-dependence, which is the enhanced growth of the cells when they attach to and spread over a suitable surface (4).

A great deal of work has been done with tissue cells in culture; the accumulated experience has shown that, while the history and environment of the cell and surface are important, certain surfaces are much more suitable than others (5).

Thus, glass (5,6,7) and polystyrene which has been processed by sulfonation (8) or exposure to electrical discharge (4,7) have become standard surfaces for tissue culture, while the use of unprocessed plastics generally has not been successful. The higher surface energy of glass and processed plastic (which may also be described as increased "wettability" (9, 10)) seems to be associated with successful anchorange and growth of tissue cells (7).

Many species of marine organisms undergoing the transition from free-swimming to attached forms appear capable of "selecting" an attachment site (1). Evidence concerning the effects of surface physico-chemical properties on settlement, however, is still sparse. Glass has been found to be relatively unattractive to barnacle larvae, but leaching of alkali and extreme smoothness have been considered probable reasons (1). Plastics, on the other hand, have been successfully used for settlement of polychaetes, (11 and unpublished results of K. Parks), barnacles (personal communication of D.J. Crisp), marine bacteria (12,13), and bryozoans (14,15, 16). Certain marine algae have been shown to develop adhesive structures on plastic substrata, whereas such structures are usually not formed at glass surfaces (17). Thus, lower-energy materials may be favored for settlement by certain marine organisms. We have investigated this possibility using a marine bacterium in the present work.

II. EXPERIMENTAL MATERIALS AND METHODS

A. The Organism and Its Culture

The bacterium used is a marine pseudomonad, (*Pseudomonas* sp. NCMB2021) and has been previously described (12). The medium contained 0.1% (w/v) peptone (Oxoid) and 0.1% (w/v) yeast extract (Oxoid) in filtered (0.2 nm porosity) seawater, pH 7.6, and was sterilized by autoclaving before use. 250 ml. portions of the medium were inoculated from a stationary phase culture, and incubated at $18^{o}C$ for 22 hr. with aeration by bubbling. After the incubation period, at which time the cells were in the late logarithmic growth phase or early stationary phase, the bacteria were collected by centrifugation and re-suspended in sterile filtered seawater to 60% of the original culture volume. The optical density of the bacterial suspension was measured and related to bacterial density via total cell counts on aliguots of the suspension. Cell densities used ranged from $2.5x10^9$ to $5x10^9$ cells/ml.

B. Substratum Materials

The following substances were used as substrata for attachment: glass microscope slides (Chance); freshly cleaved mica, (Mica & Micanite Supplies Ltd); polystyrene petri dishes (Falcon); poly (ethylene terphthalate) (DuPont Mylar film;) polyethylene film, PTFE Teflon film, and glass slides coated with polymerized m-phenylene-diamine following Kennedy, Barker, & White (18). The polystyrene petri dishes, tissue culture grade, were supplied with polystyrene covers. The dishes were processed to make them more wettable, with water and so suitable for cell tissue culture work, while the covers were unprocessed and so non-wettable. Since scanning electron microscopic examination of the polystyrene dishes and covers revealed differences in surface texture, we treated a number of samples in the following way to obtain paired samples of similar texture but different surface energy. Halves of polystyrene petri dish covers (Falcon) and of cut pieces of Mylar film (0.125 mm) were washed with detergent solution and air-dried. At this stage the surfaces did not wet with water. Half of the samples were treated for ten minutes at maximum power at 250×10^{-3} Torr residual air pressure in a radio-frequency plasma cleaning device (Harrick Scientific Corp.). The treated samples then wetted with water, and on draining the water film remained continuous even when sufficiently thin that interference colors were seen. Scanning electron microscopic examination of paired treated and untreated samples cut from the same piece did not reveal differences in surface texture, and such pairs were compared in this work.

Samples of glass, mica, and Teflon were cleaned in hot nitric acid after detergent washing. Other samples were detergent-washed only; all were exhaustively rinsed with sterile distilled water.

C. Bacterial Attachment

Thirty ml. portions of bacterial suspension were placed in petri dishes containing the test materials. Two hours were allowed for bacterial attachment. The test substrata were then rinsed gently with sterile water to remove any residual medium and unattached, or weakly sorbed, bacteria. The adhering bacteria were fixed with Bouin's fixative, and stained with crystal violet. Supplementary observations indicated that the rinsing, fixing, and staining procedures did not influence the results. Two replicates of each test material were used in each experiment, and all experiments were repeated. The number of bacteria per 100 μm^2 was determined for 25 areas on each sample, by direct counting.

III. RESULTS

The mean number of bacteria per 100 μm^2 and standard deviation are given in Table 1, with the critical surface tension (γ_c) of the substrata where known.

TABLE 1
Number of Bacteria Attached after Exposure of Various Substrata for Two Hours

Material	γ_c (a)	Mean Number Attached/ 100 μm^2	S.D
Teflon	19	31	4
Polyethylene	31	34	4
Polystyrene	33	40.5	4
Poly (ethylene terphthlate)	43	38.4	4
Glass coated with phenyline diamine	-	38	4
Processed Polystyrene (commercial)	56 (b)	4.7	6
(R.F. Plasma) (treatment)		2.8	5
Poly (ethylene terphthalate) (R.F. Plasma) (treatment)		2.4	4
Mica	-	2.6	±6
Glass	~700 (c)	0.8	±2

(a) Ref. 19 (b) Ref. 4 (c) Ref. 20

Histograms, indicating the distributions of bacteria on the various materials, are shown in Fig. 1, while Fig. 2 indicates the appearance of polystyrene samples in the low and high surface energy condition after exposure to bacteria and staining.

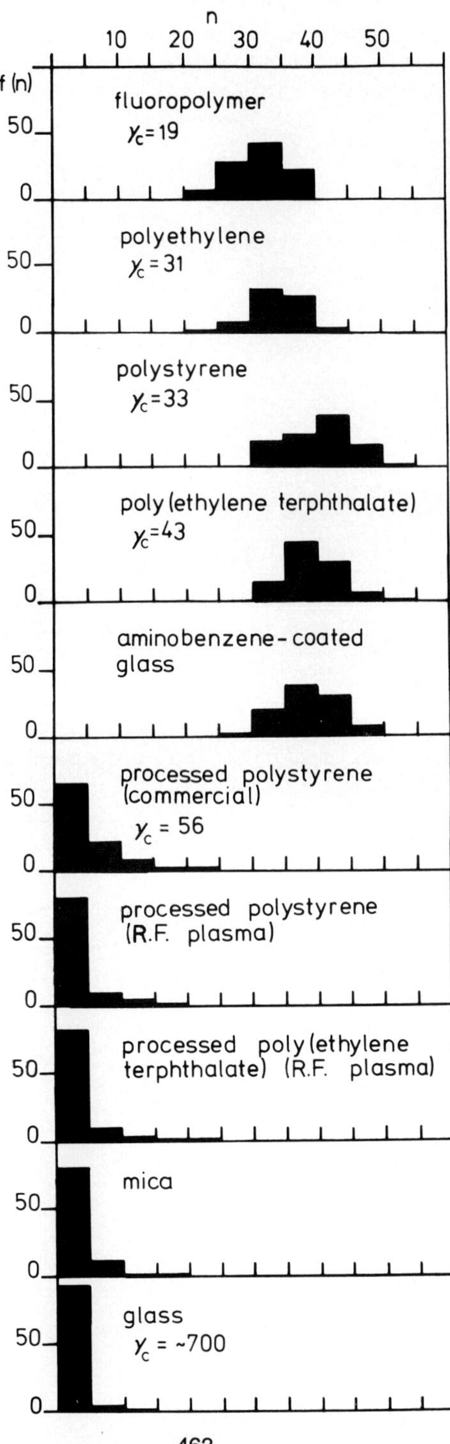
n
10
20
30
40
50
f (n)
50
0
fluoropolymer
γ_c = 19
polyethylene
γ_c = 31
polystyrene
γ_c = 33
poly (ethylene terphthalate)
γ_c = 43
aminobenzene-coated glass
processed polystyrene (commercial)
γ_c = 56
processed polystyrene (R.F. plasma)
processed poly (ethylene terphthalate) (R.F. plasma)
mica
glass
γ_c = ~700

Fig. 1. Histograms representing the distribution of attached bacteria to test substrata. n = number of bacteria (at intervals of 5) per 100 μm^2 field; f(n) = number of fields counted containing n bacteria. The critical surface tensions (γc) are given in dyne/cm, and indicate the relative surface free energies of the substrata.

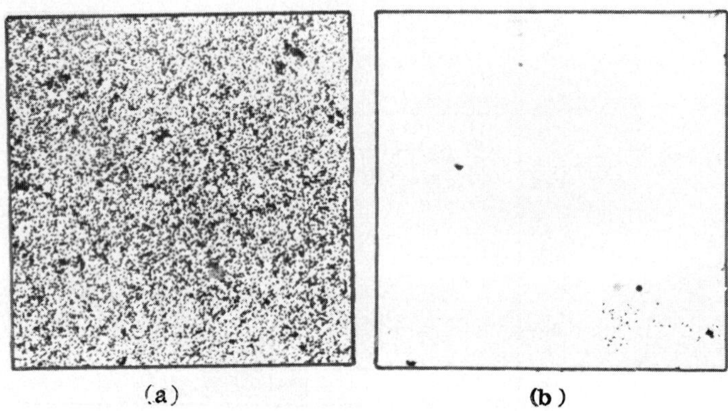

Fig. 2. Bacteria attached to (a) untreated polystyrene, and (b) R. F. plasma processed polystyrene. Large numbers of evenly distributed bacteria attached to the positively charged surface and to the low energy surfaces (a), whereas attachment to negatively charged higher energy surfaces is sparse and uneven.(b).

The data presented in Table 1 and Fig. 1 show that the lower surface energy non-water-wettable materials bear a much larger number of attached bacteria after the 2-hr exposure than the higher energy materials which are well wetted with water. The exception is the phenyline diamine coated glass slide, which bears approximately the same number of bacteria as the low energy materials although it is well wetted with water. As shown in Fig. 2, the materials bearing 30 to 40 bacteria per 100 μm^2 are quite uniformly covered; this degree of coverage is almost a monolayer, although small areas separate most of the bacteria from each other so that very little over-lap exists. The other materials, on which small numbers of bacteria are present, have a patchy appearance, with clumping of the cells.

IV. DISCUSSION

The results presented here indicate a decided preference on the part of the bacterium for attachment to low-energy, low negative charge surfaces. In contrast, most previous work on cell adhesion using cultured tissue cells, indicated attachment and spreading were most successfully achieved on high-energy materials (4,5,6). The relative importance of the various parameters affecting interaction between an attachment surface and cells is not yet clear. Polystyrene and other plastics processed to be wettable by sulfuric acid (8) or electric discharge (4), and glass (6) are most often used for tissue cell culture. These materials, however, bear a negative surface charge in addition to being of high surface energy (4,6), and indeed Rappaport has suggested that the ion exchange properties of the substratum are required for culture of such cells. In addition surface roughness, which influences adhesion and spreading, may be quite different on different substrata. The surface roughness of the treated and untreated mylar and polystyrene cut from the same stocks and used in this work is not sufficantly different insofar as is detectable by scanning electron microscopy and visual inspection of specular reflections, therefore other factors must be considered.

Attraction and repulsion among immersed bodies have been analyzed in terms of electrostatic interaction and London-van der Waals interactions. Using this DLVO (Derjagin-Landau-Verwey-Overbeek) approach, London-van der Waals attractive interactions must be dominant over electrostatic repulsion for bodies of like charge to adhere. This model has been applied to the early reversible phase of bacterial sorption, when they were removable by rinsing (21) and to tissue cells (22), and Taylor has shown that a positively charged surface is quite attractive for cell attachment. It was soon realized, however, that enhanced attachment and spreading of negatively charged tissue cells on processed polystyrene, which carries a negative charge higher than that of the unprocessed material, was inconsistent with this model (7,10,23), and other shortcomings of the model as applied to tissue cells have been described by Weiss (24). Carter (10) and Harris (23) have more recently described tissue cell attachment in terms of the surface energy of the substratum: that is, the tendency of the high energy surfaces to be wetted, and have shown that tissue cells will migrate from a low-energy surface to a high-energy surface if given the opportunity. The difference in surface energy, or some other property related to surface energy, apparently is sufficient to overcome the increased charge repulsion of the higher energy substrata. Electrostatic repulsion, therefore, is not the dominant feature of such systems, according to this hypothesis.

If other interactions in biological systems are considered, the range of possible bioadhesive phenomena is extended. Thus, hydrophobic interactions with low-energy substrates, as well as decreased charge repulsion, may be invoked to rationalize our results in this work.

Attachment to the amine-coated glass is about the same as is found with the lower surface energy materials, although this surface is wetted by water. Since the amine groups are expected to confer a positive charge to the surface, electrostatic attraction between the positive substrate and the negatively charged bacterium compensates any higher surface energy effect. There is also considerable evidence that polymeric material secreted by bacterial (26,27) and other cells (5,7) is involved in adhesion, while adsorption of components of media (8,9) may be expected to influence all of these interactions.

The clumping of cells on surfaces of low attachment, implies relatively greater cell-cell attraction than cell-surface attraction, while formation of a uniform monolayer upon the more attractive surfaces, with very little cell-cell contact, indicates a stronger cell-substrate attraction.

In preliminary tests, a number of other marine bacteria have also been found to attach preferentially to low energy polystyrene rather than glass, or processed polystyrene, and O'Neill and Wilcox have observed a preference for acrylic plastic over glass by marine bacteria in the Pacific (13).

However, since bacterial surface polymers may vary widely (28,29) and surfaces of immersed material (30) and bacteria (6) can be modified by adsorption of local organic matter, similar trends may not hold for all bacterial strains in all ecological situations (31). However, these and similar studies should help define the range of adhesive mechanisms and adaptations in aquatic environments.

V. ACKNOWLEDGMENTS

We thank Dr. N.G. Maroudas for fluoropolymer samples and helpful discussions, and Dr. R.L. Jones for help with scanning electron microscopy.

REFERENCES

1. Crisp, D.J. Chemoreception in Marine Organisms (edit by Grant,P.T. and Mackie, E.M.) 177-265 (Academic Press, London, 1974).
2. ZoBell, C.E. J. Bact. 46 39-56 (1943).
3. McHugh, W.D. ed Dental Plaque (E&S. Livingstone, Edinburgh, London 1970).
4. Maroudas, N.G., New Techniques in Biophysics and Cell Biology (Edit by Pain, R.H. and Smith, B.J.) Vol. 1 67-86 (Wiley Interscience, London 1973).
5. Taylor, A.C. Adhesion in Biol. Systems (Edit by Manly, R.S.) 51-71 (Academic Press, London 1970).
6. Rappaport, C. The Chemistry of Biosurfaces Vol. 2 (Edit by Hair, M.L.) 449-487 (Marcel Dekker, Inc. New York 1972).
7. Maroudas, N.C. J. Theor. Biol. 49 417-424 (1975).
8. Martin, G.R. and Rubin, H. Exptl. Cell Res. 85 319-333 (1974).
9. Baier, R.E. Adhesion in Biol. Systems (Edit by Manly, R.S.) 15-49 (Academic Press, London 1970).
10. Carter, S.B. Nature 213 256-260 (1967).
11. Williams, G.B. J. Mar. Biol. Ass. U.K. 44 397-414 (1964).
12. Fletcher, M. J. Gen. Microbiol. (in press).
13. O'Neill, T.B. and Wilcox, G.L. Pacific Sci. 25 1-12 (1971).
14. Crisp, D.J. and Williams, G.B. Nature 188 1206-7 (1960).
15. Ryland, J.S. Thalassia Jugoslavica (in press).
16. Harvey, P.H., Ryland, J.S. and Hayward, P.J. J. Exp. Mar. Biol. Ecol. (in press)
17. Fletcher, R. Proc. 3rd Int'l. Biodegradation Symp. Kingston, Rhode Island (Biodeterioration Soc. (in press)
18. Kennedy, J.F., Barker, S.A. and White, C.A. Car. Res. 38 13-23 (1974).
19. Shafrin, E.G. in Polymer Handbook (J. Brandrup & E.H. Immevgot, eds) 111-113 Wiley Interscience, N.Y. (1967).
20. Sisman, W.A. Adv. Chem. Ser. 43 1 (1963).
21. Marshall, K.C., Stout, R. and Mitchell, R. J. Gen. Microbiol. 68 337-348 (1971).
22. Curtis, A.S.G. The Cell Surface, (Academic Press, London 1967).
23. Harris, A. Exptl. Cell Res. 77 285-297 (1973).
24. Weiss, L. The Chemistry of Biosurfaces Vol. 2 (Edit by Hair, M.L.) 377-447 (Marcel Dekker, Inc. N.Y. 1972).
25. Taylor, A.C. Expt' Cell Res. Suppl. 8 154-173 (1961).
26. Corpe, W.A. Adhesion in Biol. Systems (Edit by Manly, R.S.) 74-87 (Academic Press, London 1970).

27. Fletcher, M. and Floodgate, G.D. J. Gen. Microbiol. 74 325-334 (1973).
28. Stacey, M. and Barker, S.A. Polysaccharides of Micro-organisms (Oxford U. Press 1960).
29. Sutherland, I.W. Adv. in Microbiol Physiol. (Edit by Rose, A.H. and Tempest, D.W.) Vol. 8. (Academic Press, London 1972).
30. Loeb, G.I. and Neihof, R.A. Adv. Chem. Series 145 (Edit by Baier, R.E.) 319-335 (Am. Chem. Soc. Wash,D.C. 1975).
31. Dexter, S.C., Sullivan, J.D. Jr., Williams, J. III and Watson, S.W. Appl. Microbiol. 30 298-308 (1975).

THE STATE OF WATER ADJACENT TO GLASS[1]

E. Nyilas, T-H. Chiu and D. M. Lederman
Avco Everett Research Laboratory, Inc.
and
F. J. Micale
Center for Surface and Coatings Research, Lehigh University

The structure of water attached to glass can affect the thermodynamics as well as the mechanism of the solution adsorption of flexible macromolecules, such as native proteins, capable of undergoing surface-induced conformational changes. Although both components of the glass/water adsorption system are quite common, even the modern literature has little, if any, information applicable to a quantitative characterization of the state of sorbed water. From a soda-lime type glass containing TiO_2 *and BaO, particles were prepared measuring* $\leq 1\mu$ *in the SEM. After exhaustive acid leaching and washings with water, the powder obtained had a B.E.T. specific surface area of 9.85* m^2/g*. Using this adsorbent, its heats of immersion into water, which did not vary with surface area, were measured at 25° and 37°C in a thermistorized isothermal-jacketed microcalorimeter capable of resolving* $\pm 1 \times 10^{-5}$°C *in 100 ml of liquid volume. The 0°, 25° and 37°C water vapor adsorption isotherms of the powder determined at relative pressures* $\geq 5 \times 10^{-5}$ *indicated no hysteresis and the absence of porosity. The vapor adsorption data have been analyzed according to the B.E.T. and the deBoer-Zwikker-Bradley methods. Consistent with the standard thermodynamic functions computed from the adsorption isotherms, the heat of immersion data indicate that, on this particular glass, interfacial water becomes ordered to a depth of 8-10 monolayers and its state is about 5-6 e.u. below that of bulk water at identical temperatures.*

---

[1]This research was performed with support from the Biomaterials Program, Devices and Technology Branch, Division of Heart and Vascular Diseases, National Heart and Lung Institute, under Contract NIH-N01-HB-3-2917.

I. INTRODUCTION

The thermodynamic quantities characterizing the state of water in the interfacial layer represent significant components in the energy balance of solution adsorption onto glass. Depending upon the depth and the extent to which that layer is structured, the entropy terms can assume varying contributions affecting the mechanism of molecular attachment as well as the results of the overall adsorption process. This is particularly true for the adsorption of flexible macromolecules, such as native proteins whose specific secondary and tertiary structures determining their functionality can be subject to adsorption-induced conformational changes. It can be expected that, at the physiologic pH of 7.2, where both the glass surface and native proteins carry a net negative charge, electrostatic forces would exert a significant influence on the adsorptive properties of these proteins. In contrast to this, evidence has been presented that the adsorption of native human plasma proteins, such as γ-(7S)-globulin[1] and fibrinogen[2] on glass is completed within periods in the order of seconds and leads to irreversible conformational alterations, at least in those molecules directly attached to the surface in the first monolayer. While the formation of protein-to-surface bonds is exothermic, the displacement of the solvent by adsorbing protein molecules and particularly, the endothermic process entailing the breakage and/or rearrangement of noncovalent intramolecular bonds which are involved in the conformational changes, would be thermodynamically unfavored without the substantial contributions of entropy factors. Thus, as an integral part of the work conducted to elucidate the adsorption mechanism of native plasma proteins on glass, the structure of water on the microparticulate glass adsorbent used in these studies has been investigated.

The first studies of the adsorption of water on glass were performed by Faraday[3] as early as 1830. His observations are consistent with those made later by several others that glass, without acid treatment, displays hysteresis in its water vapor adsorption isotherms. The data available from the old literature reflect a wide variety of contradictory views which have been summarized by Razouk and Salem.[4] It is surprising, however, that the modern literature, generated since the advent of sensitive measuring techniques and the development of the statistical mechanical theory of adsorption, contains less than a fair amount of information from which the thermodynamic quantities characterizing the state of water sorbed onto glass could be calculated. The exception to this is, of course, the set of publications which either advocated or refuted the formation of "polywater" induced in glass capillaries. To characterize quantitatively the state of water on the glass powder, the water vapor adsorption isotherms, determined to relative pressures as small as 5×10^{-5}, were analyzed in terms of the B.E.T., the deBoer-Zwikker-Bradley (D.Z.B.) methods as well as the standard thermodynamic functions. These analyses were corroborated by determining the heats of immersion of the glass powder adsorbent into water by direct microcalorimetric measurements.

II. EXPERIMENTAL MATERIALS AND METHODS

A. Glass Powder Adsorbent

This material was prepared by the continuous ball milling of commercially available glass microbeads.[2] After the separation of particles $\leq 1\mu$ by fractionation, the crude product was repeatedly extracted with 6N.HCl at room temperature until no Fe^{++} ion was detectable by SCN^- in the extract. This was followed by exhaustive washings with dist. H_2O over a period of 2 weeks, until the pH of effluents increased into the range of 6.0-7.0. An SEM of the powder shown elsewhere(1) reveals irregularly shaped particles having jagged surfaces. Despite the presence of crevices, and other surface irregularities in the order of 0.05μ or less, no hysteresis between the water vapor adsorption and desorption isotherms of the material has been observed indicating that the particles are nonporous. The major constituents of the glass were determined to be: 75% SiO_2, 2% Na_2O, 5% CaO, 3% B_2O_3 and approximately 10% as TiO_2 and BaO. B.E.T. multipoint N_2 adsorption at -195^oC gave the specific surface area of the adsorbent, $\Sigma = 9.85\ m^2/g$.

B. Determination of Water Vapor Adsorption Isotherms

These experiments were performed within the linear distension range of a vacuum quartz spring balance mounted on a vibration-free pillow, exhibiting a sensitivity of 0.1002 mg/mm, which was found to be adequate for measuring adsorption isotherms on samples with Σ's < $2\ m^2/g$. Instrumentation associated with the microbalance allowed for the measurement of manifold pressures in the range of $1 \times 10^{-6} - 1 \times 10^2$ mm Hg. The apparatus was calibrated for errors attributable to buoyancy on the spring assembly and thermomolecular flow. All isotherms given below have been accordingly corrected.

---

[2] Type A, Class V, "Close Sized Unispheres," Microbeads Div., Cataphote Corp., Jackson, Miss.

C. Determination of Heats of Immersion

All of these measurements were performed with a custom-built isothermal-jacketed microcalorimeter system designed according to some of the principles described in the literature.[5,6] A functional block diagram is shown in Fig. 1. The instrument can be operated at more than one temperature. The calorimeter cell proper is immersed into a continuously agitated 200 ℓ heat sink bath whose temperature is maintained to better than 0.01°C by a thermistor-controlled proportional heater. To detect temperature changes in the cell, two highly sensitive and matched thermistors are positioned in the opposite arms of a conventional bridge. The offset of the null bridge is fed into a chopper stabilized amplifier which drives a strip-chart recorder. By optimizing circuit parameters, resolutions of 1×10^{-5}°C in 100 ml of aqueous sample can be routinely attained. Resolutions as high as 3×10^{-6}°C have been achieved in an environment having a relatively low electromagnetic noise level. A detailed description of the design, specifications and operation of this instrument has been published elsewhere.[7]

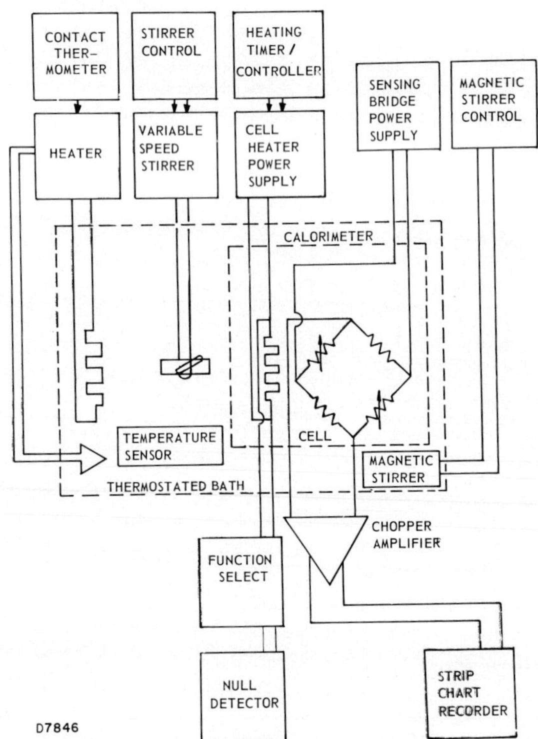

Fig. 1. Functional block diagram of the microcalorimeter system.

In a typical experiment, 0.1 g of the adsorbent is placed in a custom-made thin-walled spherical bulb (1.6 cm dia.), outgassed at either 25° or 37°C and sealed under vacuum. The cell, containing a 100 ml liquid sample and the bulb, is placed into the heat sink where its temperature is allowed to approach that of the bath to within 0.05 - 0.1°C. After two successive determinations of the nominal heat capacity of the cell and its contents by well-controlled inputs of electrical energy via a calibrating coil, the bulb is imploded by slowly lowering it against the spiked stainless steel axle of the magnetic stirrer. This effectively flushes the adsorbent into the vortex of the liquid stirred at a constant 120 rpm, achieving instantaneous mixing. The rupturing of an evacuated bulb having the size specified above, as determined in separate experiments, releases 0.0597 ±0.0030 and 0.0847 ±0.0076 cal at 25° and 37°C, respectively. All heats of immersion reported here have been corrected for the energy of bulb breaking.

III. RESULTS AND DISCUSSION

With respect to its composition given above, the glass powder adsorbent used can be characterized as a soda-lime glass, which generally contains 79% SiO_2 and 12.5% Na_2O, having most of the Na_2O substituted by TiO_2 and BaO. The water vapor adsorption isotherms of this material, measured at 0°, 25° and 37°C, are shown in Fig. 2. At any of these temperatures, the isotherms

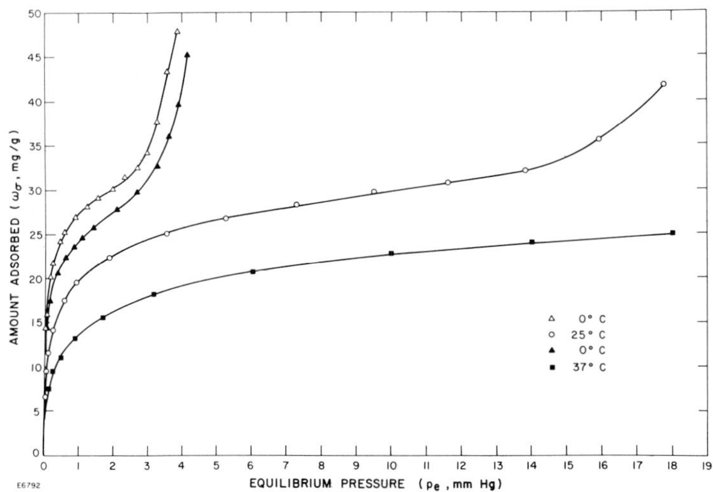

Fig. 2. Water vapor adsorption on glass powder (9.85 m^2/g) activated at 25° (open symbols) and 37°C (filled symbols).

did not display any hysteresis indicating the absence of porosity. For adsorption at either 25° or 37°C, the powder was activated at the corresponding temperatures. As indicated in this diagram, the difference between the activation temperatures has only a very small, if any, effect on the two isotherms obtained at 0°C.

The B.E.T. plots of the same isotherms, shown in Fig. 3, display deviations from linearity at relative pressures >0.2, which is not unexpected. However, after subjecting **only the** linear portions of the B.E.T. isotherms to a least squares fit, on the glass activated at 37°C, the adsorption of water vapor at the same temperature yields a first statistical monolayer with an adsorbance, $w_m(37°)$ = 17.8 mg H_2O/g, corresponding to about 60.4 water molecules per 100 A^2. The B.E.T. isotherm of water adsorbed at 0°C on the powder activated at 37°C gives similar results with w_m = 19.8 mg/g. Assuming the tightest possible packing of adsorbed water molecules, whose mean cross-sectional area derived from all possible orientations is 10.5 A^2, one theoretical monolayer has to correspond to w_m^* = 2.8 mg/g on the adsorbent having a Σ = 9.85 m^2/g. Thus, according to these data, there are 6.4 and 7.1 theoretical monolayers of water at 37° and 0°C, respectively, which appear to be energetically in the same state in terms of the B.E.T. theory, and each is bound

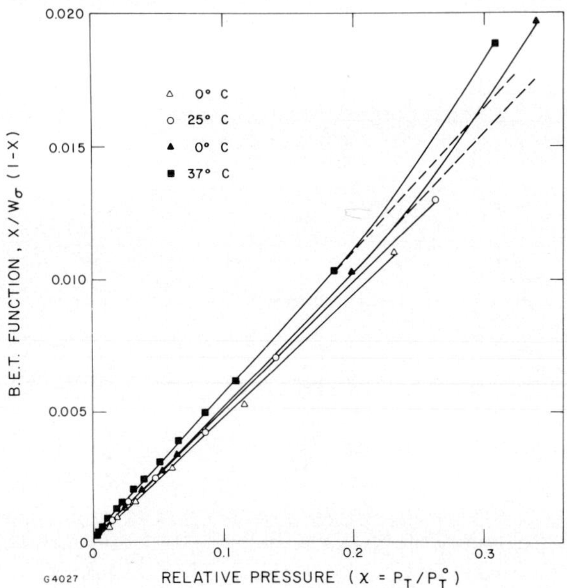

Fig. 3. B.E.T. plot of water vapor adsorption on glass powder (9.85 m^2/g) activated at 25° (open symbols) and 37°C (filled symbols).

by a mean energy of 2.97×10^3 cal/mole at 37°C or 3.05×10^3 cal/mole at 0°C. Similarly, for the 25° and 0°C adsorptions of water on a sample activated at 25°C, $w_m(25°) = 21.6$ mg/g and $w_m(0°) = 22.7$ mg/g, which convert, respectively, to 7.7 and 8.1 theoretical monolayers. The mean binding energies of these layers are 3.16×10^3 cal/mole at 25°C and 3.11×10^3 cal/mole at 0°C. The mean binding energies computed by the B.E.T. method for the first "statistical" monolayers attained at 0°C are practically equal, confirming that activation at the 2 different temperatures indicated has no appreciable effect on the glass surface.

According to the B.E.T. theory, sorbed water can exist only in two possible states, one of which is that of the molecules in the first monolayer and the other state being equivalent to that of the bulk liquid. Since the experimental data obtained are inconsistent with what can be expected from the postulates of this theory, it can be inferred that water on glass may be distributed between more than those two states. Because water has a relatively large permanent dipole, the applicability of the D.Z.B. method is well justified since this theory assumes that (a) the surface can induce and/or orient dipoles in adsorbate molecules and (b) the orientational effect can propagate through several of the sorbed layers.

The 37°C D.Z.B. isotherm of the glass powder activated at the same temperature is shown in Fig. 4. As an extension of the D.Z.B. method, it can be shown(8) that a change in the slope of these isotherms, which occurs prior to the completion of the first monolayer coverage, can be consistently attributed to a change in force field propagation and hence, in the mechanism of adsorption. Before reaching its long linear segment at about $w_\sigma > 6$ mg/g, the D.Z.B. isotherm shown in Fig. 4 displays two definite changes in slope. It may not be coincidental that these slope changes occur at about 3 and 6 mg/g, which are roughly equal to coverages corresponding to one and 2 theoretical monolayers. The intercepts of each of the 3 isotherm segments having different slopes convert to $\varepsilon' = 7.51 \times 10^3$, $\varepsilon'' = 6.96 \times 10^3$ and $\varepsilon''' = 6.44 \times 10^3$ cal/mole, respectively. The differences, $(\varepsilon' - \varepsilon'') \simeq (\varepsilon'' - \varepsilon''') \simeq RT$, implying that the binding energy of subsequent layers of water decreases by an amount about equal to the energy associated with the gain of 2 degrees of freedom. The 25° and 0°C D.Z.B. isotherms of water vapor on the glass powder adsorbent display characteristics identical to that shown in Fig.4. At all of the temperatures used, the experimental data fit reasonably the D.Z.B. isotherm up to the surface coverages which have been derived by the B.E.T. method as the first "statistical" monolayer.

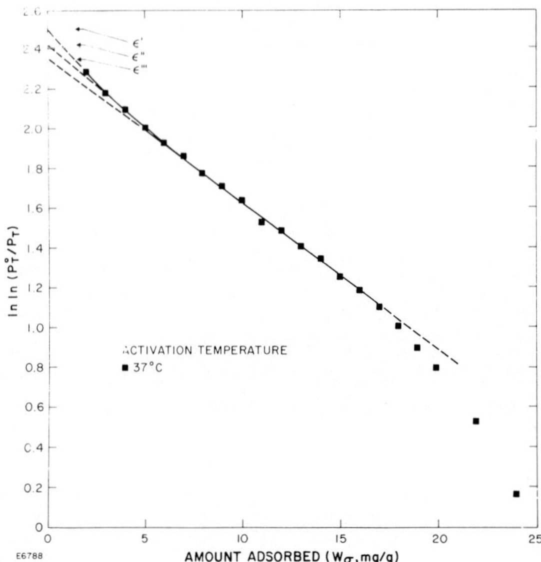

Fig. 4. deBoer-Zwikker-Bradley isotherm for the adsorption of water onto glass powder (9.85 m^2/g) at 37°C.

The differential and integral isosteric heats of the 25° and 37°C adsorption of water vapor on the glass powder, which was activated at the corresponding temperatures, are shown in Figs. 5 and 6, respectively. In the range of up to about 1.5 theoretical monolayers, $\Delta\bar{H}_T \leq Q_{V,T}$ at both temperatures, which can be attributed to the conversion of some of the translational energy lost due to adsorption into vibrational modes and the configurational entropy up to, at least, the completion of the first theoretical monolayers. In contrast to this, the release of heat at coverages >1.5 w_m^* indicates that, as a result of the attachment of further quantities of sorbed water, the adsorbate undergoes a rearrangement which, in the force field of the glass surface, leads to an energetically more stable structure. At these coverages, the formation of new water-to-surface bonds is not likely and hence, the heat released can be attributed to enthalpy changes which are associated with the formation of intermolecular bonds and the ordering of adsorbate molecules. It is noteworthy that $\Delta\bar{H}(37°) \simeq Q_{V,37}$ at a coverage where the D.Z.B. isotherm, shown in Fig. 4, displays its second change in slope.

The differential and integral absolute entropies of water, adsorbed at 25° and 37°C on samples activated at 25° or 37°C, are shown in Fig. 7. These entropy values are presented in terms of the difference which can be computed between the

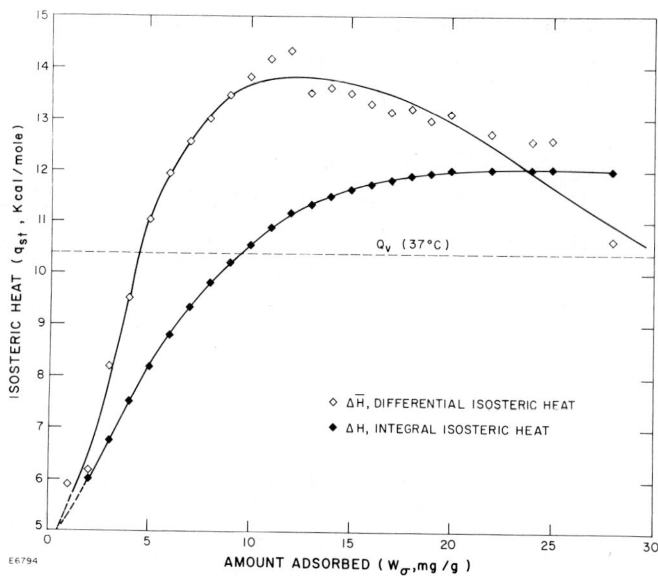

Fig. 5. Isosteric heat of water vapor adsorption at 37°C on glass powder (9.85 m²/g) activated at 37°C.

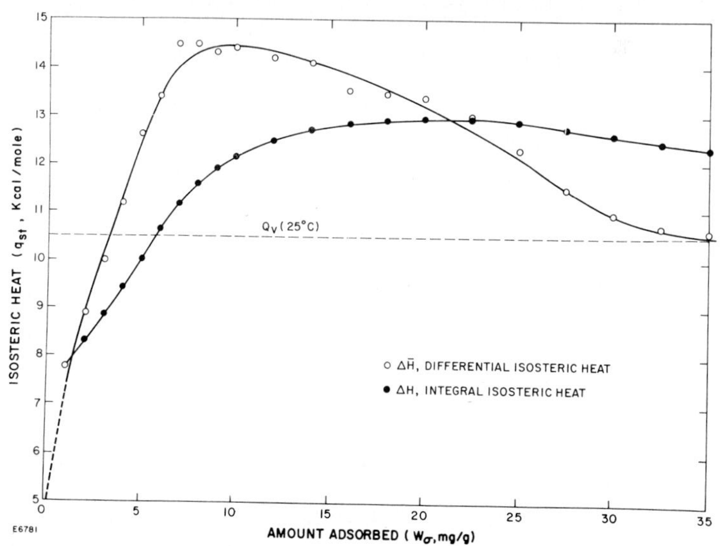

Fig. 6. Isosteric heat of water vapor adsorption at 25°C on glass powder (9.85 m²/g) activated at 25°C.

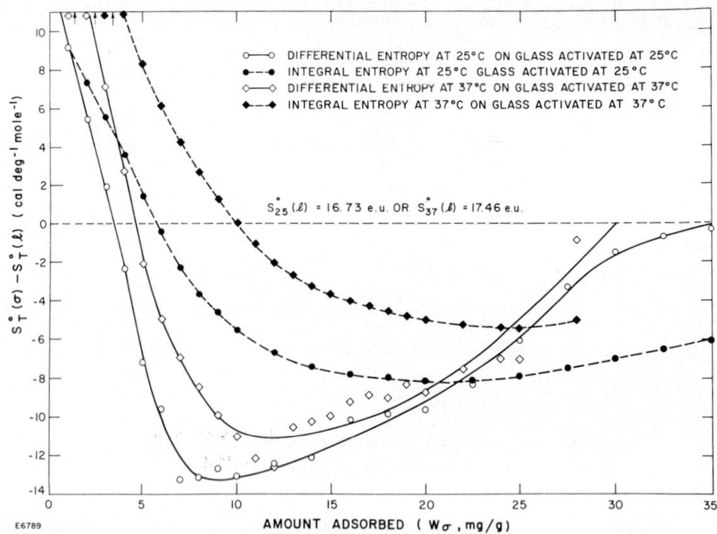

Fig. 7. Differential and integral entropy of water sorbed at 25° and 37°C onto glass powder activated at the respective temperatures.

entropy of the sorbed layer and that of bulk water at the same temperature, i.e., $[S^\circ_T(\sigma) - S^\circ_T(\ell)]$.

In the computation of the integral isosteric heats and entropies, the integration of the corresponding differential quantities has been carried out to an upper limit, w^*_σ, defined according to standard methods (9,10) as the coverage where $\Delta\bar{H}_T \to Q_{V,T}$ and $S_T(\sigma) \to S^\circ_T(\ell)$. As seen from Figs. 5 through 8, w^*_σ is equal to about 35 and 30 mg/g at 25° and 37°C, respectively. Considering the fact that the basic premises of either the B.E.T. or the D.Z.B. theory have some implicit limitations, the values of w_m, i.e., the coverages corresponding to a first "statistical" monolayer obtainable by either one of these independent treatments, are not significantly different from the value of w^*_σ taken where $\Delta\bar{H}_T \to Q_{V,T}$ and $S_T(\sigma) \to S^\circ_T(\ell)$. Thus, in terms of the characteristic quantities of both the B.E.T. and D.Z.B. analyses, the results obtained are consistent with those derivable from the standard thermodynamic functions indicating the formation of structured vicinal layers of water on the particular glass adsorbent used. This can be further substantiated by plotting the experimental data according to the Freundlich isotherm. Although this isotherm is semiempirical, the slope of this curve, n, is a constant representing the number of surface binding sites per sorbed molecule. However, none of the Freundlich isotherms which can be obtained from the water vapor adsorption data measured at either one of the temperatures is linear beyond the surface coverage that is equal or close to one theoretical monolayer. Since the nonlinearity of these isotherms indicates that n is also a variable, the values of this parameter have been incrementally computed and are shown as a function of the coverage in Fig. 8. At adsorptivities smaller than one theoretical monolayer, the values of n are close to unity implying roughly one binding site per adsorbed water molecule. With increasing coverages, at all 3 temperatures, n goes through maxima in the range of 5-6 binding sites per molecule. These maxima are attained at coverages at which both the B.E.T. and D.Z.B. analyses, and the standard thermodynamic functions implied the outer boundary of the formation of ordered structures. At coverages in excess of the completion of these layers, the values of n sharply decline to about 1.2-1.5.

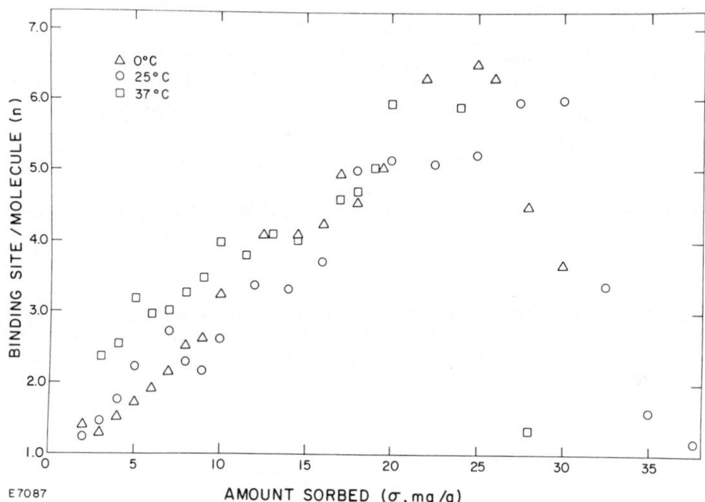

Fig. 8. Plot of binding sites derived from the Freundlich isotherms of water vapor adsorption on glass powder (9.85 m^2/g).

The heats of immersion of the glass powder adsorbent into pure water, $h_I(sLW)_T$'s, have been determined by direct microcalorimetric measurements at 25° and 37°C, and are equal to -203 ±5 and -286 ±9 erg/cm^2, respectively. Under these conditions, the heats released correspond to the enthalpy change associated with the attachment of a χ quantity of water, from the liquid phase, per unit area of adsorbent. For nonporous solids, s, which do not give rise to capillary condensation, the free energy change associated with the immersion into a pure liquid, W, is given by[11-14]

$$\Delta F_{W/s} = -\phi_{(sO/sW)} = (\gamma_{sO} - \gamma_{sW}) \tag{1}$$

where $\phi_{(sO/sW)}$ is the surface free energy change occurring when a unit area of solid/vacuum (sO) is replaced by a unit area of solid/liquid (sW) interface, and the γ's signify the respective interfacial tensions. It has been shown[11-14] that $\Delta F_{W/s}$ can be related to the adsorption of the vapor of the pure liquid used in the immersion, by the expression

$$\Delta F_{W/s} = RT/M\Sigma \int_{x_T=0}^{x_T=1} w_\sigma d\ln x + \gamma_o - T(\partial\gamma_o/\partial T)_V \tag{2}$$

where M is the adsorbate molecular weight, $x_T = p_e/p_T^\circ$, the relative equilibrium vapor pressure at temperature T^oK, w_σ is the amount of sorbed vapor per unit weight of adsorbent having specific surface area of Σ, and γ_o is the surface tension of the pure liquid. The entropy of the sorbed layer that became attached to the adsorbent upon immersion into the pure liquid, $S'_T(\sigma)$, can be computed from

$$S'_T(\sigma) = S^\circ_T(\ell) - \Delta S_{W/s} \tag{3}$$

where $\Delta S_{W/s}$ is the entropy change associated with the formation of the sorbed layer. Expressing $\Delta S_{W/s}$ in terms of $\Delta F_{W/s}$ and the experimentally determined enthalpy of immersion, $h_I(sLW)_T$, gives after rearrangement

$$S'_T(\sigma) - S^\circ_T(\ell) = -h_I(sLW)_T/T + R/M\Sigma \int_{x_T=0}^{x_T=1} w_\sigma d\ln x + \gamma_o/T - (\partial\gamma_o/\partial T)_V \tag{4}$$

The integral term on the right-hand side of Eq. (4) can be evaluated from the area under the curve obtained by plotting w_σ vs ln x. Using the water vapor adsorption data presented earlier for the glass powder adsorbent at 25° and 37°C, the curves shown in Fig. 9 are obtained. The area under the 25°C curve corresponds to 1,970 erg/cm^2 and that under the 37°C curve to 1,810 erg/cm^2. The substitution into Eq. (4) of the numerical values of γ_o and $(\partial\gamma_o/\partial T)_V$ for water, which are known quantities, and the value of $h_I(sLW)_{25}$ given earlier, gives the entropy change which is associated with the attachment of water onto the glass adsorbent upon immersion at 25°C, $\Delta S_{W/s}(25°) = -1.3 \times 10^{-7}$ cal cm^{-2} deg^{-1}. Taking $\chi = w^*_\sigma$ and recalling that $w^*_\sigma(25°) = 35$ mg/g, $\Delta S_{W/s}(25°) = -6.7$ cal mole^{-1} deg^{-1}. At the same coverage, this value can be compared to $[S'_{25}(\sigma) - S^\circ_{25}(\ell)] = -6.1$ cal mole^{-1} deg^{-1}, as indicated in Fig. 7. Similarly, at 37°C, $\Delta S_{W/s}(37°) = -6.4$ cal mole^{-1} deg^{-1}, with $\chi = w^*_\sigma(37°) = 30$ mg/g. At this

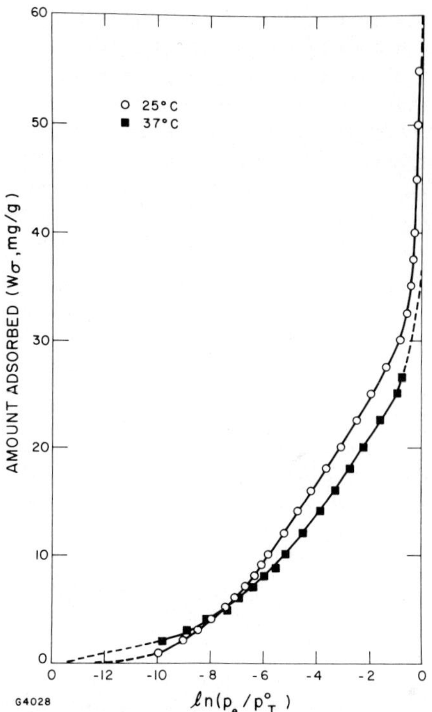

Fig. 9. Water vapor adsorption at 25° and 37°C on glass powder (9.85 m^2/g) activated at 25° (open circles) and 37°C (filled squares), respectively.

coverage, $[S'_{37}(\sigma) - S^{\circ}_{37}(\ell)] = -5.1$ cal $mole^{-1}$ deg^{-1}, as also indicated in Fig. 7. In view of the fact that the foregoing data have been derived from independent experiments, the agreement is satisfactory between (a) the entropy determined from the standard thermodynamic functions of the water layer adsorbed from the vapor, and (b) the entropy of the layer attached as a result of immersion into the pure liquid. The agreement between these values also verifies the formation of structured water layers on the surface of this particular glass adsorbent, which extend to about 10-11 monolayers at 25°C and 8-9 monolayers at 37°C.

In conclusion, the experimental data and their analyses given above definitely establish that the state of water in the near vicinity of the glass powder adsorbent is significantly more ordered than that of the free liquid. In the absence of other supporting experimental evidence, however, it is a matter of conjecture whether the structure of this vicinal water approximates that of ice or that of the supercooled liquid. On the other hand, the established orderedness of this layer can contribute significant entropy terms to the energy balance when the adsorption of a flexible macromolecule, such as a native protein capable of undergoing conformational changes, perturbs and/or displaces several thousands of structured water molecules. The entropy gain of this process, coupled with a potential increase in entropy resulting from the conformational alterations in the protein, can apparently allow for thermodynamic conditions favoring the attachment of those macromolecules despite the presence of net negative charges on both the protein and the glass surface.

IV. REFERENCES

1. Nyilas, E., Chiu, T-H., and Lederman, D. M., "Thermodynamics of Native Protein/Foreign Surface Interactions. II. Calorimetric and Electrophoretic Mobility Studies on the Human γ-Globulin/Glass System," Paper No. B8 presented at the 50th Colloid and Surface Science Symposium, San Juan, PR, 1976; and in press, "Recent Advances in Colloid and Interface Science," Academic Press, New York.
2. Chiu, T-H., Nyilas, E., and Lederman, D. M., Trans. Amer. Soc. Artif. Int. Organs 22, 498 (1976).
3. Faraday, M., Phil. Trans. Roy. Soc. 1, 49 (1830).
4. Razouk, R. I., and Salem, A. S., J. Phys. Colloid Chem. 52, 1208 (1948).
5. Zettlemoyer, A. C., Young, G. J., Chessick, J. J., and Healy, F. H., J. Phys. Chem. 57, 649 (1953).
6. Skewis, J. D., and Zettlemoyer, A. C., in "Proceedings of the 3rd International Congress of Surface Activity," Vol. No. 2, Sec. B/III/1, p. 401, Cologne, W. Germany, 1960.
7. Nyilas, E., Chiu, T-H., Herzlinger, G. A., and Federico, A., "Microcalorimetric Study of the Interaction of Plasma Proteins with Synthetic Surfaces. I. Construction of the Microcalorimeter and Preliminary Results," February 1974, NTIS Public Document PB 231 776.
8. deBoer, J. H., and Zwikker, C., Z. Physik. Chem. B3, 407 (1929).
9. Hill, T. L., J. Chem. Phys. 17, 520 (1949).
10. Hill, T. L., J. Chem. Phys. 18, 246 (1950).
11. Boyd, G. E., and Livingston, H. K., J. Am. Chem. Soc. 64, 2383 (1942).
12. Jura, G., and Harkins, W. D., J. Am. Chem. Soc. 66, 1356 (1944).
13. Loeser, E. H., Harkins, W. D., and Twiss, S. B., J. Phys. Chem. 57, 251 (1953).
14. Loeser, E. H., Harkins, W. D., and Twiss, S. B., J. Phys. Chem. 57, 591 (1953).

THERMAL EXPANSION OF INTERLAYER WATER IN CLAY SYSTEMS.

I. EFFECT OF WATER CONTENT.

David M. Clementz and Philip F. Low
Department of Agronomy, Purdue University

The specific expansibility $(\partial v_w/\partial T)_P$ *of the interlayer water in Na-saturated montmorillonite was determined as a function of temperature, T, at several water contents by a dilatometric technique. It was found that, at any water content,* $(\partial v_w/\partial T)_P$ *was linearly related to* $1/T^2$ *and that, at any temperature,* $(\partial v_w/\partial T)_P$ *decreased exponentially with increasing water content until it became equal to the specific expansibility of pure bulk water at that temperature. Also, it was found that the temperature of maximum density decreased with decreasing water content. Calculations showed that the difference between* $(\partial v_w/\partial T)_P$ *and the specific expansibility of pure bulk water could not be ascribed entirely to the effect of the exchangeable cations. Hence, the particle surfaces must have contributed to the enhancement of* $(\partial v_w/\partial T)_P$.

I. INTRODUCTION

Clay minerals are probably the most abundant of all natural colloids. They are widely distributed through the earth's crust where they are in intimate contact with water. Hence, it is important to know whether or not their surfaces affect the structure and properties of the water. Most of the available evidence is affirmative (eg., 16, 20, 24). However, additional evidence is warranted. It was for this reason that the present study was undertaken.

The coefficient of thermal expansion and the temperature of maximum density are structure-sensitive properties of water that are affected by dissolved electrolytes (6, 10, 29), nonelectrolytes (11, 28) and macromolecules (1, 2, 26) and by solid surfaces (3, 14, 22, 27, 30). Therefore, we decided to study these properties of the interlayer water in montmorillonite, a common clay mineral.

The montmorillonite crystal is composed of flat, extended aluminosilicate layers about 10 Å thick that are stacked one above the other in an ordered sequence. The layers have a net negative charge because of a partial replacement of Al^{3+} in octahedral sites and Si^{4+} in tetrahedral sites by cations of smaller valence. This negative charge is neutralized by exchangeable, interlayer cations.

When the crystal is exposed to water, the water penetrates between its superimposed layers and forces them apart, ie., swelling occurs. The total surface area of these layers amounts to about 800 m^2/gm. Consequently, the interlayer water exists in thin films even when the water/montmorillonite ratio is high. It is this feature of the montmorillonite-water system that makes it well-suited to the present study.

II. MATERIALS AND METHODS

Two Na-saturated montmorillonites were used in this study. One of them was essentially free of hydrous aluminum oxide. It will be referred to as Na-clay. The other contained 0.37 meq. of hydrous aluminum oxide per gm of montmorillonite, probably as a surface coating. It will be referred to as Na/Al-clay. Both clays had a cation exchange capacity of about 1.0 meq/gm. They were prepared from the < 2-μm fraction of Wyoming montmorillonite (Volclay 200 from the American Colloid Co.) by Davey and Low (7) and Kay and Low (15), respectively. Both were stored in the freeze-dried state.

Two sensitive dilatometers were constructed to measure the volume changes of aqueous suspensions of the two clays with changing temperature. In each dilatometer, shown diagrammatically in Fig. 1, the capillary (a) was made of precision-bore borosilicate glass tubing having an internal diameter of 0.6 ± 0.001 mm. One end of the capillary

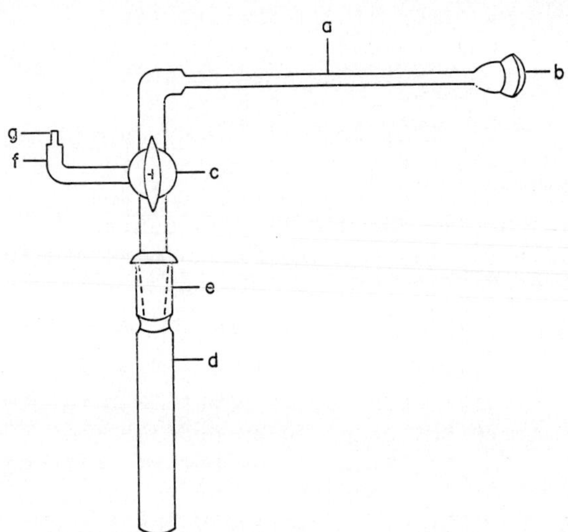

Fig. 1. Diagram of the dilatometer.

terminated in the ball (b) of a ball-and-socket joint. The other end was fused to a three-way stopcock (c) which was connected to a sample tube (d) by means of a standard taper joint (e). The stopcock had a side-arm (f) with a ground-glass Leur-Lok fitting (g) at its tip.

Calibration of the capillary was accomplished in an air thermostat at 25°C by weighing increments of mercury dispensed from it, converting the weights to the corresponding volumes and relating these volumes to the respective changes in level of the mercury in the capillary. The changes in level were measured to within 0.001 cm by a cathetometer.

To prepare a clay-water sample in tube (d), a predetermined amount of freeze-dried clay was weighed into it, degassed thoroughly and then allowed to swell in degassed, deionized water. The water was admitted to the tube under vacuum. Both the evacuation of the gas and the admission of the water occurred through a porous Teflon filter which was sealed into the lower end of a standard taper joint inserted in the tube's mouth. The filter served to confine the swollen clay to the tube.

After allowing time for equilibration, the standard taper joint was replaced by the upper part of the dilatometer which had been filled, under vacuum, with degassed, deionized water. Stopcock (c) was adjusted so that tube (d), sidearm (f) and capillary (a) were interconnected and the dilatometer was immersed in a non-freezing bath at 25 ± 0.005°C. To maintain the inside of the dilatometer at atmospheric pressure, ball (b) was joined to a socket from which a tube extended into the atmosphere above the bath. When thermal equilibrium was reached, the air-water meniscus in the capillary was adjusted to a convenient position by moving a plunger in a syringe connected to fitting (g). Then side-arm (f) was closed off, a cap was placed over fitting (g) and the position of the meniscus in capillary (a) was measured with the cathetometer. Additional measurements of the position of the meniscus were made after equilibration at the following successive temperatures: 20°, 15°, 10°, 5°, 4°, 3°, 2°, 1° and 0° C. By reference to the prior calibration, each displacement of the meniscus was converted to the corresponding volume change, ie., the observed volume change. After the measurement at 0°C was taken, the temperature was restored to 25° and the entire process was repeated. Thus there were duplicate determinations for each sample. These determinations never differed by more than 0.1 μl from the average.

Following the final measurement at 0°C, the dilatometer was removed from the bath, dried carefully and weighed. During this time, capillary (a) was sealed off to prevent evaporation. Next, stopcock (c) was closed and sample tube (d) was removed and weighed separately. Then both parts of the dilatometer were emptied [excepting side-arm (f)] and weighed. Thus, the weights of the clay-water sample and the supernatant water above it were determined. By subtracting the known weight of the clay from that of the clay-water sample, the weight of the interlayer water in the sample was also determined.

Since the observed volume change of the sample had to be corrected for the volume change of the dilatometer itself, the latter quantity had to be determined. For this purpose, the dilatometer was filled with a known weight of deionized water and the displacement of the meniscus with decreasing temperature was observed and converted to the corresponding change in volume, as before. Then the observed change in volume was subtracted from the true change in volume, as determined from handbook data, to obtain the corresponding change in volume of the dilatometer. The results, which were obtained in duplicate, were used to construct a curve from which the change in volume of the dilatometer could be determined for any drop in temperature.

If $(\partial v_w/\partial T)_P$ and $(\partial v_c/\partial T)_P$, the specific expansibility of the interlayer water and the specific expansibility of the clay, respectively, at the pressure, P, are constant over small intervals of temperature, ΔT, we have

$$\begin{aligned} \Delta V_t &= \Delta V_{obs} + \Delta V_{dil} \\ &= [m_c(\partial v_c/\partial T)_P + m_w(\partial v_w/\partial T)_P]\Delta T + m_w^\circ \Delta v_w^\circ \end{aligned} \qquad [1]$$

where ΔV_t is the true volume change; ΔV_{obs} is the observed volume change; ΔV_{dil} is the volume change of the dilatometer; and m_c, m_w and m_w° are the masses and v_c, v_w and v_w° are the specific volumes of the clay, interlayer water and pure supernatant water, respectively. Values of ΔV_{obs}, ΔV_{dil}, m_c, m_w, m_w° and ΔT were determined experimentally as described earlier and values of Δv_w° were determined from handbook data. To obtain corresponding values of $(\partial v_w/\partial T)_P$, appropriate values of $(\partial v_c/\partial T)_P$ were needed. They were determined as follows.

It is known (12) that a good approximation for β, the coefficient of volume expansion, is given by

$$\beta = \bar{\alpha}_a + \bar{\alpha}_b + \bar{\alpha}_c \qquad [2]$$

where $\bar{\alpha}_a$, $\bar{\alpha}_b$ and $\bar{\alpha}_c$ are the coefficients of linear expansion along the a, b and c axes of a mineral, respectively. Values of $\bar{\alpha}_b$ and $\bar{\alpha}_c$ for several clay minerals, not including montmorillonite, were measured by McKinstry (21) by X-ray diffraction. Such minerals have hexagonal symmetry and so $\bar{\alpha}_a = \bar{\alpha}_b$. Consequently, it was possible to calculate their respective values of β. These values ranged from 2.2×10^{-5} to 4.6×10^{-5} cm^3/°C/cm^3. Undoubtedly, the value of β for montmorillonite is within this range. However, to be specific, we assumed that it equalled 3.93×10^{-5} cm^3/°C/cm^3, which is the value of β for pyrophyllite, the prototype for montmorillonite. Then we calculated $(\partial v_c/\partial T)_P$ from the relation $(\partial v_c/\partial T)_P = \beta v_c$ in which $v_c = 0.3571$ cm^3/gm (4, 19). Its value was found to be 1.4×10^{-5} cm^3/gm/deg. Finally, having appropriate values for all the relevant variables, we calculated $(\partial v_w/\partial T)_P$ for each temperature interval by means of Eq. [1].

III. RESULTS AND DISCUSSION

Accumulated in Table 1 are values of $(\partial v_w/\partial T)_P$ in the different montmorillonite-water systems for several temperature intervals between 5° and 25°C. This table also includes values of $(\partial v_w/\partial T)_P$ for a Na-clay which had been prepared earlier and exposed to dimethyldichlorosilane vapor by Hemwall and Low (13). The exposure to dimethyldichlorosilane vapor was supposed to render the particle surfaces hydrophobic but subsequent X-ray examination indicated that the vapor had not penetrated between the superimposed layers. Therefore, only the external surfaces, constituting a small fraction of the total surface area, were hydrophobic.

TABLE 1

Specific expansibilities of interstitial water in montmorillonite-water systems from 5° to 25°C.

m_w/m_c	$(\partial v_w/\partial T)_P$			
	25-20°C	20-15°C	15-10°C	10-5°C
(gm/gm)	($cm^3/gm/°C \times 10^4$)			
Na-clay:				
52.5	2.31	1.87	1.17	0.60
27.8	2.33	1.88	1.23	0.58
10.8	2.45	1.91	1.35	0.79
5.0	2.55	2.15	1.51	0.93
3.5	2.72	2.24	1.65	1.09
3.5	2.70	2.27	1.61	1.06
Na-clay (silaned):				
1.7	3.17	2.73	2.11	1.57
Na/Al-clay				
10.3	2.57	1.83	1.33	0.79
5.6	2.56	1.95	1.41	0.82
4.0	2.70	2.15	1.47	1.02
3.9	2.70	2.20	1.58	1.02
Pure H_2O:				
	2.34	1.80	1.20	0.52

Figure 2 presents the normalized specific volume of water in Na-clay systems of different water content, m_w/m_c, as a function of temperature. To obtain the normalized specific volume at any given temperature, the sum of the values of

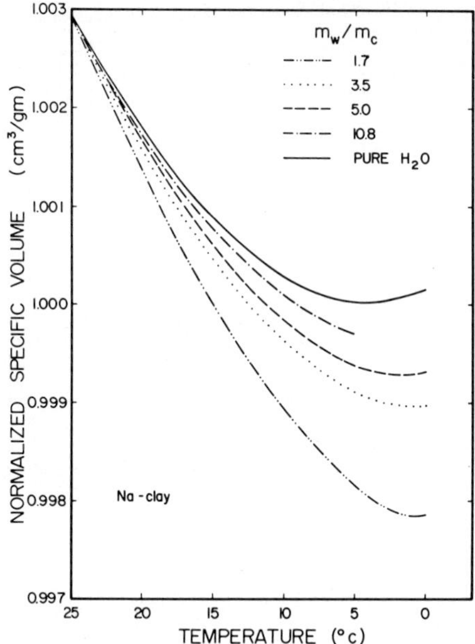

Fig. 2. The dependence of the normalized specific volume of interlayer water in Na-clay on temperature at various ratios of the mass of water, m_w, to the mass of clay, m_c.

$(\partial v_w/\partial T)_P \Delta T$ for all temperature intervals between 25°C and the given temperature was substracted from the specific volume of pure bulk water at 25°C, namely, 1.00296 cm^3/gm. Note that, if the true specific volume of the water in any system does not equal that of pure bulk water at 25°C, the curve for that system would be displaced from the one presented in the figure by an amount equal to the difference in specific volumes. Therefore, if the specific volume of the interstitial water is the same as that of bulk water at any specific temperature, T_v, it cannot be the same at any other temperature. It would have to be larger than that of bulk water at all temperatures above T_v and smaller at all temperatures below T_v. From Fig. 2 it appears that the difference in specific volumes can be appreciable, especially if the water content of the clay is low.

Figure 3 shows the relative specific volume of water in Na-clay systems of different water content as a function of temperature in the range 0-5°C. The relative specific volume

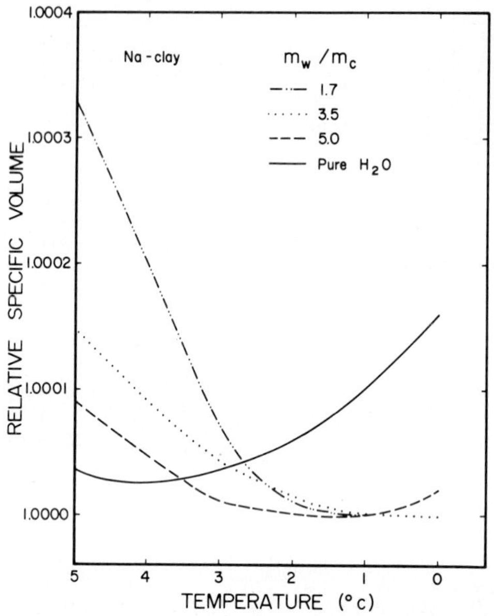

Fig. 3. The dependence of the relative specific volume of interlayer water in Na-clay on temperature at various ratios of the mass of water, m_w, to the mass of clay, m_c.

is the volume at any temperature if the specific volume at T_{MD}, the temperature of maximum density, equals 1.0 cm^3/gm. It was obtained from the data in Fig. 2 by dividing the normalized specific volume at any temperature by the normalized specific volume at T_{MD}. Note from this figure that T_{MD} for the water in the clay systems is lower than that for pure water and tends to shift downwards with decreasing water content.

The data on normalized and relative specific volumes for the Na/Al-clay were similar to those in Figs. 2 and 3.

If the tabulated values of $(\partial v_w/\partial T)_P$ at any value of m_w/m_c are plotted against $1/T^2$, where T is the mid-temperature of each five-degree interval, straight lines are obtained. Representative plots for the Na-clay and Na/Al-clay are shown in Figs. 4 and 5, respectively.

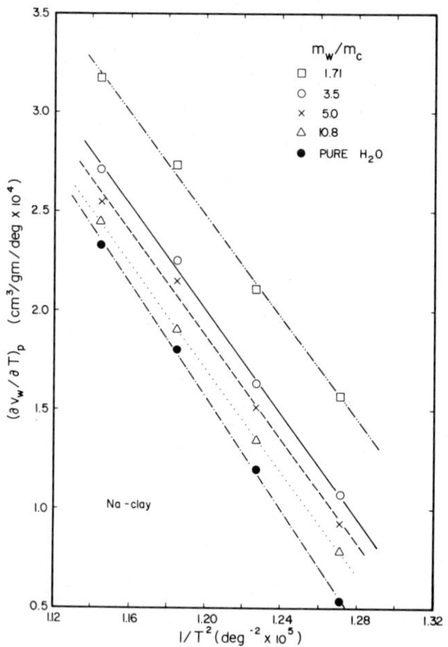

Fig. 4. The relation between the specific expansibility of interlayer water $(\partial v_w/\partial T)_P$ in Na-clay and the reciprocal of the temperature squared, $1/T^2$, at various ratios of the mass of water, m_w, to the mass of clay, m_c.

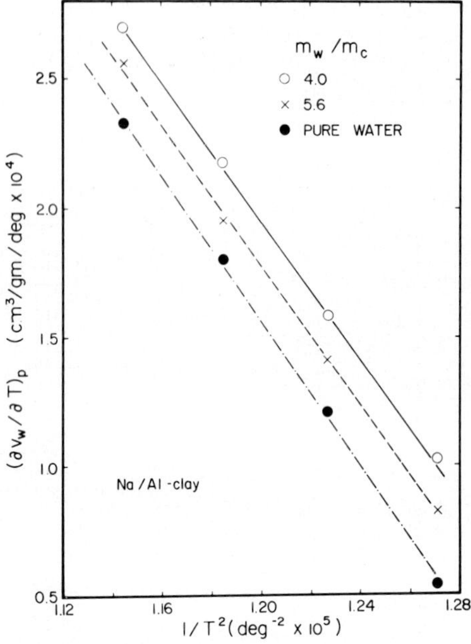

Fig. 5. The relation between the specific expansibility of interlayer water $(\partial v_w/\partial T)_P$ in Na/Al-clay and the reciprocal of the temperature squared, $1/T^2$, at various ratios of the mass of water, m_w, to the mass of clay, m_c.

Hence,

$$(\partial v_w/\partial T)_P = A + \frac{B}{T^2} \qquad [3]$$

where A and B are constants that depend on water content. The values of these constants at different water contents, determined by the method of least squares, are reported in Table 2.

TABLE 2.

The values of A, B, and B/T^2 for Na-clay and Na/Al-clay at different ratios of the mass of water, m_w, to the mass of clay, m_c.

m_w/m_c (gm/gm)	A ($cm^3/gm/deg \times 10^4$)	B (cm^3 deg/gm)	B/T^2 ($cm^3/gm/deg \times 10^4$)
Na-clay			
1.7	17.977	-129.10	-14.52[a]
3.5	17.842	-131.96	-14.84
5.0	17.608	-131.09	-14.74
10.8	17.541	-131.86	-14.83
27.8	18.032	-137.20	-15.43
Na/Al-clay:			
4.0	18.209	-135.50	-15.24
5.6	18.227	-137.05	-15.42
10.3	18.378	-138.75	-15.61
Pure H_2O:			
	18.719	-142.95	-16.10

a. *Values in this column are for* T = 298.16 K.

By utilizing the values of A and B from Table 2 in Eq. [3] and setting $(\partial v_w/\partial T)_p = 0$, the respective values of T_{MD} may be obtained. The results are presented in Table 3.

TABLE 3.

The calculated temperature of maximum density, T_{MD}, of the interlayer water in Na-clay and Na/Al-clay at different ratios of the mass of water, m_w, to the mass of clay, m_c.

Na-clay		Na/Al-clay	
m_w/m_c (gm/gm)	T_{MD} (°C)	m_w/m_c (gm/gm)	T_{MD} (°C)
1.7	-5.01	4.0	-0.21
3.5	-1.04	5.6	1.21
5.0	-0.15	10.3	1.77
10.8	1.18		
17.8	2.84		

Calculated T_{MD} for pure water = 3.35°C.

Obviously, the calculated value of T_{MD} decreases with decreasing water content. Comparison of the calculated values of T_{MD} with those for corresponding water contents in Fig. 3 indicates that the former are 0.5 - 1.5°C lower than the latter. However, the trends are the same.

Shown in Fig. 6 are plots of $(\partial v_w/\partial T)_P$ versus m_w/m_c for both clay systems at 298°K. The data in the figure were obtained by substituting into Eq. [3] the appropriate values of A and B from Table 2 and letting T = 298°K. The solid line is the best-fitting curve. The equation for this curve is

$$(\partial v_w/\partial T)_P = 2.58 \times 10^{-4} \exp[0.5/(m_w/m_c)] \quad . \qquad [4]$$

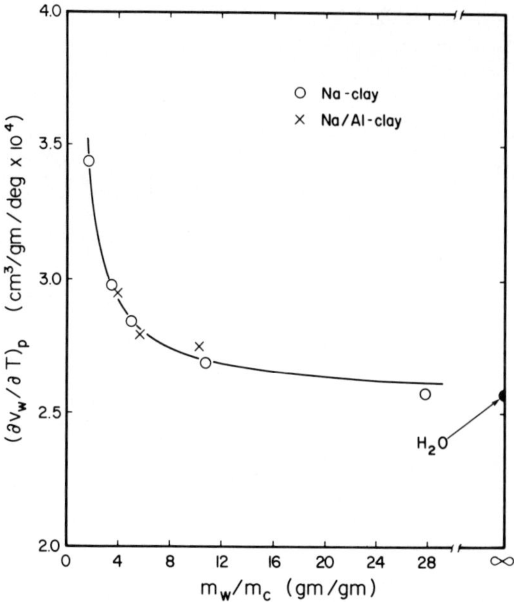

Fig. 6. The relation between the specific expansibility of the interlayer water $(\partial v_w/\partial T)_P$ as determined from dilatometric data and the ratio of the mass of water, m_w, to the mass of clay, m_c, at 25°C (interlayer distance approximately 25 m_w/m_c).

It was obtained by plotting $\ln(\partial v_w/\partial T)_P$ against m_c/m_w and determining the slope and intercept of the resulting straight line by the method of least squares. From the figure, it is apparent that the value of $(\partial v_w/\partial T)_P$ for the interlayer water decreases exponentially with increasing water content until, at a relatively high water content, it becomes equal to that of pure bulk water.

The results in Fig. 6 are supported by the work of Kay and Low (16) who measured the maximum electromotive force, E_m, generated by the thermopiles of a Calvet differential calorimeter (5) when a quantity of heat, Q, was released from a clay-water mixture in the reaction cell of the calorimeter in response to the application of a pressure, P. They showed that, under the conditions of their experiment,

$$(\partial E_m/\partial P)_T/m_w = -\gamma(\partial Q/\partial P)_T/m_w = \gamma T(\partial V/\partial T)_P/m_w \qquad [5]$$

in which V is the volume of the clay-water system and γ is a constant that is characteristic of the calorimeter. From the information they provided, we calculated that γ = 262 μv/cal = 6.35 μv/cm³ atmos. Also, the specific expansibility of the clay is small relative to that of the water and so, as long as m_w/m_c is greater than about 2.0 gm/gm, $(\partial V/\partial T)_P/m_w$ approximately equals $(\partial v_w/\partial T)_P$. Therefore,

$$(\partial E_m/\partial P)_T/m_w \simeq 6.35\, T(\partial v_w/\partial T)_P \quad . \qquad [6]$$

Their data were obtained on a Na-clay, a Na/Al-clay and a Na/Al*-clay. The former two clays corresponded to the ones that we used. The latter contained more hydrous aluminum oxide, namely, 0.7 meq/gm. We substituted their measured values of $(\partial E_m/\partial P)_T/m_w$ into Eq. [6] with T = 298°K to obtain the data in Fig. 7. The equation of the best-fitting curve in this figure was obtained as described previously.

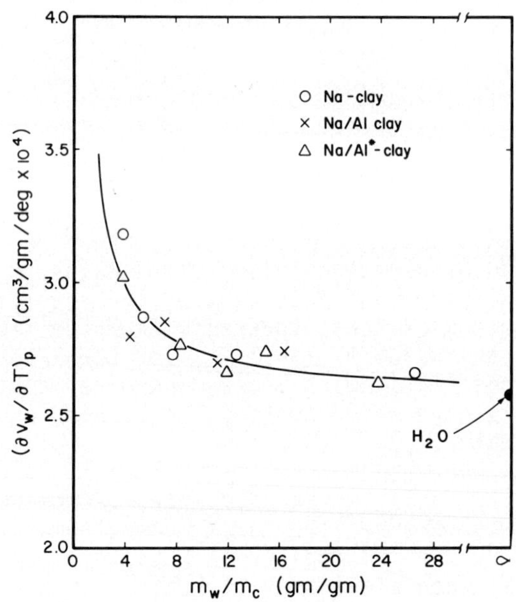

Fig. 7. The relation between the specific expansibility of the interlayer water $(\partial v_w/\partial T)_P$ as determined from calorimetric data and the ratio of the mass of water, m_w, to the mass of clay, m_c, at 25°C (interlayer distance approximately equal to 25 m_w/m_c).

It is

$$(\partial v_w/\partial T)_P = 2.58 \times 10^{-4} \exp\ [0.6/(m_w/m_c)] \quad [7]$$

which is within experimental error of Eq. [4]. Attention is called to the fact that Eq. [7] is a first-order approximation and does not take into account the slight minimum in $(\partial v_w/\partial T)_P$ that appears to exist at m_w/m_c approximately equal to 9.0 gm/gm. A minimum in the apparent specific heat capacity of the interlayer water occurs at the same water content (7). This is the water content at which there is a sudden rearrangement of the clay particles (18). Attention is also called to the fact that the presence of hydrous aluminum oxide had little effect on the specific expansibility of the interlayer water.

The enhancement of the specific expansibility of the interlayer water could be due to the effect of the particle surfaces or to the effect of the exchangeable cations. To estimate the effect of the exchangeable cations, we relate the volume, V, of the system to the volumes of its constituents in the following way

$$V = m_w v_w^\circ + m_- v_-^\circ + m_+ \phi_+ \quad [8]$$

in which v_w° and v_-° are the specific volumes of pure water and pure clay anion, respectively, ϕ_+ is the apparent specific volume of the exchangeable cation, and m_w, m_- and m_+ are the masses of the interlayer water, clay anion and exchangeable cation, respectively. We define the clay anion as being the bare negatively charged clay particle. Thus, all the effects on V of interaction between the constituents are arbitrarily included in ϕ_+. When Eq. [8] is differentiated with respect to temperature at constant pressure, the result is

$$(\partial V/\partial T)_P = m_w(\partial v_w^\circ/\partial T)_P + m_-(\partial v_-^\circ/\partial T)_P + m_+(\partial \phi_+/\partial T)_P \ . \quad [9]$$

In a clay-water system containing, for example, two gm of water and one gm of Na-clay with a cation exchange capacity of one meq/gm, the values of m_w, m_- and m_+ are 2.0, 0.977 and 0.023 gm, respectively. At 25°C and one atm pressure,

$(\partial v_w^\circ/\partial T)_P$ can be calculated from handbook data to be 2.58×10^{-4} cm^3/gm/deg and $(\partial v_-^\circ/\partial T)_P$ can be regarded as being equal to the specific expansibility of the dry Na-clay, ie., 1.4×10^{-5} cm^3/gm/deg. Now, assume that the particle surfaces have no effect on the vicinal water and that, therefore, the interlayer solution is identical to a bulk solution of Na^+ in water at the same average concentration. Then $(\partial \phi_+/\partial T)_P$ can be approximated closely by dividing the partial molal expansibility of Na^+ in bulk aqueous solution at infinite dilution by its atomic weight. The value obtained in this way from the data of Millero (23) is 1.39×10^{-3} cm^3/gm/deg. As we now have values for all the independent variables in Eq. [9], the value of $(\partial V/\partial T)_P$ can be calculated. It is 5.6×10^{-4} cm^3/deg. Hence, $(\partial V/\partial T)_P/m_w$ approximately equals $(\partial v_w/\partial T)_P = 2.8 \times 10^{-4}$ cm^3/gm/deg. If we compare the latter value with those reported for the same water content in Figs. 6 and 7, we are led to the conclusion that the exchangeable cations alone cannot account for the enhanced specific expansibility of the interlayer water; the particle surfaces must also have a significant influence. This conclusion is reinforced by the fact that the calculated value of $(\partial v_w/\partial T)_P$ is a maximum. It was obtained by assuming that all of the exchangeable cations are dissolved in the interlayer solution, whereas, only a small fraction of them are so dissolved. Most of them remain in the adsorbed state.

It is probable that the specific expansibility of the interlayer water is composed of two components, namely, that due to rapid changes in lattice spacing, designated hereafter by the subscript ∞, and that due to structural or relaxational rearrangement, designated hereafter by the subscript r. Therefore, we write

$$(\partial v_w/\partial T)_P = [(\partial v_w/\partial T)_\infty + (\partial v_w/\partial T)_r]_P \,. \qquad [10]$$

Since an increase in thermal energy causes a lengthening of hydrogen bonds, the lattice component is positive. An increase in thermal energy also causes these bonds to bend or break. In the "continuum" theory of water (17) they are supposed to bend, whereas, in the "mixture" theory (9) they are supposed to break. In either case, a closer packing of water molecules results and so the relaxational component is negative. There is no quantitative treatment of thermal expansion based on the "continuum" theory. However, according to the "mixture" theory as developed by Davis and Litovitz (8) and Davis and Jarzynski (9), the lattice and relaxational components in Eq. [10] can be identified with A and B/T^2, respectively, in Eq. [3]. Reference to Table 2 shows that the former becomes less positive and the latter becomes less negative as the water content decreases, ie., as the particles come closer together. Hence, if the "mixture" model is valid, we can conclude that the presence of clay particles enhances the thermal stability of the water. The reason for this enhancement is discussed in the following paper by Ruiz and Low (25).

Acknowledgment: *This paper is published with the permission of the Purdue University Agricultural Experiment Station as Journal Paper No.* 6335

IV. REFERENCES

1. Ahsanullah, A.K.M. and Mian, M.E., Pakistan J. Sci. Ind. Res. 14, 29 (1971).
2. Aizawa, M. and Suzuki, S., Bull. Chem. Soc. Japan, 44, 2967 (1971).
3. Bijl, A.J., Rec. Trav. Chim. Pays-Bas, 46, 763 (1927).
4. Bradley, W.F., Nature, 183, 1614 (1959).
5. Calvet, E. and Prat, H., "Recent Progress in Microcalorimetry," The Macmillan Co., New York, 1963.
6. Darnell, A.J. and Greyson, J., J. Phys. Chem., 72, 3021 (1968).
7. Davey, B.G. and Low, P.F. Soil Sci. Soc. Amer. Proc., 35, 230 (1971).
8. Davis, C.M. and Litovitz, T.A., J. Chem. Phys., 42, 2563 (1965).
9. Davis, C.M. and Jarzynski, J., in "Water and Aqueous Solutions" (R.A. Horne, Ed.), p. 377, Wiley-Interscience, New York, 1972.
10. Despretz, M.C., Ann. Chim. Phys., 70, 49 (1839).
11. Franks, F. and Watson, B., Trans. Faraday Soc., 63, 329 (1967).
12. Gorton, A.T., Bitsianes, G. and Joseph, T.L. Trans. Metallurgical Soc. A.I.M.E., 233(8), 1519 (1965).
13. Hemwall, J.B. and Low, P.F., Soil Sci., 82, 135 (1956).
14. Karasev, V.V., Deryagin, B.V. and Efremova, E.N. Colloid J. (English translation), 24, 404 (1962).
15. Kay, B.D. and Low, P.F. J. Colloid Interface Sci., 40, 337 (1972).
16. Kay, B.D. and Low, P.F., Clays and Clay Minerals, 23, 266 (1975).

17. Kell, G.S., in "Water and Aqueous Solutions" (R.A. Horne, Ed.), p. 331, Wiley-Interscience, New York, 1972.
18. Lerot, L. and Low, P.F., Clays and Clay Minerals, in press.
19. Low, P.F. and Anderson, D.M., Soil Sci. Soc. Amer. Proc., 22, 22 (1958).
20. Low, P.F. and White, J.L., Clays and Clay Minerals, 18, 63 (1970).
21. McKinstry, H.A., Amer. Mineral., 50, 212 (1965).
22. Metzik, M.S., Perevertaev, V.D., Liopo, V.A., Timoshtchenko, G.T. and Kiselev, A.B. J. Colloid Interface Sci., 43, 662 (1973).
23. Millero, F.J., J. Phys. Chem., 72, 4589 (1968).
24. Ravina, I. and Low, P.F., Clays and Clay Minerals, 20, 109 (1972).
25. Ruiz, H.A. and Low, P.F., This volume. in press.
26. Sakurai, M., Komatsu, T. and Nakagawa, T. Bull. Chem. Soc. Japan, 45, 1038 (1972).
27. Schufle, J.A. and Venugopalan, M., J. Geophys. Res., 72, 3271 (1967).
28. Wada, C. and Umeda, S., Bull. Chem. Soc. Japan, 35, 646, 1797 (1962).
29. Wright, R., J. Chem. Soc., 115, 119 (1919).
30. Zheleznyi, B.V. and Sobolev, V.D., Kolloidnyi Zhurnal, 34, 696 (1972).

THERMAL EXPANSION OF INTERLAYER WATER IN CLAY SYSTEMS.

II. EFFECT OF CLAY COMPOSITION.

Hugo A. Ruiz and Philip F. Low
Department of Agronomy, Purdue University

The specific expansibility $(\partial v_w/\partial T)_P$ of the interlayer water in five Na-saturated montmorillonites of different composition was determined as a function of temperature, T, at a water content of 4.0 gm/gm of montmorillonite. It was found that $(\partial v_w/\partial T)_P = A + B/T^2$ where A and B depend on the composition of the montmorillonite. It necessarily follows that the temperature of maximum density, at which $(\partial v_w/\partial T)_P = 0$, also depends on composition. Since the composition of montmorillonite is altered by ionic substitutions which affect its cation exchange capacity and/or the b-dimension of its unit cell, a multiple linear regression analysis was performed with $(\partial v_w/\partial T)_P$, cation exchange capacity and b-dimension as variables. As a result, $(\partial v_w/\partial T)_P$ was found to be a linear function of the other two variables. Since $(\partial v_w/\partial T)_P$ should be a function of the b-dimension of the montmorillonite only if an epitaxial association exists between the montmorillonite and water, this finding was regarded as evidence in favor of such an association.

In Part I of this study (1), it was shown that, in Na-saturated Volclay 200[a], the specific expansibility of the interlayer water at constant pressure, $(\partial v_w/\partial T)_P$, increased exponentially as the water content decreased. In addition, it was shown that the enhancement of $(\partial v_w/\partial T)_P$ was not due exclusively to the exchangeable cations. The surfaces of the montmorillonite layers were also responsible. How these surfaces react with the water is unknown. However, Davidtz and Low (2) and Ravina and Low (4) proposed that, because of hydrogen bonding between surface oxygens and neighboring water molecules, the structure of the water is strained to match that of the montmorillonite up to an appreciable thickness. In other words, they proposed that epitaxy exists between the two structures. If this proposal is valid, the magnitude of $(\partial v_w/\partial T)_P$, which is a structure-sensitive property of the water, should be related to the lattice dimensions of the montmorillonite in the plane of the surface, ie., to the a- and b-dimension. These dimensions are related approximately by the expression

---

[a] Volclay 200 is a composite of similar montmorillonites from the Black Hills region of South Dakota and Wyoming.

$b = \sqrt{3}\ a$ because of the existing crystal symmetry. Therefore, we decided to determine $(\partial v_w/\partial T)_P$ in several Na-saturated montmorillonites having different b-dimensions.

I. MATERIALS AND METHODS

The sources of the five montmorillonites used in this study, and their A.P.I. numbers (in brackets) were: Belle Fourche, S.D. (No. 27); Upton, Wyo. (No. 25); Polkville, Miss. (No. 21); Otay, Calif. (No. 24) and Cameron, Ariz. (No. 31). All of them were obtained from Wards National Science Establishment.

Two hundred grams of each raw montmorillonite were made up to 20 liters with a 0.1% solution of technical grade Na-hexametaphosphate. The resulting suspension was stirred mechanically for 48 hours, after which the particles above 2 μm in equivalent spherical diameter were allowed to settle out. Then the supernatant suspension was syphoned off and enough NaCl was added to it to bring its concentration to 2N. At this concentration, the < 2-μm montmorillonite became Na-saturated. Also, it flocculated and settled out.

The flocculated, Na-saturated montmorillonite was washed three times by centrifugation and decantation with a 2N NaCl solution and then washed once with deionized water in the same way. To remove the remaining NaCl, the montmorillonite was put in Visking dialysis tubes which were immersed in flowing deionized water. Dialysis was continued until the water flowing out of the dialyzer gave a negative chloride test with silver nitrate.

Water was removed from the dialyzed, Na-saturated montmorillonites by evaporation followed by freeze-drying. Thereafter, they were dispersed by ball-milling them briefly.

The entire procedure for wetting the dry montmorillonite samples in the sample tubes and for determining $(\partial v_w/\partial T)_P$ was the same as that described in Part I (1). However, in the present experiment, every sample was wet to essentially the same water content, namely, 4.0 gm of water per gm of montmorillonite. Also all determinations of $(\partial v_w/\partial T)_P$ were made in duplicate.

The cation exchange capacities of the different < 2-μm montmorillonites were determined by saturating them with

Mg^{++}, displacing the Mg^{++} with Ca^{++} and measuring the amount of Mg^{++} displaced. The b-dimensions were determined by X-ray analysis using powdered silicon as an internal standard. Both determinations were made by J. Margheim. Details of his procedures will be published later.

II. RESULTS AND DISCUSSION

Table 1 presents the relevant cation exchange capacities and b-dimensions. Table 2 presents the measured values of $(\partial v_w/\partial T)_P$ in the various temperature intervals between 0.1° and 25°C.

TABLE 1.

Cation exchange capacities and b-dimensions of the Na-montmorillonites used in this study.

Origin of montmorillonite	*Cation Exchange Capacity* (meq/gm)	*b-dimension of unit cell* (Å)
Belle Fourche, S.D.	0.85	8.9815
Upton, Wyo.	0.84	8.9810
Polkville, Miss.	0.90	8.9875
Otay, Calif.	1.12	8.9945
Cameron, Ariz.	0.68	9.0195

TABLE 2

Specific expansibility, $(\partial v_w/\partial T)_P$, of interlayer water in different Na-montmorillonites[a] *at temperatures between 25° and 0.1°C.*

Origin of Montmorillonite	$(\partial v_w/\partial T)_P$				
	25--20 °C	20--15 °C	15--10 °C	10--5 °C	5--0.1[b] °C
	($cm^3/gm/deg \times 10^4$)				
Belle Fourche, S.D.	2.68	----	1.69	1.07	----
	2.67	2.19	1.66	1.04	0.38
Avg.	2.67	2.19	1.67	1.05	0.38
Upton, Wyo.	----	2.13	1.62	0.96	0.32
	2.64	2.15	1.57	0.96	0.31
Avg.	2.64	2.14	1.59	0.96	0.31
Polkville, Miss.	2.64	2.13	1.61	0.95	----
	----	2.17	1.64	0.97	0.33
Avg.	2.64	2.15	1.62	0.96	0.33
Otay, Calif.	----	2.17	1.64	1.00	0.37
	2.64	2.19	1.64	1.01	0.37
Avg.	2.64	2.18	1.64	1.00	0.37
Cameron, Ariz.	2.57	2.10	1.52	0.86	----
	2.50	2.05	1.48	0.85	----
Avg.	2.53	2.07	1.50	0.85	-----

a. Water content = 4.0 gm/gm montmorillonite

b. The specific expansibility of the interlayer water was measured between 5° and 1°C in one Polkville sample and in both Cameron samples. The resulting values were 0.37, 0.30 and 0.26 $cm^3/gm/deg \times 10^4$, respectively.

Figure 1 compares the normalized specific volumes of the interlayer water in the Belle Fourche and Cameron montmorillonites with the specific volume of pure bulk water at different temperatures. The normalized specific volume at

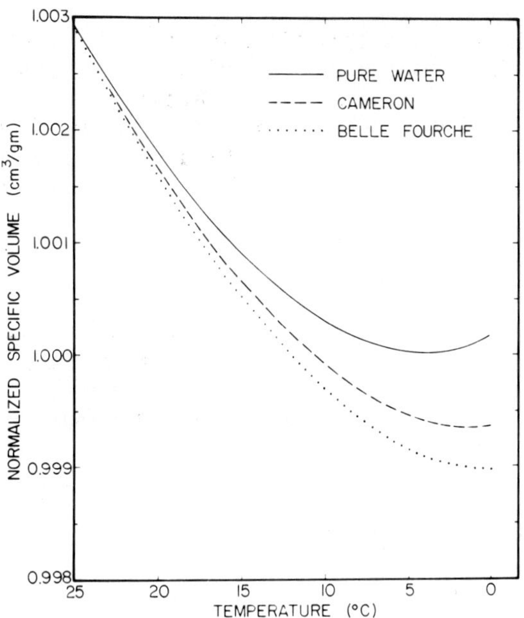

Fig. 1. Comparison of the normalized specific volume of the interlayer water in Belle Fourche and Cameron montmorillonites (water content = 4.0 gm/gm) with the specific volume of pure water between 25° and 0°C.

any given temperature was obtained by summing the values of $(\partial v_w/\partial T)_P\ \Delta T$ for all temperature intervals, ΔT, between 25° and that temperature and subtracting the result from 1.00296 cm^3/gm, the specific volume of pure bulk water at 25°C. The normalized specific volumes of the interlayer water in the other montmorillonites were intermediate between those shown for Belle Fourche and Cameron. Note that the nature of the montmorillonite has a significant effect on the normalized specific volume. Now, the curve for the true specific volume of the interlayer water would be parallel to the corresponding curve for the normalized specific volume. Therefore, we conclude that, between 0° and 25°C, the true specific volume of the interlayer water in the Belle Fourche montmorillonite cannot be the same as that in the Cameron montmorillonite (or as that of pure water) at more than one temperature. The same conclusion applies to the other montmorillonites as well because the respective curves of normalized specific volume versus temperature are not coincident with each other.

If each value of $(\partial v_w/\partial T)_P$ in Table 2 is identified with the median temperature of the corresponding five-degree interval and $(\partial v_w/\partial T)_P$ is plotted against $1/T^2$, where T is the absolute temperature, a straight line is obtained. This is illustrated in Fig. 2 by the data for the Belle Fourche and Cameron montmorillonites. The data for water are also

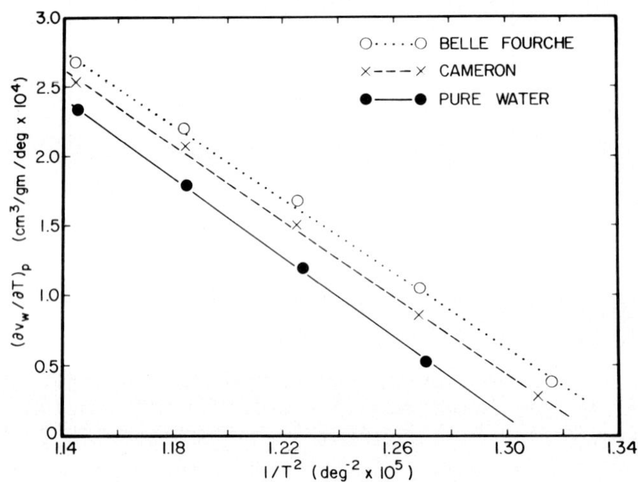

Fig. 2. Relation between $(\partial v_w/\partial T)_P$, the specific expansibility of the interlayer water, in Belle Fourche and Cameron montmorillonites and $1/T^2$, the reciprocal of the temperature squared.

presented for the purpose of comparison. Hence, we can write

$$(\partial v_w/\partial T)_P = A + \frac{B}{T^2} \qquad [1]$$

where A and B are constants. The values of A and B for all montmorillonites are presented in Table 3.

TABLE 3.

The values of A and B in equation (1) for the different Na-montmorillonites[a] used in this study.

Origin of montmorillonite	A	B
	($cm^3gm^{-1}deg^{-1} \times 10^4$)	($cm^3deg\ gm^{-1}$)
Belle Fourche, S.D.	17.980	-133.50
Upton, Wyo.	18.250	-136.22
Polkville, Miss.	18.313	-136.68
Otay, Calif.	17.950	-133.46
Cameron, Ariz.	18.201	-136.60
Pure water	18.719	-142.95

a. *Water content = 4.0 gm/gm of montmorillonite.*

By using the data in Table 3, it can be shown that A and B are negatively correlated with a correlation coefficient of -0.987. It was noted earlier (1) that, according to the "mixture" theory of water, A and B/T^2 are identifiable with the lattice and relaxational components, respectively, of the specific expansibility. Hence, if this theory is correct, we can conclude that structural relaxation increases as lattice expansibility increases. This is reasonable because the strength of a hydrogen bond in any species of water should decrease as its length increases. Evidently, the species composing the interlayer water undergo less lattice expansion and less structural relaxation than those composing pure bulk water.

If the values of A and B in Table 3 are inserted into Eq. [1] with $(\partial v_w/\partial T)_P = 0$, the respective values of T_{MD}, the temperature of maximum density, are obtained. The results are reported in Table 4. Also reported in this table are the observed values of T_{MD} which were obtained by plotting the average values of $(\partial v_w/\partial T)_P$ in Table 2 against the corresponding median temperatures and extrapolating the resulting curves to $(\partial v_w/\partial T)_P = 0$.

TABLE 4

Values of T_{MD}, the temperature of maximum density, of the interlayer water in the Na-montmorillonites[a] used in this study.

Origin of Montmorillonite	T_{MD}	
	Calculated	*Observed*
	(°C)	(°C)
Belle Fourche, S.D.	-0.67	-0.33
Upton, Wyo.	0.04	0.17
Polkville, Miss.	0.03	0.17
Otay, Calif.	-0.49	-0.17
Cameron, Ariz.	0.79	0.92
Pure water	3.184	4.0

a. Water content = 4.0 gm/gm montmorillonite.

The calculated values of T_{MD} in Table 4 are slightly smaller than the observed values but the two are highly correlated. Regardless of which values are correct, it is obvious that the T_{MD} of interlayer water is affected by the nature of the montmorillonite and is smaller than that of pure bulk water.

The specific expansibility of the interlayer water should depend on its structure as affected by dissolved cations and, if epitaxy exists, by the geometry of the oxygens in the montmorillonite surface. Therefore, we decided to determine the dependence of $(\partial v_w/\partial T)_P$ at 25°C on the cation exchange capacity and b-dimension by multiple linear regression analysis. Values of $(\partial v_w/\partial T)_P$ at 25°C were obtained by substituting the appropriate values of A and B from Table 3 into Eq. [1] and setting $T = 298.16°K$. Values of the cation exchange capacity and b-dimension were obtained from Table 1. The resulting regression equation is

$$(\partial v_w/\partial T)_P = [22.460 + 0.107\ c - 2.183\ b]\ 10^{-4} \qquad [2]$$

in which c represents the cation exchange capacity in meq/gm and b represents the b-dimension in Å. The linear multiple correlation coefficient, r, is 0.951 and $r^2 = 0.904$. Consequently, 90.4% of the total variation in $(\partial v_w/\partial T)_P$ can be explained by means of the regression equation. Values calculated by using it and the data in Table 1 are presented in Table 5. Note that these values are in fair

TABLE 5

Specific expansibility, $(\partial v_w/\partial T)_P$, of the interlayer water in several Na-montmorillonites[a] at 25°C as calculated by different equations.

Origin of Montmorillonite	$(\partial v_w/\partial T)_P$		
	Eq. [1]	*Eq. [2]*	*Eq. [3]*
		($cm^3/gm/deg \times 10^4$)	
Belle Fourche, S.D.	2.96	2.94	2.96
Upton, Wyo.	2.92	2.94	----
Polkville, Miss.	2.94	2.93	2.94
Otay, Calif.	2.94	2.94	2.94
Cameron, Ariz.	2.84	2.84	2.84

a. *Water content = 4.0 gm/gm of montmorillonite.*

agreement with the ones calculated by means of Eq. [1]. The latter values are considered to be more reliable because, in effect, they were obtained by a slight extrapolation of the experimental data.

If the data for the Upton montmorillonite are omitted in the regression analysis, the regression equation becomes

$$(\partial v_w/\partial T)_P = [28.006 + 0.067\ c - 2.795\ b]\ 10^{-4} \qquad [3]$$

for which $r = 0.999$ and $r^2 = 0.998$. Evidently, nearly 100% of the total variation in $(\partial v_w/\partial T)_P$ can be explained by this regression equation. The results obtained by using it and the data in Table 1 are also included in Table 5. They are in excellent agreement with those obtained by means of Eq. [1]. Thus, omission of the data for Upton montmorillonite improves the regression equation.

In a paper being prepared for publication, data will be presented on the maximal water contents attained by the five montmorillonites used in the present study when they are brought into contact with pure water and allowed to swell spontaneously. It will be shown that these water contents are related to the cation exchange capacity and b-dimension by an equation of the form of Eqs. [2] and [3] with r^2 = 0.996. If they are plotted against the respective values of $(\partial v_w/\partial T)_P$ determined by Eq. [1], all of the data points fall close to a straight line except that for Upton montmorillonite. Hence, two different analyses indicate that the value of $(\partial v_w/\partial T)_P$ for the Upton montmorillonite may be somewhat discrepant. For this reason, Eq. [3] is considered to be more reliable than Eq. [2].

In the previous paper of this series (1), we showed that the relation between $(\partial v_w/\partial T)_P$ and m_w/m_c, the mass ratio of water to montmorillonite, was

$$(\partial v_w/\partial T)_P = 2.58 \times 10^{-4} \exp [0.5/(m_w/m_c)]\ . \qquad [4]$$

It is reasonable to suppose that the functional relation between $(\partial v_w/\partial T)_P$ and m_w/m_c would have the same form for all montmorillonites. Therefore, we will write

$$(\partial v_w/\partial T)_P = 2.58 \times 10^{-4} \exp [\lambda/(m_w/m_c)] \qquad [5]$$

in which λ is a constant that is characteristic of the montmorillonite. Attention is called to the fact that the factor preceding the exponential term remains the same because, when $m_w/m_c = \infty$, the value of $(\partial v_w/\partial T)_P$ must equal that of pure water, namely, 2.58×10^{-4} cm^3/gm/°C. Now, when $\lambda/(m_w/m_c) \ll 1.0$, equation [5] becomes

$$(\partial v_w/\partial T)_P = 2.58 \times 10^{-4}\{1 + [\lambda/(m_w/m_c)]\} \ . \qquad [6]$$

If m_w/m_c in this equation is assigned a value of 4.0 gm/gm, the water content of the systems under investigation, and the resulting equation is compared with Eqs. [2] and [3], it is evident that λ is a linear function of the cation exchange capacity and b-dimension. In fact, an approximate expression for λ in terms of these variables can be obtained by such a comparison. However, a more exact expression (not involving series expansion and elimination of higher order terms) can be obtained by using values of λ, calculated by substituting values of $(\partial v_w/\partial T)_P$ from Table 5 (column 4) into Eq. [5], in a multiple linear regression analysis with cation exchange capacity and b-dimension as independent variables. The result is

$$\lambda = 34.968 + 0.093\, c - 3.841\, b \qquad [7]$$

for which $r^2 = 0.998$. Combination of this equation with Eq. [5] yields

$$(\partial v_w/\partial T)_P = 2.58 \times 10^{-4} \exp[34.968 + 0.093c - 3.841b/(m_w/m_c)] \qquad [8]$$

which expresses $(\partial v_w/\partial T)_P$ as a function of all the independent variables studied in this experiment except temperature.

Figure 3 presents $(\partial v_w/\partial T)_P$ as a function of m_w/m_c for the Belle Fourche and Cameron montmorillonites. The data for this figure were calculated by means of Eq. [8]. As before, the corresponding curves for the other montmorillonites occupy intermediate positions. Obviously, the difference between the values of $(\partial v_w/\partial T)_P$ in the two montmorillonites increases with decreasing water content.

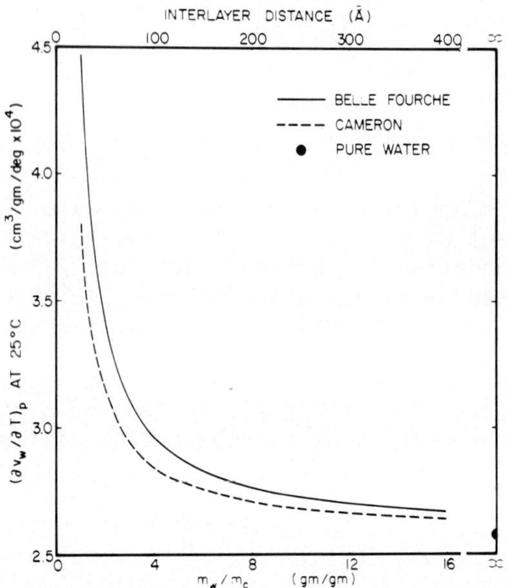

Fig. 3. Relation between $(\partial v_w/\partial T)_P$, the specific expansibility of the interlayer water, in Belle Fourche and Cameron montmorillonites and m_w/m_c, the ratio of the mass of water to the mass of montmorillonite.

We have shown that, at any water content, the magnitude of $(\partial v_w/\partial T)_P$ depends on the cation exchange capacity and b-dimension. Since ions are known to affect the specific expansibility of bulk water (3), the dependence of $(\partial v_w/\partial T)_P$ on the cation exchange capacity is to be expected. However, there is no obvious reason why $(\partial v_w/\partial T)_P$ should depend on the lattice parameters of the montmorillonite unless the structure of the water is constrained to match that of the montmorillonite, ie., unless epitaxy occurs. Therefore, we regard the evidence presented in this paper as being in favor of an epitaxial association between the water and montmorillonite. In view of the data in Fig. 3, this association must perturb the structure of the water to an appreciable depth.

*Acknowledgment: This paper is published with the permission of Purdue University Agricultural Experiment Station as Journal Paper No.*6336

III. REFERENCES

1. Clementz, D.M. and Low, P.F. This volume.
2. Davidtz, J.C. and Low, P.F., Clays and Clay Minerals, 18, 325 (1970).
3. Millero, F.J., J. Phys. Chem., 72, 4589 (1968).
4. Ravina, I. and Low, P.F., Clays and Clay Minerals, 20, 109 (1972).

phase transitions - near 15, 30, 45 and 60°. Peschel and Adlfinger (1,2) demonstrated that both viscosity and disjoining pressure showed anomalous behavior near the above temperatures. These authors also showed that the influence of modified vicinal water structures may be observed over distances of more than 900 Å (≃ .1 μm) from the surface. Such distances certainly cover the interfacial distances in biological cells. Wiggins (3) showed the presence of highly temperature dependent selectivity for certain ions by the vicinally structured water in the pores of a silica gel. She observed specifically that the pore water in the silica gel selected K+ over Na+ and, in the same study, also established qualitatively similar behavior for the rat renal cortex.

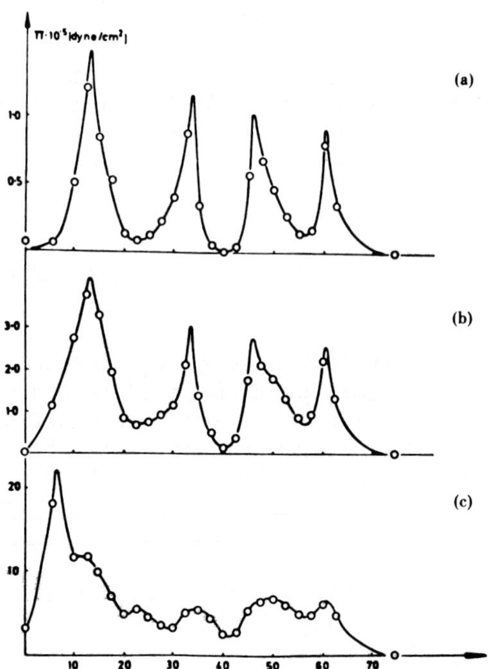

Fig 1. Disjoining pressure of water between two highly polished quartz surfaces for various plate separations as a function of temperature. (Data by Peschel and Adlfinger, 2). Note the maxima near 15°, 30°, 45°C.

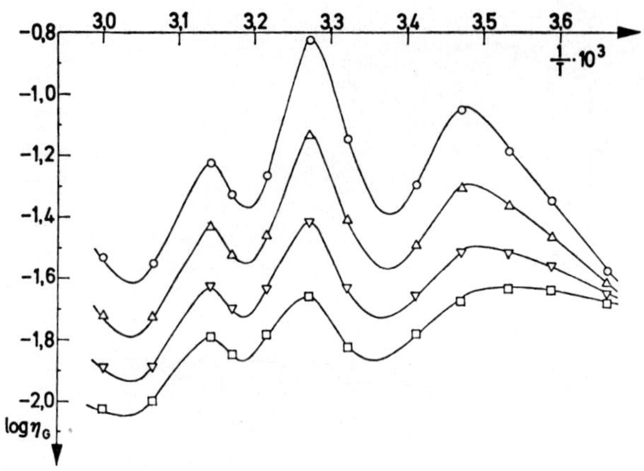

Fig. 2. Arrhenius plots of the viscosity of water in thin films between quartz plates as a function of temperature. (Data by Peschel and Adlfinger, 1). Note maxima and minima observed for all distances measured.

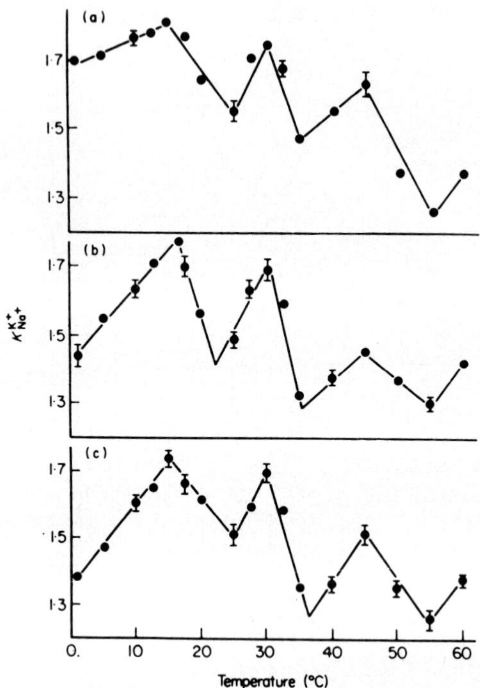

Fig. 3a. Variations in the selectivity coefficient (K^{k+}_{Na+}) in a silica gel with temperature. Shown, from top to bottom, for solutions containing $SO_4^{=}$, I^-, and Cl^- as the anion. (Data by Wiggins, 3). Note maxima near 15°, 30° and 45°.

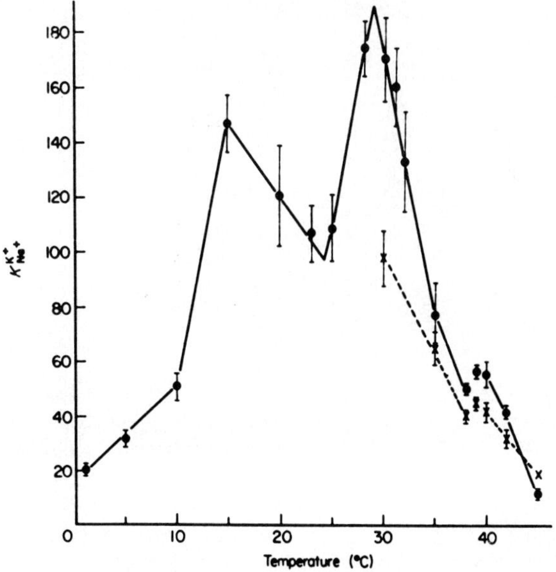

Fig. 3b. Variation in selectivity coefficient (K^{K+}_{Na+}) in rat renal cortex with temperature. (Data by Wiggins, 3). Note similarity to Figure 3a.

It is our contention that vicinal water in cells is structured notably differently from bulk water and that these structures play a prominent role in some biophysical systems. As an example of the role of vicinal water, multiple growth optima for various organisms have been reported in the literature and by the present authors as well (6). Examples of organisms exhibiting this feature include *Streptococcus faecalis* (7), *E. coli* (9), *Aerobacter aerogenes* (W. Drost-Hansen, unpublished) and *Neurospora crassa* (8).

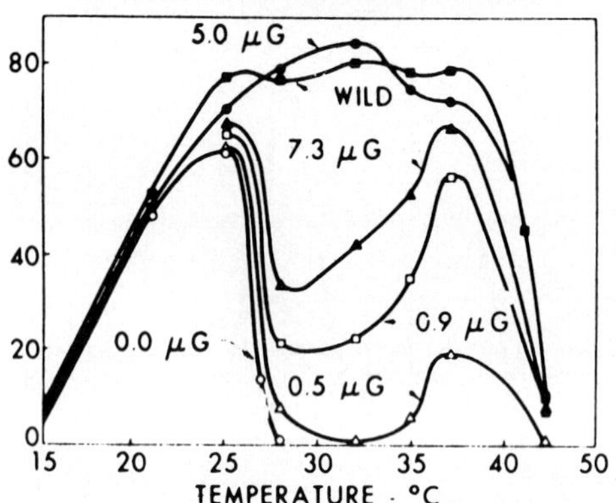

Fig. 4. Growth of a mold (a mutant of Neurospora crassa) as a function of temperature with various amounts of riboflavin supplied to the nutrient broth. Riboflavin appears to be a required nutrient for growth above 30°C. The present authors suggest that the change in metabolism is due to changes in vicinal water structure. (Data by Mitchell and Houlahan, 8).

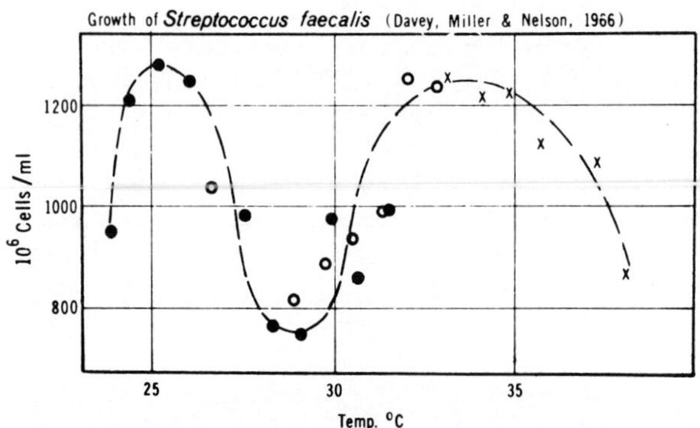

Fig. 5. The growth of Streptococcus faecalis as a function of temperature. Note the minimum near 30°C. (Data by Davy, et al., 7).

It was observed that for the mutant of Neurospora crassa, riboflavin was required for growth above 30°; however, it was not required for growth below 30°. This suggested to Oppenheimer and Drost-Hansen (6) that different metabolic pathways are utilized by organisms above and below the transition temperatures of the vicinal water. Hence, it is the contention of the present authors that this water exerts an extremely important - and sometimes controlling - role in determining which metabolic processes are utilized.

The study of seed germination suggested itself to the present authors because it was felt that if carried out relatively carefully, such a study could shed light on the nature of the influence of vicinal water on an enzymatic system. In general, protein activity is very sensitive to the nature of hydration. Fig. 6 shows the temperature dependence of D-amino acid oxidase activity (10). We consider

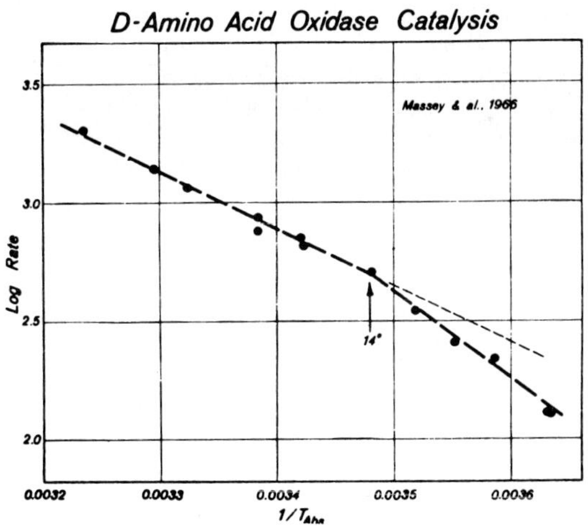

Fig. 6. Log rate of catalyses for D-amino acid oxidase as a function of reciprocal, absolute temperature. (Data by Massey et al., 10). Note the apparent, relatively abrupt change in activation energy at 14°C.

the abrupt change in slope at 14°C to be a manifestation of the change in the hydration of the protein, reflecting the change in the nature of vicinal water at this temperature. This aspect will be discussed below together with our rate data and some ideas related to the independent and concurrent studies of Hageseth and Joyner (11) as well as those of Goloff and Bazzaz (12). The present authors suggest a hypothesis for incorporating structural effects of vicinal water into expressions for the rate processes in germination of seeds.

II. MATERIALS AND METHODS

The germination studies have been carried out using turnip seeds, obtained from W. Atlee Burpe Co. (Philadelphia, Pa.). Approximately 60 to 120 seeds were placed in 18 x 150 mm Pyrex test tubes, containing two layers of Whatman #1 filter paper. At the bottom of each tube were placed absorbent cotton balls. The cotton and filter paper were saturated with 4 ml of H_2O: this provided for saturation of the cotton and filter paper without allowing excess water to flow freely in the test tubes. The tubes were stoppered loosely with a cork, covered with paraffin film. Thirty such tubes were prepared and placed into the wells of a Model TN-3 Temperature Gradient Incubator manufactured by Toyo Kagaku Sangyo Co., Ltd., Japan. The tubes were examined at two-hour intervals beginning the 10th hour after the start of the experiment. The emergence of the radicle was used as the criterion for germination.

III. DISCUSSION AND RESULTS

A typical germination curve is shown in Fig. 7. The abscissa is the time in hours; the ordinate is the percent germination (i.e., number of seeds germinated, divided by total numbers of seeds x 100).

The germination data obtained have been analyzed in different manners. In Fig. 8 is shown an Arrhenius plot of maximum germination rate vs. reciprocal absolute temperature. The maximum germination rate is defined as the maximum slope (α) of the sigmoid-shaped growth curve (see Fig. 7). Figs. 9 and 10 show log $(1/t_{50})$ and log $(1/t_{16})$, as a function of reciprocal absolute temperature, where t_{50} and t_{16} are, respectively, the time (in hours) for 50% and 16% germination.

In Figs. 9 and 10 straight line segments through the data points have been suggested by the present authors. This approach resembles that utilized by Hageseth and Joyner (11) and by Goloff and Bazzaz (12). However, this approach differs radically from the "classical" growth curve suggested by an expression such as derived by Johnson and Eyring (13, 14).

$$G = \frac{cT\exp(\Delta H^{\neq}/RT)}{1 + \exp(\Delta H/RT)\ \exp(\Delta S/R)} \tag{1}$$

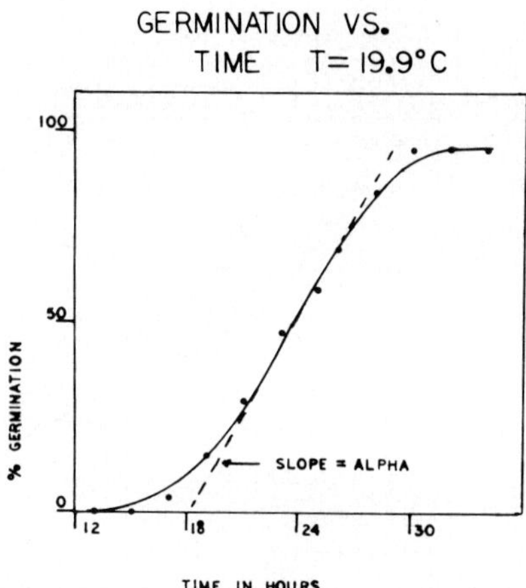

Fig. 7. The percent germination of turnip seeds as a function of time (in hours) at 19.9°C. The maximum germination rate (α) is defined as the maximum slope indicated by the dotted line. This parameter is similar to the intrinsic germination rate defined by the logistics equation generally used in population studies.

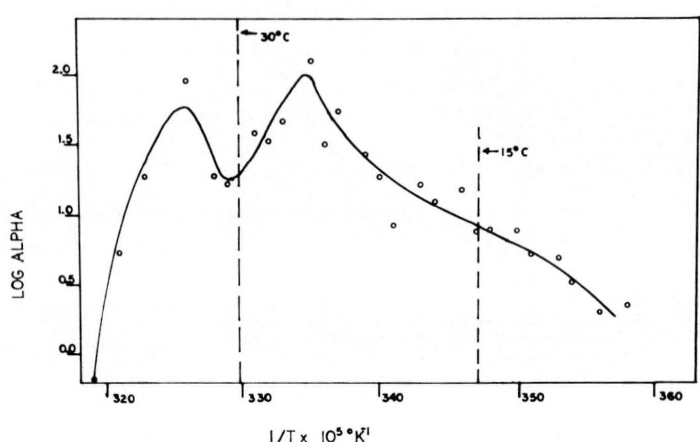

Fig. 8. Log α (i.e., maximum rate of germination) as a function of reciprocal absolute temperature. Note the minimum near 30°C. This trend in the data has been found to be reproducible both qualitatively and quantitatively, at least within the seed population studied here. In the temperature region above 30°, no chlorophyll appears to develop in the seedlings in the early stages of growth. Chlorophyll is, however, produced below 30°.

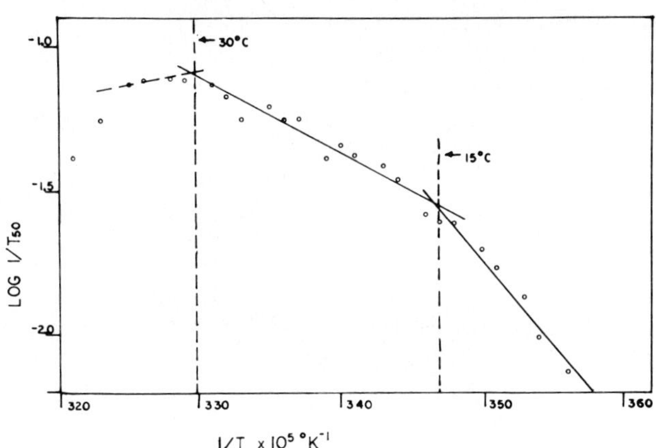

Fig. 9. Log $(1/t_{50})$ (i.e., reciprocal time in hours to 50% germination) as a function of reciprocal, absolute temperature. Note the apparent changes in slope near 15° and 30°.

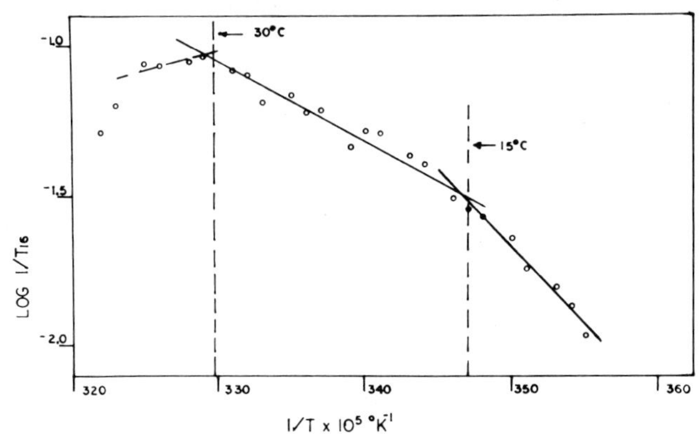

Fig. 10. Log $(1/t_{16})$ (i.e., reciprocal time in hours to 16% germination) as a function of reciprocal, absolute temperature. Note apparent changes in slope near 15° and 30°.

In this equation, G represents the rate of "growth" and c is a constant. It should be noted that the constant, c, contains reference to a variety of parameters, introduced in the derivation of equation 1. See discussion below regarding the validity of this derivation in the present treatment. ΔH and ΔS are, respectively, the enthalpy and entropy of the germination process, while $\Delta H^{\neq}$ is the enthalpy of activation.

Equation 1 is derived on the basis of a number of assumptions, some of which may not be appropriate except as a first approximation. Johnson and Eyring (14) point out that the curve, represented by Equation 1, does not provide a substantially better fit than may be achieved using several line segments. Many examples have been reported in the literature which utilize either the Johnson and Eyring expression or individual curve segments, often straight lines. In the derivation of Eq. 1, the thermodynamic and activation parameters are assumed to be temperature independent. This assumption may not be entirely correct in view of the existence of different vicinal water structures. Thus, if vicinal water undergoes structural transitions, these may impose conformational constraints upon the proteins, giving rise to temperature dependent activation parameters. Hence, in many respects, the interpretation of the present authors resemble the ideas of Crozier and co-workers; (see Johnson, Eyring and Stover, 14) - developed in the mid twenties - and, in fact, resemble the proposal by Blackman (1905) and Putter (1914). The difference between those early authors and the present authors is that we can now propose a molecular interpretation - namely in terms of a controlling role played by the thermal stability limits of vicinal water.

The thermodynamic functions (ΔH; ΔS) may be also temperature dependent. Hageseth and Joyner (11) in fact find the enthalpy and entropy change abruptly over the temperature range studies. However, the thermodynamic functions may be far less sensitive to vicinal water structure changes as compared to the activation parameters. Such insensitivity may exist for at least two reasons. First, enzymes operate as catalysts for chemical processes by lowering the activation parameters, but without altering the thermodynamic functions of the process. Secondly, there is a tendency for enthalpies and entropies to compensate (ΔH/ΔS compensation) which in turn may make the rate data less sensitive to the superimposed effects of vicinal water structures. Enthalpy/Entropy compensation has been observed for a large variety of systems (see, for instance, Lumry and Rajender, 15). Hence, in light of entropy/enthalpy compensation, the denominator of Eq. 1 may remain fairly constant with temperature even though ΔH and ΔS are temperature dependent. In this connection, it would be of considerable interest to study systematically the "characteristic temperatures", i.e., the proportionality factor to the entropy change. The "characteristic temperatures" have frequently been reported to fall in narrow temperature ranges and may possibly reflect the influence of the solvent.

It is our contention that many biological data, particularly where sufficiently closely spaced measurements have been made, show the occurrence of thermal anomalies at the temperatures of vicinal water structure changes. The turnip seeds used in the present study similarly exhibit this behavior, i.e., unexpected thermal anomalies in the rate data. Hageseth and Joyner (11), as well as Goloff and Bazzaz (12) have measured the rate of germination of seeds as a function of temperature. In both studies, the authors reported changes in the thermodynamic and kinetic parameters for the germination process. The temperatures at which the anomalies were observed are consistent with the temperatures for one of the vicinal water transitions - namely, the one occurring in the vicinity of 30°C. The data obtained by the present authors have also confirmed an anomaly near 30° in the germination data for turnip seeds. In addition, our own data suggest an anomaly near 15°. Unfortunately, this temperature range was not studied by the previous authors. As shown in the figures 9 and 10, we have fitted straight line segments to the rate of germination data vs. reciprocal, absolute temperature plots, thus dividing the region studied in a segment below 15°; between 15 and 30°; and a third segment above 30°. Again, the transition near 15° may possibly be most readily explained in terms of a conformational change in one or more of the enzymes involved in seed germination. Note also, however, that vicinal water may as well interact with the genetic apparatus directly in some manner and this, then, in turn could affect the germination rate. (Discussion of this aspect is presently largely unexplored and beyond the scope of this paper; this process may still be influenced by proteins). It appears that the anomaly near 30° is more complex than the one observed at 15°C. Two phenomena may be involved simultaneously - a "simple" heat denaturation, as implied by the treatment by Johnson and Eyring (14), together with a vicinal water structure induced conformational change. Finally, note also that independently, Dau and Laborieu (16) have measured the rate of germination for a species of cactus as a function of temperature. In these data, both the anomalies at 15 and 30° are present.

Fig. 8 shows a plot of log maximum germination rate vs. 1/T. The maximum growth rate is again simply the maximum germination rate taken from the sigmoidal growth curves. (Compared Fig.7). The log maximum germination rate vs. 1/T shows interesting trends: the two optima correspond to temperatures where seeds germinate and later develop stem tissue to approximately the same degree; however, the plants grown at the optimum above 30° do not develop chlorophyll in the early stages. This observation is not inconsistent with the contention of the present authors that organisms utilize different metabolic pathways in various temperature intervals and that these differences are caused by changes in vicinal water structures and, in turn, are manifested in different cellular processes. In conclusion, it is seen that vicinal water structure appears to play an important role in the germination of seeds. It would seem reasonable to consider carefully the thermal conditions of germination to optimize germination rates where these may be of critical economic importance.

IV. SUMMARY

The present authors have demonstrated that vicinal water appears to have a great impact upon the germination of seeds. Thus, turnip seeds show anomalous behavior at the temperatures of vicinal water transitions (15° and 30°). Additionally, it appears important in general to include vicinal water transitions in any theory describing the kinetic or thermodynamic behavior of seed germination. It is envisioned that changes in vicinal water structure may confer conformational transitions upon proteins such that changes in activation parameters result. Obviously, the present authors, however, do not wish to imply that only proteins are affected by vicinal water structure nor that various lipids and macromolecules do not have intrinsic transitions.

V. ACKNOWLEDGEMENTS

The authors wish to thank Toyo Kagaku Sangyo Co., Ltd., Japan for the donation of the Temperature Gradient Incubator with which these experiments were made. The authors also wish to thank Mrs. Bee Drost-Hansen who initiated the seed germination experiments and for continued interest in this work.

REFERENCES

1. Peschel, G., Adlfinger, K. H., J. Colloid Interface Sci., 25, 131 (1970).
2. Peschel. G., Adlfinger, K. H., Z. Naturforschung, 26a, 707 (1971).
3. Wiggins, P. M., Clinical and Experimental Pharmacology and Physiology, 2, 171 (1975).

4. Drost-Hansen, W., in: "Chemistry of the Cell Interface", Part B (H. D. Brown, ed.). Academic Press (1971).
5. Drost-Hansen, Ind. Eng. Chem. , 61(11), 10 (1969).
6. Oppenheimer, C. H., Drost-Hansen, W. , J. Bact., 80, 21 (1960).
7. Davy, C. B., Miller, R. J. and Nelson, L. A., J. Bact., 91, 1827 (1966).
8. Mitchell, H. K. and Houlahan, M. B., Am. J. Bot., 33, 31 (1946).
9. Schmidt, M. G. and Drost-Hansen, W. Paper presented at ACS meeting, Chicago, Ill., Sept. 1961.
10. Massy, V., Curti, B. and Ganther, H., J. Biol. Chem., 241, 2347 (1966).
11. Hageseth, G. T. and Joyner, R. D., J. Theor. Biol., 53, 51 (1975).
12. Goloff, A. A. and Bazzaz, F. A., J. Theor. Biol., 52, 259 (1975).
13. Johnson, F. H., Eyring, H. and Williams, R. W., J. Cell. Comp. Physiol., 20, 247 (1942).
14. Johnson, F. H., Eyring, H. and Stover, B. J., "Theory of Rate Processes in Biology and Medicine". John Wiley and Sons, N. Y. (1974).
15. Lumry, R., Rajender, S. Biopolymers, 9, 1125 (1970).
16. Dau, L. and Laboreau, L. G., Un. Acad. Brasil Cienc., 46, 311 (1974).

"A DSC Study of the Heat Capacity of Vicinal Water in Porous Materials"

Chester V. Braun, Jr. and W. Drost-Hansen
University of Miami

Vicinal water is believed to possess structures which differ notably from the bulk structure. If this is indeed true, then measureable differences must be expected in the thermodynamic properties. Ling and Drost-Hansen (6) attempted a DTA study of water in porous glass, and the data obtainted were suggestive of differences in the properties of bulk water and water in narrow pores (range: 20 nm to 200 nm). We have continued the study of thermodynamic properties employing a DSC method utilizing different types of porous materials. The results to date have not provided firm additional evidence for the occurrence of thermal anomalies. However, the data do suggest that the heat capacity of water in porous glass, activated charcoal, a zeolite, and diamond powder is significantly larger than the value obtained for bulk water (measured by the same technique). Assuming <u>*all*</u> *the water present possesses a modified structure, the apparent heat capacity is at least 20% larger than that for bulk water; and, to a first approximation, independent of the specific chemical nature of the solid material. This observation rules out the possibility that the anomalous values obtained in porous glass are due to some type of silicic acid formation. Furthermore, the substrate-independence agrees qualitatively with the "paradoxial effect" observed for the temperatures of vicinal water transitions (Drost-Hansen, 3).*

I. THERMAL PROPERTIES OF VICINAL WATER

Water adjacent to many (solid) interfaces possess properties which differ from the bulk and such water must therefore also be structured differently from the bulk. Water thus modified will be referred to as <u>vicinal</u> water. The properties of such water appear to exhibit anomalous temperature dependencies; for reviews of the properties of interfacial water, see, for instance, Clifford (1) or

Drost-Hansen (2,3,4). Earlier, Lumry and Rajender (5) discussed one aspect of vicinal water structuring: they noted that if many properties of interfacial water exhibit more or less abrupt changes in their temperature dependencies, over narrow temperature intervals, such abrupt changes must be associated with large entropy changes, as the result of cooperative behavior of a rather large number of water molecules. Lumry and Rajender continue: "However, such behaviour must reveal itself by heat capacity spikes which have thus far defied detection". On this basis, Ling and Drost-Hansen began a study of the thermal properties of water in porous glass by Differential Thermal Analysis (DTA). The study, unfortunately, proved inconclusive, although suggestive. For this reason, the present authors have initiated a re-examination of the thermal properties of interfacial water by a Differential Scanning Calorimetry (DSC) approach. However, before addressing ourselves to the possible existence of abrupt changes in vicinal water structuring over narrow thermal transition ranges, it was decided to attempt to obtain values (at 25°C) for the heat capacity of interfacially modified water.

The two most obvious questions regarding vicinal water are A) what types of structures may be induced by proximity to a solid?, and B) over what distances from the surface are these structures stabilized? Obviously, it is not likely that an answer to question 'A' will be achieved much before the structure of bulk water and bulk aqueous solutions are determined - a question which may not be resolved for many years, judging from the rate with which progress is presently taking place. (For some speculative reviews of possible types of vicinal water structures, see Drost-Hansen, 2,3,4). On the other hand, it is far easier to state how vicinal water properties differ from bulk properties and to determine the distances over which such differences can be observed.

A unique feature of vicinal water is the relative insensitivity of the temperatures of the thermal anomalies to the detailed chemical nature of the substrate. This phenomenon has been referred to as the "paradoxical effect" (1,4,7). Recently, a suggestion has been made to explain such "substrate independent" effects (8); this tentative explanation is outlined below: Clifford (1) (and independently, Bee Drost-Hansen, unpublished) suggested that in sufficiently "narrow" spaces, bulk water structures may not be able to develop because of geometric constraints on what must then be assumed to be rather large (cooperative) structures, possibly "clusters", for instance, in the sense of

---

Contribution #23: LABORATORY FOR WATER RESEARCH

Luck (9), Frenkel (10) and others. If indeed this is true, then it seems reasonable that, to a first approximation the detailed chemical nature of the confining surfaces play no, or only a minor, role. (For cases where anomalies are observed at the interface between a solid surface and a "semi-infinite" thickness of water, it is difficult to see how this explanation of the paradoxial effect might be applicable). The "paradoxial effect" has been introduced because the results of the heat capacity studies, to be reported in this paper, appear consistent with this interpretation.

II. METHODS AND MATERIALS

A DuPont Model 900 DTA instrument was used for the measurements of heat capacities using a DuPont DSC Cell. Stable temperatures below ambient were obtained by adding a copper cooling coil around the cell block and a heavy layer of insulation. A continuous flow of ethylene glycol and ethanol at approximately -15°C was maintained at all times. To prevent condensation, a plug containing silica gel was placed directly above the silver lid of the cell. No other modifications were made on the instrument. DuPont aluminum sample cups, which could be hermetically sealed, were used in these runs; the volume of each cup was 0.05 cm^3. An Ice-Point Calibration Standard device was used as the cold junction reference with a temperature stability at 0°C of 0.025°C. A Sartorius Model $123\text{-}360_c$ balance was used for all weighings.

A heating rate of 5.00° $\pm$ 0.07°C/minute was used; the ΔT scale sensitivity was 0.1°C/inch; the T scale sensitivity was 1.0°C/inch. The calibration coefficient (E_T) at 25°C was determined to be 169.92/mcal/°C-min.(by heat capacity measurements) using tin, sulfur, zinc, silver, and platinum.

The following substrates were used:

A) Bio Rad Laboratories Bio-Glas 200 porous glass for chromatography with a nominal exclusion limit of 200 Å. This material was used as received.

B) Breedmore Deluxe Grade Activated Carbon No. DGC-16 with a surface area of 20,000 $feet^2$/gram. The carbon was continuously washed with distilled water for 24 hours, dried at 120°C for 24 hours, and stored in a dessicator over silica gel until used.

C) Linde Molecular Seives (a Zeolite from Union Carbide), Type 4A Mesh was used with the same preparation as the carbon (B)

D) Synthetic diamond powder (obtained from a local distributor). The particle size was in the range of 5 μm to 6.3 μm with a Coulter Electronics Counter, Model B.

Except for the individual differences noted above, sample preparations and heat capacity determinations did not vary. The experimental design was based on the hypothesis that the specific chemical nature of the substrate did not affect the apparent heat capacity of the water (beyond the first few molecular layers). Thus, operationally the substrates were considered to be part of the "container". To test the hypothesis on which the experimental design was based, substrates of notably different chemical nature were used. Thus, the materials ranged from the hydrophilic silicates to the hydrophobic diamond. The test of the validity of the hypothesis is discussed in the section describing our results. For the reason discussed above, cups, substrates, and lids were weighed, placed in the DSC cell and allowed to come to equilibrium at the starting temperature.

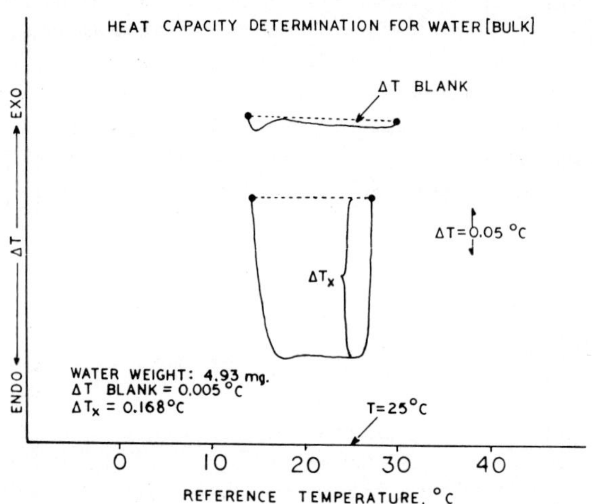

Fig. 1. A typical DSC run for heat capacity determination (for bulk H_2O, i.e., no substrate).

Empty, hermetically sealed, aluminum sample cups were used as the reference in order to partially counteract the large ΔT due to the sample cup and substrate. In this configuration the cell was heated at a rate of 5.00°C/minute. At a point slightly above 25°C the mode was changed from "heat" to "hold" to establish thermal equilibrium. The difference (at 25°C) between the trace thus obtained and a straight line drawn between the starting temperature equilibrium and the final equilibrium point corresponds to the absolute differential temperature under no sample (H_2O) condition (i.e. ΔT_{Blank} °C). Water was then added to the same substrate with a microliter syringe, and the cup sealed and weighed to determine the amount of water (M in milligrams). After equilibrium was reached at the desired starting temperature the above procedure was repeated to find the absolute differential temperature (i.e., ΔT_x °C) for the sample (H_2O). With these two values (ΔT_{blank} and ΔT) and (heating rate (Z °C/minute), and the calibration coefficient ($E_{25°C}$ mcal/°C-minute) - the heat capacity ($Cp_{25°C}$ cal/gram-°K) was determined as:

$$(Cp)_{25°C} = \frac{(\Delta T_x - \Delta T_{blank}) \cdot E_{25°C}}{M \cdot Z}$$

Note: An endothermic ΔT was given a positive value and an exothermic ΔT a negative value. This was to allow for an endothermic or exothermic ΔT_{blank}; ΔT_x was always endothermic.

In addition to the measurements on the materials listed above, James Clegg (in the authors' laboratory) has performed heat capacity determinations on Artemia salina cysts at various water contents. (The cysts are the fertilized eggs of the brine shrimp, A. salina). The measurements were carried out in the same manner as described above except that the value for $(Cp)_{25°C}$ thus obtained now refers to the water-cyst system. To arrive at the value for the heat capacity of the water in this system the following relation was used:

$$(1+w_1) \cdot Cp = \overline{Cp}_2 + \overline{Cp}_1 \cdot w_1$$

The result of $(1+w_1) \cdot Cp$ plotted against w_1 is shown in Fig. 2. The intercept represents $\overline{Cp}_2$ and the slope $\overline{Cp}_1$ where:

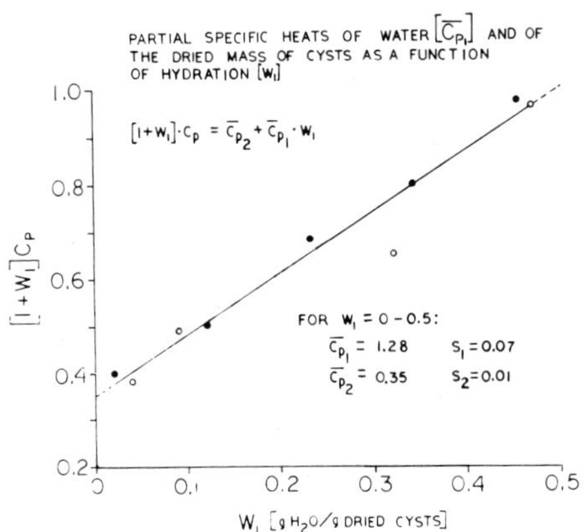

Fig 2. Plot of (1+w) . Cp versus w_1 for A. salina cysts. The slope of the line is the heat capacity of the water in the cysts ($\overline{Cp}_1$). (Data by Clegg).

w_1 = number of grams of water per gram of dried cyst.
Cp = apparent heat capacity of the whole system
$\overline{Cp}_2$ = "partial molal" heat capacity of the dried cysts.
$\overline{Cp}_1$ = partial molal heat capacity of the cellular water (see 11,12,13).

III. RESULTS AND DISCUSSION

The results of the heat capacity measurements are shown in Table 1. On the left hand side are shown the observed values and the standard deviations. The apparent value for bulk water obviously reflects a systematic difference, most likely associated with the calibration coefficient used in these experiments. (*Artemia* cysts are not included in this set because a systematic error in E_T would show up as an error in the heat capacity of the cysts (Cp_2) but the slope (Cp_1) would be unchanged). Hence, in the top right hand side of the Table are listed the "corrected values" for the heat capacities, normalized to Cp,H_2O,bulk = 1.00.

HEAT CAPACITY OF VICINAL WATER [cal./gm.-°K] at 25°C		
	OBSERVED	REDUCED TO C_p[BULK] = 1.00
BULK	1.08 ± 0.08	[1.00 ± 0.08]
POROUS GLASS	1.37 ± 0.20	[1.27 ± 0.20]
ACTIVATED CARBON	1.38 ± 0.03	[1.28 ± 0.03]
ZEOLITE	1.31 ± 0.03	[1.21 ± 0.03]
DIAMOND	1.30 ± 0.08	[1.20 ± 0.08]
COLLAGEN [Hoeve]		1.24
EGG ALBUMIN [Bull]		1.25 ± 0.02
DNA [Bull]		1.26 ± 0.06
ARTEMIA CYSTS [Clegg]		1.28 ± 0.07

Table 1. Heat capacities of some aqueous systems. Top half represents data obtained by the present authors; lower data from references 20, 12, 13.

Several aspects of these results are striking: A) All the values are notably larger than 1.00 and B) all the values center around 1.24 cal/gram-°K, i.e., well within the experimental scatter on the individual sets of heat capacity values. Finally for comparison purposes, some values are listed for the heat capacities of water in some aqueous biochemical systems and in one cellular system. The average value is 1.26 cal/gram-°K - apparently within essentially similar error limits as determined by our own DSC measurements.

To a first approximation, then, the heat capacity of vicinal water appears to be constant; approximately 1.25 cal/gram-°K. If the water involved in these widely different systems is all in the form of "vicinal water", this result is then in agreement with the paradoxial effect: the detailed chemical nature of the confining solids does not influence the specific modification of the "permissible" vicinal water structures. Unfortunately, this facile explanation must be suspect on several grounds: for the very innermost layers of interfacial water, specific water/"solid" interactions must play unique roles: ion-dipole interactions, dipole-dipole interactions, induced dipole-dipole interactions, etc. Furthermore, for sufficiently thick layers of interfacial water, water with bulk attributes must be expected beyond certain distances from the nearest solid surface. Hence, it is truly remarkable that such a narrow range of Cp values is observed. On the otherhand, if the thickness of the water layers all exceed the dimensions of the high energy, short-range order (due to ion-dipole interactions, etc.) yet all fall within what appears to be the range of "vicinal water structuring", (namely around 0.01-0.1 μm) the major contribution to the observed heat capacities may be constant, representative of this long-range, vicinal ordering. (This explanation, however, may not account for the result obtained with the zeolite which possesses only very small voids). For discussions of distances over which vicinal water structure effects in general may operate, see Peschel, et al. (14,15,16), Clifford (1) (for what appears to be minimum estimates), Drost-Hansen (3,8) and Schufle, Huang and Drost-Hansen. It should be noted that in all cases "excess" of solid was chosen. Thus, full saturation of individual pores may not have been achieved; this particularly applies to the diamond dust. Note also that spurious results due to some type of silicic acid formation can be ruled out as statistically identical results were obtained with glass, activated carbon, and diamond dust. The same type of argument may be applied to the biochemical systems: in other words, cellular water is likely also to possess some characteristics, independent of the specific proteins present. The "paradoxical effect" applied to living systems then predicts some "invariant properties" related only to vicinal water. Examples of such "invariant properties" are sharp thermal death limits and minima for metabolism at the temperatures of transitions of vicinal water - see (18,3) and Etzler and Drost-Hansen, this Volume.

We stress again that if the observed values for the heat capacity of vicinal water are constant, this finding then is in excellent agreement with the explanation for the "paradoxical effect". It should be observed that not all studies of dispersed aqueous systems have yielded constant values for Cp. Thus, values for Cp of interfacial water, exceeding the value for bulk water, but strongly dependent on total water contents have also been reported by other authors; see, for instance, Oster and Low (19) who studied clay-water mixtures. The research on the heat capacities of vicinal water is being continued in the authors' laboratory.

IV. ACKNOWLEDGEMENT

The authors wish to acknowledge partial support for this study by the U.S. Environmental Protection Agency through grant number R80 38 26 01 to the Senior Investigator. We also gratefully acknowledge many stimulating discussions with Professor James S. Clegg, and his permission to use the unpublished data for heat capacities obtained with the *Artemia* cysts.

REFERENCES

1. Clifford, F. in "Water - A Comprehensive Treatise" (F. Franks, ed.), Vol. 5, p. 75. Plenum Press, New York, 1975.
2. Drost-Hansen, W. *Ind. and Eng. Chem.*, 61 (November), 10 (1969).
3. Drost-Hansen, W. in "Chemistry of the Cell Interface" (H. D. Brown, ed.), Vol. B, p. 1, 1971.
4. Drost-Hansen, W. *Preprint*, *Div. of Petroleum Chemistry,* Advances in Petroleum Recovery (Symposium), Vol. 21, 278 (1976).
5. Lumry, R. and Rajender, S. *Biopolymers*, Vol. 9, 1125, (1970).
6. Ling, C. S. and Drost-Hansen, W. in "Adsorption at Interfaces", (K. Mittal, ed.), *ACS Symposium Series*, No. 8, 1976.
7. Drost-Hansen, W. *Annals, N. Y. Acad. Sci.*, Vol. 204, 100 (1973).
8. Drost-Hansen, W. Invited paper, International Conference on Colloids and Surfaces, San Juan, Puerto Rico, June 1976; to be published.
9. Luck, W. *Fortschritte der Chemishen Forschung*, Vol. 4 (Number 4), 653 (1964).
10. Frenkel, J. "Kinetic Theory of Liquids", Dover Publications, Inc., New York, 1955.
11. White, P. and Benson, G. C. *Journal of Physical Chemistry*, Vol. 64, 599 (1960).
12. Bull, H. B. and Breese, K. *Archives of Biochemistry and Biophysics*, Vol. 128, 497 (L968).

13. Chattoraj, D. K. and Bull, H. B. Journal of Colloid and Interface Science, Vol. 35, 220 (1971).
14. Peschel, G. and Adlfinger, K. H. Naturwissenschaften, Vol. 11, 558 (1969).
15. Peschel, G. and Adlfinger, K. H. Journal of Colloid and Interface Science, Vol. 34, 505 (1970).
16. Peschel, G. and Adlfinger, K. H. Z. Naturforsch., Vol. 26, 707 (1971).
17. Schufle, J. A. and Huang, C. T. Journal of Colloid and Interface Science, Vol. 54, 184 (1976).
18. Drost-Hansen, W. Annals, N. Y. Acad. Sci., Vol. 125, Art. 2, 471 (1965).
19. Oster, J. D. and Low, P. F. Soil Science Society of America Proceedings, Vol. 28, 605 (1964).
20. Hoeve, C. A. J. and Kakivaya, S. R. Journal of Physical Chemistry, Vol. 80, 745 (1976).

INTERFACIAL FREE ENERGY AND COMPOSITION PROFILE OF INTERPHASES BETWEEN DILUTE POLYMER SOLUTIONS

A. Vrij and G.J. Roebersen

Van 't Hoff Laboratory

Transitorium III, Padualaan 8

Utrecht, The Netherlands

ABSTRACT

When a (dilute) solution of a polymer in a poor solvent is cooled off sufficiently below its critical solution temperature, a phase separation will occur.
A model has been constructed to study the interface between the two separate phases. Calculated results will be reported of polymer concentration profiles in the interface and of interfacial free energies.

NEW VALUES OF THE LIGHT SCATTERING INTENSITY, DEPOLARIZATION RATIO AND ANISOTROPY OF WATER

R.L. Rowell and R.S. Farinato
University of Massachusetts at Amherst

We have measured a new value for the depolarization ratio of pure water at 514.5 nm using an argon ion laser light source, a high resolution photometer (0.3°) and photon counting detection. With no filter in the viewing optics we find $\rho_V = 0.026$ which is equivalent to $\rho_u = 0.051$. A full account of the work including the spectral dependence is reported elsewhere(1). Below, we combine our new measurement of the depolarization ratio with a recent measurement of the Rayleigh ratio by Pike et al.(2) to obtain new values for the anisotropic Rayleigh ratio R_{anis} and the effective optical anisotropy Δ^2.

The best available value for the Rayleigh ratio of water appears to be that due to Pike et al.(2) who found $R_{90,V,V+H} = 0.92 \times 10^{-6}\ cm^{-1}$ at 633.0 nm. The subscripts indicate vertically polarized incident intensity and total scattered intensity. We convert Pike's value to $1.66 \times 10^{-6}\ cm^{-1}$ at 514.5 nm considering the $1/\lambda^4$ dependence as being the only significant dispersion within the envelope of experimental error. Using $\rho_u = 0.051$ we readily(2) obtain $R_{u,90} = 0.872 \times 10^{-6}\ cm^{-1}$ at 514.5 nm. This is in good agreement with the Cohen and Eisenberg value(3) of $0.865 \times 10^{-6}\ cm^{-1}$. The anisotropic contribution is obtained from $R_{anis} = R_{u,90}\ (13\rho_u)/(6+6\rho_u)$ after Deželić(4) which gives $R_{anis} = 0.0917 \times 10^{-6}$ at 514.5 nm. The latter value is 3.38 times smaller than the value of $0.31 \times 10^{-6}\ cm^{-1}$ used by Kielich(5) in an earlier critical comparison between experiment and theory. The new results improve the self consistency of calculations(5) of the effective optical anisotropy of water leading to $\Delta^2 = 0.0030$ as discussed in detail elsewhere(1).

REFERENCES

1. Farinato, R.S. and Rowell, R.L., J. Chem. Phys., in press.
2. Pike, E.R., Pomeroy, W.R.M. and Vaughan, J.M., J. Chem. Phys. 62, 3188 (1975).
3. Cohen, G. and Eisenberg, H., J. Chem. Phys. 43, 3881 (1965).
4. Deželić, G., Pure and Applied Chem. 23, 327 (1970).
5. Kielich, S., J. Chem. Phys. 46, 4090 (1967).